INTRODUCTION TO DATA SCIENCE

INTRODUCTION TO DATA SCIENCE

Authored by

Gaoyan Ou
Beijing Institute of Big Data Research, China

Zhanxing Zhu
Peking University, China

Bin Dong
Peking University, China

Weinan E
Beijing Institute of Big Data Research, China

Translated by

Binyang Li
Shumin Shi

World Scientific

NEW JERSEY • LONDON • SINGAPORE • BEIJING • SHANGHAI • HONG KONG • TAIPEI • CHENNAI • TOKYO

Published by

Higher Education Press Limited Company
4 Dewai Dajie, Beijing 100120, P. R. China

and

World Scientific Publishing Co. Pte. Ltd.

5 Toh Tuck Link, Singapore 596224

USA office: 27 Warren Street, Suite 401-402, Hackensack, NJ 07601

UK office: 57 Shelton Street, Covent Garden, London WC2H 9HE

Library of Congress Cataloging-in-Publication Data
Names: Ou, Gaoyan, author. | Zhu, Zhanxing, author. | Dong, Bin (Professor), author. |E, Weinan, 1963– author. | Li, Binyang (Translator), translator. | Shi, Shumin (Translator), translator.
Title: Introduction to data science / Gaoyan Ou, Beijing Institute of Big Data Research, China, Zhanxing Zhu, Peking University, China, Bin Dong, Peking University, China, Weinan E, Beijing Institute of Big Data Research, China. Translated by: Binyang Li, Shumin Shi.
Description: New Jersey : World Scientific, 2024. | Includes bibliographical references.
Identifiers: LCCN 2023006247 | ISBN 9789811263897 (hardcover) | ISBN 9789811263903 (ebook) | ISBN 9789811263910 (ebook other)
Subjects: LCSH: Big data. | Electronic data processing. | Multivariate analysis. | Data mining. | Algorithms.
Classification: LCC QA76.9.B45 O9 2024 | DDC 005.7--dc23/eng/20230404
LC record available at https://lccn.loc.gov/2023006247

British Library Cataloguing-in-Publication Data
A catalogue record for this book is available from the British Library.

B&R Book Program

First published 2024 (hardcover)
Reprinted 2026 (in paperback edition)
ISBN 9789819832903 (pbk)

For any available supplementary material, please visit
https://www.worldscientific.com/worldscibooks/10.1142/13071#t=suppl

Desk Editors: Aanand Jayaraman/Tan Rok Ting

Typeset by Stallion Press
Email: enquiries@stallionpress.com

Preface

Data science is an emerging discipline which emphasizes the cultivation of big data talents with interdisciplinary ability. Such talents should have the following three abilities: theoretical ability (such as the ability to understand and use algorithms and models), practical ability (such as the ability to process real data) and applicative ability (such as the ability to use big data to solve practical problems in specific industries). Training such talents requires close cooperation among the disciplines of mathematics, statistics, computer science and others, as well as close cooperation with industrial partners and data owners. Data science courses also need to adopt the new methods of combining theory and practice, just like physics, chemistry and biology courses. For most data science courses, we need an experimental platform that provides practical problems, datasets, as well as basic data processing and analytical tools.

In order to meet the challenges of training teachers, constructing curricula and developing teaching resources, Beijing Institute of Big Data Research established Boya Big Data College to establish the infrastructures needed for the cultivation of big data talents. These infrastructures include new curriculum, textbooks, experimental platforms and instructional methods, etc. In recent years, Boya Big Data College has successfully established a complete big data curriculum system. Courses are divided into three modules: fundamental courses, core courses and elective courses. Fundamental courses include Mathematical Basis of Big Data Analysis, Python Basis for Big Data and Data Management, etc. Core courses include Data

Acquisition and Web Crawler, Data Cleaning, Data Visualization, Introduction to Data Science, Machine Learning and Distributed Computing, etc. Elective courses include Deep Learning, Knowledge Graph, Text Mining, Health Data Science, Transportation Data Science and Financial Data Science, etc. Boya Big Data College has built a data science experimental platform named iDataCourse, which has been widely used by universities in China.

Introduction to Data Science is the first textbook by Boya Big Data College. Its purpose is to comprehensively introduce models and algorithms in data science from a technical point of view. This book systematically introduces the basic theoretical content of data science, including data preprocessing, basic methods of data analysis, processing of special problems (such as text analysis), deep learning, and distributed systems. In addition, this book provides a large number of case studies for data analysis application practice. Students can conduct practical training and interact with data on the iDataCourse platform. Instructors can assign the online practices as homework and receive feedback reports, greatly reducing their workload. Readers can visit iDataCourse to access these practical cases and resources.

This book is the result of collaboration. Gaoyan Ou is responsible for writing the data preprocessing, classification model, ensemble model, association rule mining, dimensionality reduction, text analysis, distributed computing and appendix chapters. Zhanxing Zhu is responsible for writing the regression model, clustering model, feature selection, EM algorithm, probabilistic graphical model and deep learning chapters. Bin Dong reviewed the textbook and put forward many valuable suggestions. Weinan E is in charge of planning and organization the book, as well as revising the chapters, and for writing the introduction. Lei Zou (from the Wangxuan Institute of Computer Technology of Peking University) and Peng Peng (from the School of Information Science and Engineering of Hunan University) are responsible for writing the first draft of the graph and network analysis chapter. Jiahao Yao, Xinhang Yu, Jia Chen and Wenjia Wang (students majoring in data science from Peking University) are responsible for reviewing the first draft of the book. Xiaodong Yan and Yang Gao (data analysts of Boya Big Data College) are responsible for writing the practical cases. Xiaotong Dai

(UI designer of Boya Big Data College) is responsible for producing most of the original figures in the book.

Gaoyan Ou, Researcher
Beijing Institute of Big Data Research

Zhanxing Zhu, Senior Research Professor
Changping National Lab

Bin Dong, Assistant Professor
Peking University

Weinan E, Professor
Peking University

About the Authors

Gaoyan Ou is a researcher at Beijing Institute of Big Data Research, Vice President of Beijing Big Data Association, member of the Professional Committee of Database of China Computer Federation, and Alibaba Cloud MVP (the most valuable expert). He has participated in digital transformation projects of many banks and operators in China. The main work at present includes the construction of smart city's new generation infrastructure and the development of data element platform.

Zhanxing Zhu is a senior research professor at Changping National Lab, Beijing, China, leading a group on machine learning and its application for biology. He previously was an assistant professor at Peking University and Beijing Institute of Big Data Research. He obtained his Ph.D. degree in machine learning from the University of Edinburgh in 2016. His research interests cover machine learning and its applications in various domains. Currently, he mainly focuses on theoretical and methodological foundation of deep learning, and AI for computational biology. He has co-authored more than 60 papers in top machine learning journals and conferences (with Google Scholar citation $\sim$5000), such as NeurIPS, ICML, and ICLR. He has been serving as area chair for top machine learning conferences, such as AISTATS, IJCAI, and AAAI. He has been recognized as the 2023 AI 2000 Most Influential Scholar Honorable Mention in AAAI/IJCAI for his outstanding and vibrant contributions to this field between 2013 and 2022. He was awarded "2019 Alibaba Damo Young Fellow" and obtained "Best Paper Finalist" from the top computer security conference CCS 2018.

Bin Dong is a professor at the Beijing International Center for Mathematical Research, Peking University. He is also the deputy director of the Center for Machine Learning Research at Peking University, the deputy director of the Peking University Institute for Computing and Digital Economy, and an affiliated faculty member of the National Biomedical Imaging Center and the National Engineering Laboratory for Big Data Analysis and Applications. He received his B.S. from Peking University in 2003, M.Sc from the National University of Singapore in 2005, and Ph.D. from the University of California Los Angeles in 2009. Bin Dong's research interest is in mathematical analysis, modeling, and computations in computational imaging, scientific computing, and machine learning. He currently serves on the editorial board of Inverse Problems and Imaging, CSIAM Transactions on Applied Mathematics, *Journal of Computational Mathematics*, and *Journal of Machine Learning*. He received the Qiu Shi Outstanding Young Scholar Award in 2014 and was invited to deliver a 45-minute sectional lecture at the International Congress of Mathematicians (ICM) 2022.

Weinan E is a professor at the Center for Machine Learning Research (CMLR) and the School of Mathematical Sciences at Peking University. He is also a professor at the Department of Mathematics and Program in Applied and Computational Mathematics at Princeton University. His main research interest is numerical algorithms, machine learning, and multi-scale modeling, with applications to chemistry, material sciences, and fluid mechanics. Weinan E was awarded the ICIAM Collatz Prize in 2003, the SIAM Kleinman Prize in 2009, the SIAM von Karman Prize in 2014, the SIAM-ETH Peter Henrici Prize in 2019, and the ACM Gordon Bell Prize in 2020. He is a member of the Chinese Academy of Sciences and a fellow of SIAM, AMS, and IOP. Weinan is an invited plenary speaker at ICM 2022. He has also been an invited speaker at ICM 2002, ICIAM 2007, as well as the AMS National Meeting in 2003. In addition, he has been an invited speaker at APS, ACS, AIChe annual meetings, the World Congress of Computational Mechanics, and the American Conference of Theoretical Chemistry.

About the Translators

Binyang Li graduated from the Chinese University of Hong Kong. Currently, he is a professor and master's supervisor in the School of Cyberspace Science and Engineering at the University of International Relations. He is engaging in natural language processing and social computing. As PI, Dr. Li has completed several National Natural Science Foundation projects, National key research and development plans, as well as provincial or ministerial research projects, and published more than 50 academic papers. He won the Science and Technology Progress Award of the Ministry of Education in 2017, the Science and Technology Award of Qian Weichang Chinese Information Progress in 2022, and the Information and Communication Technology Excellence Award of Hong Kong in 2013.

Shumin Shi got her doctorate degree at Nanjing University of Science and Technology in 2008. Currently, she is an associate professor in the School of Computer Science and Technology at BIT (Beijing Institute of Technology) and is also serving as the deputy secretary-general of the board of governors at CAAI (Chinese Association for Artificial Intelligence). Since joining BIT in 2009, she has been engaging in research on natural language processing and teaching computer science as a master's supervisor. Her research focuses on semantic computing, machine translation, and sentiment analysis.

From 2015 to 2016, as a visiting scholar in EECS at the University of California, Berkeley, she cooperated with Prof. Lotfi A. Zadeh in computational theory of perceptions and precision natural language. She has undertaken a dozen of projects from the National Natural

Science Foundation, National Key R&D Program, and 863 and 973 Program of China and received several awards including Progress Prize in Scientific from the People's Government of Beijing Municipality and CIE (Chinese Institute of Electronics). She published over 30 papers and five textbooks (one was awarded National Excellent Textbook) and also owns more than 10 patents.

Contents

Chapter 1

Introduction

1.1. The Fundamental Contents of Data Science

What is data science? What is the difference between data science and similar disciplines such as information science, statistics and mathematics? As a new discipline, data science is based on two essential properties: the diversity of data and the universality of data research. In modern society, data pervades all walks of life. There are many types of data, including structured data and unstructured data, such as web pages, text, image, video and audio. The objective of data analysis is to solve inverse problems, usually that of stochastic models. So the research on these problems shares a lot in common. For example, both natural language processing and biomacromolecule model employ hidden Markov model (HMM) and dynamic programming method. The fundamental reason is that they both deal with one-dimensional random signals. Regularization, which is widely used in image processing and statistical learning, is also one of the most commonly used technique in mathematical models to solve inverse problems.

Data science mainly consists of two aspects: using data to study scientific problems and using scientific methods to study data. The former includes bioinformatics, astroinformatics, digital earth, etc.; and the latter includes statistics, machine learning, data mining, database, etc. These are all essential to data science and when put together, they form the overall picture of data science.

The most typical example of scientific research is Kepler's three laws of planetary motion, which was summarized based on the data

(a) Tycho: Observation and data collection

(b) Kepler: Data analysis to create value

Figure 1.1 Example of using data to study scientific problems: Kepler's three laws of planetary motion.

observed by an astronomer, named Tycho (as shown in Figure 1.1). Table 1.1 is a typical example. Here, the dataset listed the number of orbital cycles that a planet completes per year and the average distance a planet is away from the sun. From this dataset, it is observed that the square of the orbital period is proportional to the cube of the average distance from the sun, which is Kepler's Third Law.

Although Kepler discovered the three laws of planetary motion, he did not understand the inherent mechanism behind them. It was Newton who developed a deep understanding of them. He combined Kepler's Second Law with the law of Universal Gravitation and considered it as a purely mathematical problem, i.e., a system of ordinary differential equations. Disregarding the interactions between different planets, it was a two-body problem between the planet and the sun. So it was easy to figure out the corresponding solutions and then deduce Kepler's three laws.

Newton adopted this method by seeking the fundamental principles, which is far more profound than what Kepler sought. He not only knew the results, but also knew how they were concluded. Therefore, the method of seeking basic principles became the preferred methodology for scientific research. This kind of methodology reached its peak in the early 1900s when physicists discovered

Table 1.1 Dataset about the movement of the eight planets of the solar system around the sun.

Planet	Cycle time/ year	Average distance	Cycle time^2/ average distance^3
Mercury	0.241	0.39	0.98
Venus	0.615	0.72	1.01
Earth	1.00	1.00	1.00
Mars	1.88	1.52	1.01
Jupiter	11.8	5.20	0.99
Saturn	29.5	9.54	1.00
Uranus	84.0	19.18	1.00
Neptune	165	30.06	1.00

Table 1.2 Illustration of SNP data.

	SNP_1	SNP_2	⋯	SNP_m
Volunteer 1	0	1	⋯	0
Volunteer 2	0	2	⋯	1
⋮	⋮	⋮		⋮
Volunteer n	1	9	⋯	1

quantum mechanics under its guidance. Theoretically speaking, the natural phenomena encountered in our daily life can be explained from the perspective of quantum mechanics. Moreover, quantum mechanics also provides fundamental principles for researching chemistry, materials science, engineering science, life science, and almost all the other natural sciences and engineering fields. Quantum mechanics has been successful, but is not without complications. As early as 1928, the famous British theoretical physicist, Dirac, pointed out that if these problems were solved based on the fundamental principles of quantum mechanics, then the math problem would be too difficult.[1] So, in order to make progresses, certain compromise is necessary, that is to make an approximation of the fundamental principles.

Insightful as the *Newton's* methodology is, the *Kepler's* methodology, however, is more effective in dealing with more complex problems. For instance, Table 1.2 illustrates a group of data about the

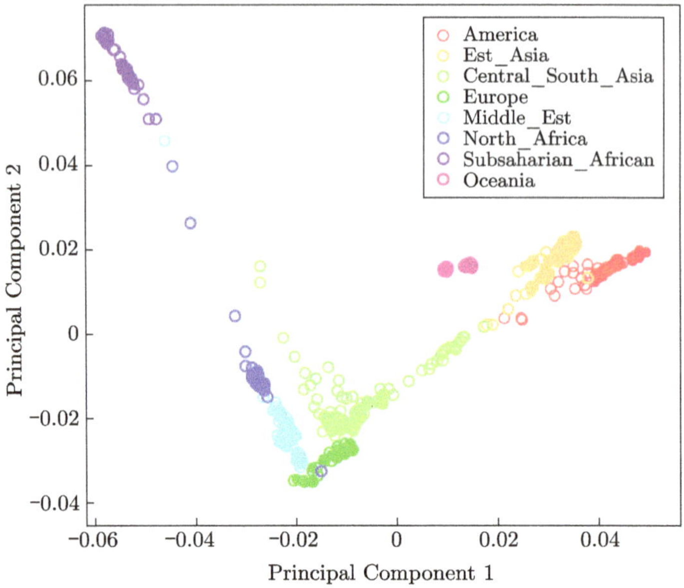

Figure 1.2 Principal component analysis to the SNP data.

single nucleotide polymorphism (SNP). 1,064 volunteers across the world were chosen, and then their SNPs were digitalized, i.e., the 10 possible base pairs at each position are represented by numbers. Then, principal component analysis was applied to the data in order to obtain the results shown in Figure 1.2.[2] The horizontal and vertical axes stand for the vectors of the first and second singular value, respectively. All these vectors consist of 1,064 components corresponding to the 1,064 volunteers. We could see that human evolution is demonstrated by these data through the common statistical analysis method — principal component analysis (PCA). PCA is a statistical analysis method that analyzes the spectral decomposition of the covariance matrix.

The above problem is hard to solve by Newton's methodology, which is based on using the fundamental principle to understand the underlying mechanism. On the contrary, Kepler's data-driven methodology will work. Some good examples of Kepler's methodology are bioinformatics and human genome projects. Because of their success, similar projects such as materials genome engineering have also been put on the agenda. Similarly, celestial informatics and computational sociology have become popular subjects. These are examples of using data to study scientific problems. Another typical example is image processing. The success of image processing is determined by the human visual system. Therefore, to solve the problem of image processing, it is necessary to start by understanding the human visual system, including how different quality of images affects the human visual system. Of course, such understanding is profound, and this may be what we ultimately need. But it is too difficult and complicated from the current point of view. Simple mathematical models are used for solving practical problems without a full understanding of their basic principles.

Using data-driven methods to study scientific problems does not mean that there is no need for models. They are different starting points, either from the data-driven approach or from the basic principles approach. Taking image processing as an example, a model based on basic principles needs to describe the human visual system as well as its relationship with the image while the data-driven method can be based on simple mathematical models, such as function approximation.

What is the scientific approach of studying data? The answer lies in data collection, data storage and data analysis. We will focus on data analysis in the rest of this section.

1.1.1. *Core Problem of Data Analysis*

Before introducing data analysis, we should first understand data type. As shown in Table 1.3, some common data types are listed as follows:

1. **Table:** It is the most typical data type. In general, each row refers to a sample, while each column refers to an attribute or feature.
2. **Point set:** Data can be considered as a set of points in a certain space.
3. **Time series:** Text, conversation and DNA sequences can be included in this category, which can also be thought of as a function of time variable.
4. **Image:** It can be treated as a function of two variables.
5. **Video:** It is a function of both the time variable and space variables.

Table 1.3 Mathematical models for common data types.

Data type	Model
Point set	Probability distribution
Time series	Stochastic process (such as hidden Markov model)
Image	Random field (Gibbs random field)
Network	Graphical model, Bayesian model

6. **Web page and newspaper:** Though each article on the webpage or in the newspaper is a part of time sequence, the web page and newspaper have a space structure as well.
7. **Network data:** Essentially, network is a graph, consisting of the vertices and the edges linking them.

Apart from the above basic data types, other data of higher order are also taken into consideration, such as the set of images, the set of time sequences, table sequences and so on. Note that a basic assumption of data analysis is that the observed data are generated by a model. The core problem of data analysis is to find out this model. Since noises are inevitable during the process of collecting data, all these models are usually stochastic models.

Of course, in most cases, we are not interested in the whole model. Instead, we just want to find out some important contents. For example, we use correlation analysis to determine whether two datasets are correlated or not and we use classification and clustering techniques to divide data into groups.

Usually, we also need to approximate the stochastic models. One approach is to approximate the stochastic model with a deterministic model. Regression and image processing models based on variational principle belong to this type. Another approach is to simulate the distribution of data, for instance, by assuming that data follows certain distribution, or the time sequence satisfies the assumptions of a Markov chain.

1.1.2. *Mathematical Structure of Data*

Before performing data analysis, we should introduce the mathematical structure of data, including measure structure, network structure, and algebraic structure.

1. **Measure structure:** Some metrics (e.g., distance) are introduced to embed the dataset into a metric space. The cosine distance function is a typical example in text processing.
2. **Network structure:** Some data, including social network, has network structure, while some other data without a network structures can be incorporated into a network. For a point set in a metric space, for example, we can decide whether there exists an edge between two vertices or not based on the distance between them, from which we can get a network structure. The PageRank algorithm is a typical example of network structure.
3. **Algebraic structure:** Data can be regarded as vectors, matrices or higher order tensors. Some datasets with hidden symmetry can also be expressed by algebraic structures.

Advanced structures can be constructed based on the aforesaid mathematical structures, such as the topological structure and the functional structure.

1. **Topological structure:** For the same dataset, the topological structure may be different from different perspectives. The most famous example is that the two-dimensional Klein Bottle is hidden in a 3×3 natural image dataset.[3]
2. **Functional structure:** A fundamental statistical problem is discovering the functional structure of a given point set. These functional structures include linear functions for linear regression, piecewise constant functions for clustering or classification, piecewise polynomials such as spline function, and other functions such as wavelet expansion.

1.1.3. *Major Difficulties in Data Analysis*

The data we used usually have the following characteristics. Firstly, the data volume is very large. For a better idea of the data that you are going to encounter, think about how many pages there are on the World Wide Web, and how much data there is on those pages. Because of the large volume, stochastic algorithms and distributed computing will become important. Secondly, the data dimension is very high. For example, the SNP data mentioned above has 640,000 dimensions. Thirdly, the data format is complex. Data can be generated in the format of web pages, newspapers, images or videos. So, it is difficult to implement data fusion due to the different formats of data. Fourthly, there is a great amount of noisy data. The process of data generation, data collection and data transformation will result in noisy data. Such noise will bring more challenges for data cleaning and data analysis. So, models with the function of modification are required, such as regularization in image processing and denoising autoencoder in machine learning.

Among the above difficulties, the major one is the high dimension, which will lead to the curse of dimensionality problem, i.e., the complexity and the computational cost of the model increase exponentially with the increase of the dimension.

To address this problem, there are usually two methods. One is to confine the mathematical models to a smaller special class, such as the linear model. Another one is to use special structures of data, such as sparseness, low-dimension or low-rank and smoothing, etc. These characteristics can be implement by regularizing the model, or by reducing the dimension.

Essentially, data analysis is an inverse problem. Therefore, many methods that deal with inverse problems, such as regularization, will play a very important role in data analysis. This is precisely the difference between statistics and statistical mechanics. Statistical mechanics deals with direct problems, while statistics deals with inverse problems.

1.1.4. *The Importance of Algorithms*

Complementary to model are algorithms and their implementation on computers. Especially when processing large-scale data, algorithms are particularly important. From the perspective of algorithm, there are two main methods for processing big data.

The first is to reduce the complexity of the algorithm, i.e., the amount of computation. For extremely large datasets, such as data on the World Wide Web or social network data, we can adopt a sampling method. The most typical example is Stochastic Gradient Descent (SGD). The other one is distributed computing. The basic idea is to use the divide and conquer approach, break down a large computational job into several small tasks. The famous MapReduce framework is one such example.

At present, research on algorithm divides into two areas that hardly have connections with each other: computational mathematics and computer science. The algorithm research in computational mathematics basically aims at continuous structures such as functions, and its main applications are for differential equations. The algorithm research in computer science mainly aims at discrete structures, such as networks. However, the characteristics of real data are in between: the data itself is discrete, while there is usually a continuous model behind it. Therefore, to develop an algorithm for data, it is necessary to effectively combine algorithms in computational mathematics and computer science.

1.2. Impact on the Development of Discipline

Data science provides unprecedented opportunities and challenges for the development of discipline. To make use of these opportunities, we must establish a new discipline and education system. At the level of college and university, we must put data science in the rightful place and establish a platform for its research. Meanwhile, we need to be more innovative to coopperate with industry, to cultivate an environment for a young generation of data scientists that meet both the needs of the academic community and the industrial community.

Data science will also exert tremendous influence on the development of many traditional disciplines. Mathematics comes first, whose development is mainly driven by two aspects: one is from the internal. Its development is the result of the promotion of itself. Another comes from the outside. Its development is influenced by the needs of other disciplines, society, or industry. For the actuality of study, the influence of the first aspect on mathematics far exceeds that of the second aspect. The result is that, on the one hand, mathematics, as a discipline, has been widely recognized for its importance. On the other hand, the influence of mathematicians as a group on the overall development of society and science is relatively indirect. Unfortunately, in many schools, even in the scientific community, mathematicians are becoming more and more isolated. It not only greatly affects the development of mathematics, but also affects the development of other disciplines, technologies and even society. Almost a hundred years ago, Dirac pointed out that at least in terms of theoretical research, the bottleneck problem in many disciplines is the mathematical problem. As a result, in many disciplines, we find that scientists who do not specialize in mathematics, are doing mathematical research. For mathematics itself, this may not be problem. But for mathematicians, they lost the role that they should be playing.

The starting point of natural sciences comes from the basic principles of physics. For mathematicians, this is a basic obstacle. Data science is different that its basic principles come from mathematics. It builds a direct bridge between mathematics and practical applications, which are the most active part of society in the era of intelligence. Mathematics is the theoretical foundation of these disciplines. Mathematicians have not exhausted their strength, but it cannot be denied that data science brings a golden opportunity to them. What's more, data analysis covers almost all aspects of modern mathematics. Even extremely abstract topics like representation theory have room in the field of data science. In a word, data science's requirements and promotion of mathematics are comprehensive rather than limited to just a few fields. Data should be one of the basic objects of mathematics research beyond numbers, graphs, and equations.

Data science will also have a great influence on the development of computer science. John Hopcroft, a Turing Award winner, once pointed out that the research of computer science has mainly focused on the computer itself in the past few decades, including hardware and software. Afterwards, the development of computer science mainly focuses on the application. However, as for computer science itself, the main research object provided by these applications is data. Although computer science always pays much attention to the study of data, the status of data will greatly improve in the future.

Statistics, as a subject of researching data, is also one of the most central parts of data science. However, within the framework of data science, the development of statistics will be greatly impacted from at least two aspects. One is that the data model will jump out of the traditional framework of statistical model. More general mathematical concepts, such as topology, geometry and random fields, will play an important role in data analysis. Another is that algorithms and distributed computation will become part of the central research topics.

1.2.1. *The Impact on Traditional Disciplines*

We have two examples, one of which is sociology. As a branch of social science, sociology is always a data-based discipline, whose basic data is from the national and social levels to familial and personal levels. From this perspective, the relation between sociology and data is not a new phenomenon. But even so, the rise of data science still has a huge impact on sociological research, which is at least reflected in the following aspects.

Firstly, the emergence of networks and the study of network science have brought a new research level, the mesoscopic level. It not only provides new directions but also new practical values for sociological research, such as information dissemination, advertising placement, and hotspot analysis.

Secondly, new data sources and new methods of data analysis will further quantify the research of sociology and avoid experience-based research.

Thirdly, more rigorous and systematic scientific methods are introduced into sociological research, such as the methods of data collection. The home visit survey launched by the China Social Survey Center of Peking University is a good example: they not only paid attention to the results in the process of this survey, but also recorded the data during the investigation. Such a rigorous scientific method will certainly make a great influence on sociological research.

In the past, sociology was not counted as a technology-based or utility-based subject for most people. However, with the quantification of this subject, people have changed their mind. The near future will witness the great influence made by sociological study in product marketing, information spreading and public opinion warning.

Linguistics is the second example. Linguistics, just like sociology, is another subject that was historically far from utility and technology. Yet in recent years, the rapid development of machine translation (MT), natural language processing (NLP), speech recognition, text analysis, etc., provides a great opportunity for the practical application of this subject. Noticeably, the validity of processing method based on probabilistic model largely exceeds that of processing method based on grammar.

Professor Chomsky, the founder of contemporary linguistics, expressed his opinions on this problem on the 150th anniversary commemoration of Massachusetts Institute of Technology (MIT). He argued that there were some limits on the success of probabilistic model, and its success was only in the circle of approximating the un-analyzed data. What he meant was that the probabilistic model was just a technological success, which could not be regarded as traditional scientific success for it did not bring any new recognition to traditional linguistic problems like the grammar problem. Apparently, Chomsky is partly right, since he only considered linguistics instead of the basic sociological functions of language. The social influence of the latter is much larger than that of the former. However, we should keep in mind that the current success of probabilistic model does not mean all the linguistic study is useless. It is time for linguists to admit the effectiveness of probabilistic model.

1.2.2. *The Birth of a New Discipline: Computational Advertising*

There has been a long history of advertising, yet it cannot be considered as science. In China, in particular, the typical advertisements are commonly seen on media like TV for marketing products. In recent years, as search engines like Baidu took commercial advertising as the main strategy to make a profit, a new discipline — computational advertising emerged.

The main problem computational advertising deals with is how to deliver a targeted advertising. There are two most basic indicators in online advertising: click-through rate and conversion rate. The click-through rate is the probability of one advertisement being clicked. The conversion rate is the probability of a commodity being sold after the relevant advertisement is clicked. Since the latter is more difficult to estimate, online advertising often uses the click-through rate as a major indicator. This requires us to predict the click-through rate of different ads based on the information provided by the users, such as the keywords they input in the searching engine. This is a basic problem of computational advertising. The main idea to solve this problem is to construct a utility function to estimate the users' levels of interest in different ads.

At present, the course of computational advertising is offered at some top universities such as Stanford University and the University of California, Berkeley. Being one of the mathematical research institutes affiliated with the National Foundation of the United States, the Statistics and Applied Mathematics Institute in North Carolina also held a workshop on computational advertising.

1.3. Impact on Scientific Research

Big data has provided unprecedented opportunities and challenges for the development of science and education. It will bring great changes to the existing scientific research and educational system and exert great influence on the relationship between science and industry as well as between science and society. To sum up, the impact of big data will mainly come from the following aspects.

First of all, data science will become an important part of scientific research, and gradually acquire an equal status with the natural sciences, including physics, chemistry and biological science. The future scientific research and education system should consist of two main approaches: One is based on the fundamental principle. Most of the present disciplines, for example, physics, chemistry and mechanical engineering as well as biological science, materials science, astrophysics and geoscience, are developed along this main route. The other is based on data. It includes statistics, data mining, machine learning, bioinformatics, astroinformatics and many social sciences. It also includes some emerging disciplines, such as Computational Advertising. The rise of data science will greatly promote the development of many social science disciplines in the direction of quantification and gradually break them out of the empirical mode.

Secondly, the links between scientific research and markets and industries will become tighter, and the cycle from the development of basic principles to industrialization will be greatly. It only took a few years for Google to emerge from an abstract algorithmic breakthrough to becoming a successful search engine company in the market. Such examples are not rare in the data science and information industry. However, it often requires a long time from breakthroughs of basic principles to technology or industry in traditional natural science.

Finally, society, including the Internet, social networks, public transportation and smart cities, serves as one of the main data sources. So, it is fair to say that research on data science has a close relationship both with our daily life and society. Google and Baidu's web search algorithms, for instance, have posed great impact on our daily life. Therefore, the needs of people and society will become one of the major sources of research problems in data science.

The most important part of scientific research is to propose forward-looking research questions. It is quite difficult to propose forward-looking research problems for many disciplines which are facing limitations from practical applications. Data science is not the case. We have already been facing the most challenging and forward-looking problems because of China's huge population, culture, character, historical background and the need of social development. The key is whether we can solve these problems in a forward-looking way. If this is done well, we naturally come to the forefront of the world in the field of data science.

1.4. The Curricula of Data Science

Data science largely emphasizes on the cultivation of big data talent with interdisciplinary abilities. This major focuses on training students with the following three qualities: the first quality is at the theoretical level, mainly about the understanding and application of models; the second is at the practical level, mainly about the ability to deal with actual data; and the third is at the application level, mainly about the ability to adopt big data methods to solve specific industrial application problems. Educating such students requires close coordination among disciplines such as mathematics, statistics and computer science, as well as cooperation with industrial communities or other departments that possess data.

The curricula of data science should include the following aspects:

1. Basic knowledge of mathematics, including calculus, linear algebra, probability theory and calculation methods;
2. Basic knowledge of computer science, such as computer languages, databases, distributed systems, data mining and data visualization;

3. Basic knowledge of algorithms, including numeric algebra, function approximation, optimization, Monte Carlo method, network algorithm, computational geometry, etc.;
4. Basic methods of data analysis;
5. Professional courses, such as statistical learning, machine learning, etc.;
6. Application courses and other professional courses, such as health and medical big data, transportation data analysis, financial data analysis, bioinformatics and astroinformatics.

Part $1 \sim 2$ are professional basic courses, Part $3 \sim 4$ are professional core courses, and Part $5 \sim 6$ are professional elective courses. The setting of professional courses can also include the cooperation with the industrial community to meet the ever-changing practical needs. Cooperation with business circles is also more conducive to the supply of suitable talent to the industry.

1.5. Contents

This book is devoted to a systematic introduction of data science's basic contents, including data preprocessing, basic methods of data analysis, solutions to special problems (text analysis, for example), deep learning and distributed system. Data science is a discipline that must be closely integrated in theory and practice. For that reason, Beijing Institute of Big Data Research has established the iDataCourse platform to provide students with an environment for practical training. Most chapters of this book also cover case studies.

Given a dataset, we usually divide it into two parts: one part is for training models, named training set, the other part is for evaluating models, named test set. Different models often have different requirements of data type, data format and data distribution. Therefore, after getting a real dataset, data preprocessing is needed. This book will introduce some common data preprocessing methods in Chapter 2, including feature transformation, missing value handling, data standardization, feature discretization and outlier detection.

The core tasks of data analysis fall into two main categories: supervised learning and unsupervised learning. The former requires that samples in the dataset carry an output label, with the aim of finding out an optimum mapping from samples to label. Typical supervised learning contains regression and classification, of which the former's label is continuous, and the latter discrete. Regression models will be discussed in Chapter 3, including linear regression, regularized linear regression and nonlinear regression. And we will offer an in-depth discussion in Chapter 4 of typical classification models including logistic regression, K-nearest neighbor, decision trees, Naive Bayes and support vector machine (SVM). Ensemble is able to learn multiple base models from one training set to improve the model prediction accuracy. Chapter 5 will introduce two kinds of ensemble approaches: bagging and boosting and their representative models.

Supervised learning can only deal with data with label, while there's no such requirement for unsupervised learning, which depicts some sort of statistical properties of data according to our interest. The typical tasks of unsupervised learning include clustering and association rule mining. This book will introduce clustering models in Chapter 6, including K-means, hierarchical clustering, spectral clustering and others. Association rule mining is another kind of unsupervised learning that extracts rules from data and will be introduced in Chapter 7.

After a model is built by applying data, whether the model's performance is "good" enough or not has to do with how to evaluate the model. And model evaluation will be discussed in Appendix E, which includes dataset partition method and model evaluation metrics.

There's always noise in the training set. Overfitting is prone to occur when the model parameters are too large. Overfitting means that the model fits the training data well, while performs poorly on unknown data (for example, the test data). Overfitting is a common problem in data analysis.

Dimensionality reduction and feature selection are the two methods of reducing data dimensions. Dimensionality reduction will be introduced in Chapter 8. In Chapter 9, we will introduce some basic feature selection methods.

Chapter 10 introduces the EM algorithm, which is widely applied in model optimization and its typical case is Gaussian mixture model (GMM).

Probabilistic graphical model uses graphs to provide an intuitive and concise framework to describe dependence among features so as to represent data distribution. In Chapter 11, typical probabilistic graphical models including hidden Markov model (HMM) and conditional random fields (CRF) will be introduced.

Chapter 12 introduces text analysis, including text data representation, topic models and sentiment analysis.

Chapter 13 introduces network data. Network data can be represented as a graph, where each node represents the object and each edge represents the relationship between the objects. For example, if the web pages on the Internet are regarded as nodes, the links between them are considered as edges, the whole Internet can be represented as a large directed graph. Graph and network analysis focus on how to discover useful information from the above graph, such as how to assess the importance of a given node and how to identify community structures.

Chapter 14 introduces deep learning, which is the most popular technique in both artificial intelligence and data science at present. This framework allows computers to learn from experience and comprehend complicated data according to hierarchical concept system, including image, audio and natural language. Each concept is defined by its relationship with some relatively simple concepts. Hierarchical concepts enable computers to learn complicated concepts by constructing relatively simple concepts, so as to complete the tasks that are difficult for previous approaches. This chapter covers the basic knowledge of deep learning, the method for training a deep learning model and other widely used model structures.

In Chapter 15, we will discuss the fundamental theory and architecture of Hadoop, Spark and other well-known distributed systems and describe how to use these platforms for large-scale data processing and model building.

Chapter 2

Data Preprocessing

In real life, we obtain data from multiple channels. Examples include traditional questionnaires (surveys), web crawlers or relational databases. Data preprocessing is usually required before data analysis. For example, some respondents in the questionnaire may choose not to answer specific questions which results in missing data. Data downloaded directly from a web page includes both web page structure and web page content, so these semi-structured data are required to be structured. Data often need to be encoded in relational databases. For example, the "gender" feature values "male" and "female" are encoded as "M" and "F", respectively. When "gender" is used as a feature in the model, its values need to be encoded into some types of digital data. Preprocessing is a critical step for special types of data such as text.[a] For example, in the field of opinion analysis (also known as sentiment analysis), one of the core tasks is to judge the emotional orientation of text. For this reason, unstructured text data needs to undergo a series of preprocessing such as word segmentation, part-of-speech tagging (POS tagging), emotional word extraction (extraction of aspect and opinion terms), and structured representation.

In addition, noise may be introduced in the data during the collection and transformation steps, so outlier detection methods are needed. In some models that require distance calculations, such as

[a]See details in Section 12.1 for the preprocessing of text data.

K-means and support vector machine (SVM),[b] the orders of magnitude affect the performance of the model. Therefore, the data must be standardized before the model is trained.

In this chapter, we first introduce several typical data preprocessing methods, including feature encoding, missing value processing, data standardization, data discretization, and outlier detection. Then we summarize and introduce some other data preprocessing methods.

2.1. Feature Encoding

In practice, the original datasets may contain non-numeric features. For example, customer's profile data often include discrete features such as gender, occupation, income level, education level, and driving car brand. These discrete features are often stored in a string format, while data analysis models usually require numeric input features. Therefore, it needs to be digitally encoded, and the discrete values are represented using numbers. Simple digital encoding uses numbers to represent specific values of a feature, and a feature is still a feature after being digitally encoded. For example, for the "gender" feature, 1 and 0 are used to represent "male" and "female", respectively. In some cases, we need to use special encoding methods to represent a feature by multiple features, such as One-Hot encoding.

2.1.1. *Numeric Encoding*

A simple numeric encoding method is to assign each value of the feature a nonnegative integer. For ordinal features, performing numeric encoding from small to large values of the feature ensures that the encoded data retains the original order relationship. For example, the feature "income level" $=$ {poverty, low income, well-off, middle income, wealthy} can be converted into "income level" $=$ $\{0, 1, 2, 3, 4\}$.

[b]K-means and SVM will be introduced in Section 6.1 and Section 4.5, respectively.

However, for the nominal features, the above encoding method may cause some problems. For example, "car brand" = {Land Rover, Geely, Audi, Volkswagen, Mercedes-Benz}, can be converted into "car brand" = $\{0, 1, 2, 3, 4\}$. When using the encoded data for analysis, it is equivalent to introducing an order relationship to the "car brand" feature. This may lead to subsequent incorrect modeling and analysis results. For example, the distance between "Geely" and "Land Rover" is smaller than the distance between "Mercedes-Benz" and "Land Rover" because we encode "Land Rover" as 0, "Geely" as 1, "Mercedes-Benz" as 4. In order to avoid the above misleading results, for the discrete features (especially nominal features), another encoding method can be used: One-Hot encoding.

2.1.2. *One-Hot Encoding*

The One-Hot encoding converts a K-valued discrete feature into K binary feature (values of 0 or 1). For example, we discussed the feature "car brand" = Land Rover, Geely, Audi, Volkswagen, Mercedes-Benz in the previous section. All 5 different values are included. We can encode them as five features f_1, f_2, f_3, f_4 and f_5. The five features correspond to the values of the original car brand. The values of the converted features are shown in Table 2.1.

One-Hot encoding does not introduce ordering relationships to the nominal feature values. For example, the distance between "Geely" ($[0, 1, 0, 0, 0]$) and "Land Rover" ($[1, 0, 0, 0, 0]$) is no longer smaller than that between "Mercedes-Benz" ($[0, 0, 0, 0, 1]$) and "Land Rover"

Table 2.1 One-Hot encoding.

Original feature value	f_1	f_2	f_3	f_4	f_5
Land Rover	1	0	0	0	0
Geely	0	1	0	0	0
Audi	0	0	1	0	0
Volkswagen	0	0	0	1	0
Mercedes-Benz	0	0	0	0	1

([1, 0, 0, 0, 0]). In fact, after the One-Hot encoding, the values of different original features have the same distance. In a linear regression model,[c] One-Hot encoding on nominal features is usually better than numeric encoding. One-Hot produces very good results on classification models containing discrete features.[d]

One-Hot encoding has its drawbacks. First of all, it will significantly increase the feature dimension. Assuming there are 10 discrete features containing 100 values each, then the number of features after One-Hot encoding will become 1000. Second, it will increase the correlation between the converted features. Observe Table 2.1 and we can find that the five encoded features have the following linear relationship:

$$f_1 + f_2 + f_3 + f_4 + f_5 = 1. \tag{2.1}$$

A linear relationship between features will affect the performance of models such as linear regression, so we need to make some changes to the One-Hot encoding. For a discrete feature containing K values, we can convert it to $K - 1$ binary features. This encoding method is called "dummy encoding". For example, the feature "car brand" = {Land Rover, Geely, Audi, Volkswagen, Mercedes-Benz} can be encoded as 4 binary features. The converted feature values are shown in Table 2.2.

Table 2.2 Dummy encoding.

Original feature value	f_1	f_2	f_3	f_4
Land Rover	1	0	0	0
Geely	0	1	0	0
Audi	0	0	1	0
Volkswagen	0	0	0	1
Mercedes-Benz	0	0	0	0

[c]We will introduce the linear regression model in Section 3.1.

[d]We will discuss classification models in Chapter 4.

2.2. Missing Value Processing

Missing data refer to the incompleteness of data due to certain reasons during data collection, transformation, and processing. Table 2.3 shows a student information dataset. Each sample contains three basic features: "year of enrollment", "gender" and "age". The remaining features are students' preference for an activity such as "soccer", "basketball", "shopping" and "makeup". There are missing values in the "gender" and "age" features.

There are several reasons why a dataset contains missing values. First, the data collection process may cause missing data. For example, in the process of a statistical survey, the respondents may be unwilling to, unable to or refuse to respond to part of the survey, resulting in missing values from the survey data; secondly, when the data is transmitted through the network and other channels, data loss or error may occur, resulting in missing data. In addition, missing values may also be introduced during data integration. For example,

Table 2.3 Student information dataset.

Year of enrollment	Gender	Age	Soccer	Basketball	⋯	Shopping	Makeup
2012	M	18	0	0	⋯	0	0
2012	F	18	0	1	⋯	0	0
2010	M	20	0	1	⋯	0	0
2012	F	18	0	0	⋯	0	2
2011	F	18	0	0	⋯	1	1
2012	F		0	0	⋯	1	0
2012	F	18	0	0	⋯	0	0
2011	M	18	2	0	⋯	0	0
2011	F	19	0	0	⋯	0	0
2012		18	0	0	⋯	1	0
2012	F	18	0	0	⋯	0	0
2011		19	0	1	⋯	0	0
2012	F	18	0	0	⋯	0	0
2012	F		0	0	⋯	0	2
2012	F	18	0	0	⋯	0	1

in the financial area, suppose we need to assess the customer's value through the customer's annual credit card expenses. If a customer does not have a credit card, then the customer's annual credit card expenses will have a missing value. The student information data set shown in Table 2.3 was collected through a social network platform in which age and gender are voluntarily filled out by the user. The user may refuse to fill in this information, resulting in missing values in the dataset.

Missing values can cause problems in the statistics of the data. Suppose we need to calculate the students' average age. Due to the missing values of the "age" feature, it is difficult to calculate the average value. In addition to simple statistics, some data analysis models cannot directly process datasets with missing values. Therefore, if you want to perform further statistical analysis of the dataset, you first need to deal with the missing values.

Then, how to deal with dataset with missing values? In this section, we will introduce common methods. These methods can be divided into two categories: the first is to delete samples or features that have missing values; the second is to impute missing values, including mean imputation, stochastic imputation, and model-based imputation.

2.2.1. *Deletion Method*

The deletion method obtains a complete subset of data by deleting data containing missing values. The deletion can be performed by sample or feature:

1. **Deleting samples:** We can delete samples containing missing values. This method is suitable when some samples have several features with missing values and small proportion of samples have missing values. When there are few samples in the dataset, deleting the samples affect the performance of the data analysis. Therefore, when the proportion of missing samples is large, deleting samples may result in a significant loss of information;

2. **Deleting features:** When there are many missing values for a feature, and the feature has little effect on the data analysis objective, the feature can be deleted.

The deletion method is simple and easy to apply, but it has great limitations. While ensuring the integrity of the information by deleting the data, it may lose much of the information in the deleted data. In some cases, the data collection costs are high and the missing values cannot be avoided. The deletion method may waste a lot of resources. For example, in the field of health care, we want to assess surgical risk using the patient's preoperative diagnostic indicators. For a patient, it is difficult to ensure that every preoperative diagnostic indicator is available before surgery. For some feature indicators, multiple tests are required, and it is also difficult to ensure that each patient conducts every test on time.

2.2.2. *Mean Imputation*

For a feature with missing values, the mean imputation method first calculates the average or mode of non-missing values and then uses the average or mode to replace the missing values. For continuous features, the average is usually used for imputation; while for discrete features, the mode is used. In the dataset shown in Table 2.3, since the mode of the "gender" feature is "F", we use "F" to impute the missing values. For the "age" feature, the average of non-missing values is 18.3. Then we can use 18.3 to impute the missing values for.

Mean imputation will make the data over-concentrated on the average or the mode, resulting in an underestimation of the variance of the feature. In addition, the mean imputation method greatly lowers the correlation between features. In practice, the dataset can be divided into multiple subsets according to certain auxiliary features, and then the mean imputation method is applied for each subset. For example, in the student information dataset, the "year of *enrollment*" feature records the year in which the student enrolled. We can first divide the data into subsets according to the year of enrollment, and then use the average "age" in each subset to fill in the missing values.

2.2.3. *Stochastic Imputation*

Stochastic imputation adds random terms to the mean imputation method and overcomes the problem of over-concentration of missing values by increasing the randomness of filled values. Stochastic imputation methods include the Bayesian bootstrap method and the approximate Bayesian bootstrap method.

Assume that the dataset has n samples, and there are k non-missing values and $(n-k)$ missing values for a feature f. The Bayesian bootstrap method imputes the missing values in two steps: in the first step, it randomly draws $k-1$ random numbers from the uniform distribution of $U(0,1)$, sorting the numbers in ascending order and denoting them by $a_1, a_2, \ldots, a_{k-1}$; in the second step, it fills in the $(n-k)$ missing values by drawing numbers from the values $\{f_1, f_2, \ldots, f_k\}$ with probabilities $\{a_1, a_2 - a_1, \ldots, 1 - a_{k-1}\}$.

The approximate Bayesian bootstrap method first creates a new set F of size k from the k values drawn with replacement among the non-missing values $\{f_1, f_2, \ldots, f_k\}$. Then to impute each of the $n-k$ missing values, one value is randomly drawn from F.

2.2.4. *Model-based Imputation*

The model-based method takes the feature f with missing values as a prediction objective. The remaining features or their subsets are used as input features. The training set is constructed with the non-missing values of the feature f, and a classification or regression model is trained. The trained model is then used to predict missing values for the feature f. For example, in the student information dataset, we found that a students' interest can be a good indication of her gender. Therefore, we can use interest words as input features and use the "gender" feature as a prediction objective to build a classification model for gender prediction, and then use the model to impute missing values for the "gender" feature.

Model evaluation is needed for the imputation model. If the model is built with poor prediction performance, this method is not suitable. In addition, the model-based imputation method will increase the correlation between features.

2.2.5. *Other Missing Values Imputation Methods*

There are many different methods to impute missing data. In addition to the methods described above, dummy variable method and EM algorithm can also be applied. For a discrete feature, the missing values can be treated as a single value. This method is called the dummy variable method. For example, in the student information dataset, the missing values of the "gender" feature are given a special value "unknown", indicating that the gender is unknown. Then the "gender" feature is considered to contain three different values of "F", "M" and "unknown". The EM algorithm is a parameterized algorithm that can use incomplete information to implement a probabilistic model. The reader can refer to Chapter 10 for a detailed introduction of the EM algorithm. The EM algorithm can be used to impute missing values, and missing features are treated as implicit variables.

2.3. Data Standardization

In the data analysis and modeling process, many machine learning algorithms require the input features to be standardized. For example, in many learning algorithms (such as the RBF kernel of Support Vector Machines, the l_1 and l_2 regularizes of the linear model, etc.), the objective function often assumes that all features are centered at zero and have variances of the same order. If a feature has a variance with a larger order of magnitude, it might dominate the objective function. This may lead to a model with poor performance and failing to learn from other features.

In some scenarios, we need to calculate the similarity between samples. If the orders of magnitude between the features of the sample vary too much, the results will be affected by the features with larger orders of magnitude, resulting in a bias of the sample similarity.

Therefore, data standardization is important in the data analysis process. Common data standardization methods include Z-score standardization, Min-Max standardization, decimal scaling standardization, and Logistic standardization.

2.3.1. *Z-score Standardization*

Z-score standardization is the most commonly used standardization method. It processes each sample in the dataset and the resulting feature has a fixed mean and standard deviation.

Assume that the set of values for the feature f is $\{f_1, f_2, \ldots, f_n\}$, then the feature value f_i standardized by Z-score method is f_i' with

$$f_i' = \frac{f_i - \mu}{\sigma}, \tag{2.2}$$

where $\mu = \frac{1}{n}\sum_{i=1}^{n} f_i$ is the mean of the feature f, $\sigma = \sqrt{\frac{1}{n}\sum_{i=1}^{n}(f_i - \mu)^2}$ is its standard deviation.

Intuitively, the Z-score standardized feature can reflect the distance between each value and the feature mean, thus providing an understanding of the overall distribution of the feature. The mean of the standardized value of a sample is interpreted as the standard deviation distance of the sample from the mean.

When there are outliers in the data, the standard deviation in the Z-score standardization method can be replaced by the average absolute deviation. In this case, the average absolute deviation of the feature f is

$$s = \frac{1}{n}\sum_{i=1}^{n} |f_i - \mu|. \tag{2.3}$$

The new Z-score standardization formula is

$$f_i' = \frac{f_i - \mu}{s}. \tag{2.4}$$

The Z-score standardization can be applied when the maximum or minimum value of the feature is unknown and the sample distribution is quite wide.

2.3.2. *Min-Max Standardization*

Min-Max standardization linearly transforms a feature so that the converted feature values are within the interval $[0, 1]$. Assuming that the set of values for the feature f is $\{f_1, f_2, \ldots, f_n\}$, the feature value f_i after Min-Max standardization is f_i' with

$$f_i' = \frac{f_i - f_{min}}{f_{max} - f_{min}}, \tag{2.5}$$

where f_{min} is the minimum value of the feature f, and f_{max} is the maximum value of the feature f. Therefore, if we want the standardized feature to take values within the interval $[-1, 1]$, we can slightly modify formula (2.5) to

$$f_i' = \frac{2(f_i - f_{min})}{f_{max} - f_{min}} - 1. \tag{2.6}$$

Further, if we want to linearly map the feature f to any interval $[a, b]$, the Min-Max standardization formula is

$$f_i' = \frac{(b - a)(f_i - f_{min})}{f_{max} - f_{min}} + a. \tag{2.7}$$

Min-Max standardization is applicable when a simple linear mapping of feature values into an interval is required. The disadvantage is that the maximum or minimum value of the feature changes when new data are added to the dataset. In this case, we need to calculate the new minimum and maximum values and re-standardize the dataset. In addition, since Min-Max standardization requires the calculation of the minimum and maximum when the data have outliers, the standardization performance may be poor.

2.3.3. *Decimal Scaling Standardization*

The decimal scaling method standardizes the data by moving the decimal part of the feature values so that the absolute standardized feature values are always less than 1. The number of decimal places moved depends on the maximum absolute value of the feature values.

Assume that the set of values for the feature f is $\{f_1, f_2, \ldots, f_n\}$, then the feature value f_i is standardized by the decimal scaling to f_i' with

$$f_i' = \frac{f_i}{10^j}, \tag{2.8}$$

where j is the smallest integer that satisfies $\max\{|f_1'|, |f_2'|, \ldots, |f_n'|\} < 1$. For example, a feature takes values in the interval

$[-3075, 2187]$. The maximum absolute feature value is 3075, and then the value of j is 4.

The decimal scaling standardization method applies to features with quite wide value distribution, especially when feature values are distributed over multiple orders of magnitude. This method is simple and practical. Known the movement of decimal points, it is easy to restore the original feature values from the standardized ones.

However, the decimal scaling standardization method also has many weaknesses. If the feature value distribution is concentrated on certain orders of magnitude, the standardized feature will also be concentrated near some values, which is not convenient for sample discrimination in the subsequent data analysis. Similar to the Min-Max standardization method, when a new sample is added, the decimal scaling method needs to re-determine the number of decimal digits to move. In addition, decimal scaling standardization is also affected by outliers.

2.3.4. *Logistic Standardization*

Logistic standardization uses the Logistic function to map feature values to the interval $[0, 1]$. The Logistic function was invented by Pierre *Francois* Verhulst to study population growth. It is in the form of

$$\sigma(x) = \frac{1}{1 + e^{-x}}. \tag{2.9}$$

Figure 2.1 shows a graph of the function. The Logistic function maps continuously the real-valued data to the interval $[0, 1]$. We can use this function to standardize features. Suppose that the set of the feature f values is $\{f_1, f_2, \ldots, f_n\}$, then the feature value f_i after Logistic standardization is f_i' with

$$f_i' = \frac{1}{1 + e^{-f_i}}. \tag{2.10}$$

The Logistic standardization method is applicable to cases where the distribution of feature values is relatively concentrated on both sides of zero. If the value distribution is quite wide and most values are far away from 0, the standardized feature values will be clustered around 0 or 1, causing changes in the distribution of the original

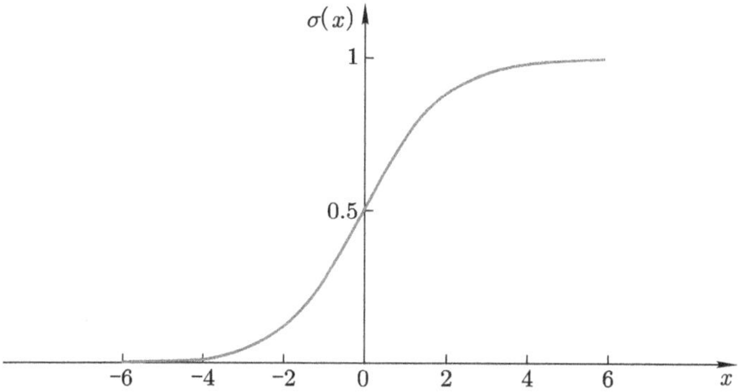

Figure 2.1 The curve of Logistic function.

feature values and the relationship among the values. We should first analyze the distribution of the original feature before performing *Logistic standardization*.

2.3.5. *Comparison of Standardization Methods*

It should be pointed out that the standardization process will change the values of the original features. Therefore, in practice, the parameters of the applied standardization method should be kept in order to standardize the subsequent data.

Z-score standardization does not require the computation of the minimum and maximum values. The mean and standard deviation of the feature need to be recorded, which is applicable when the feature maximum or minimum value is unknown and the sample distribution is relatively fragmented.

Min-Max standardization linearly transforms the data and preserves the relationship among the original data. However, when the maximum or minimum changes, it is necessary to recalculate the value for each sample. This standardization method is applicable when the relationship among the original data needs to be preserved, and the maximum and minimum value are fixed.

Decimal scaling standardization is simple and practical, and it is easy to restore the original data. However, when the maximum absolute value of the original data changes, it is necessary to recalculate the value for each sample. This standardization method is suitable for discrete feature, especially for feature distributed over multiple orders of magnitude.

Logistic standardization is simple and easy to use. It standardizes the data through a single mapping function. It does not work well for feature that is discretely distributed and far from zero. This standardization method is suitable when the feature values are relatively concentrated and evenly distributed on both sides of zero.

2.4. Data Discretization

Features can be classified into continuous and discrete based on the values they take. A continuous feature takes values in real numbers and is expressed in floating point numbers such as the sales of goods, the temperature of the weather, and the economic growth rate. A discrete feature uses concepts or symbols to describe the qualitative feature. For example, one can use "unmarried", "married" or "widowed" to describe a person's marital status. You can also use numbers to encode symbols, such as using a set of numbers $\{1, 2, 3, 4\}$ to quantify people's preference for things.

However, the real world is complex. The data sets that we encounter often contain both continuous and discrete features. In real applications, algorithms usually require the dataset to have a specific feature type. For example, the association rule mining algorithm, which will be introduced in Chapter 7 of this book, can only handle data with Boolean input features. The ID3 decision tree algorithm, which we will introduce in Section 4.3, can only handle data with discrete input features. Although most data science algorithms are capable of dealing with continuous features, studies have shown that Some models perform better through data discretization. After the features of the dataset have been discretized, the accuracy of the naive Bayes classification algorithm is 10% higher than that without discretization on average.[4]

Therefore, in order to improve the accuracy of the algorithm, discretizing the features of the data has played an important role in data preprocessing and feature engineering. The process of converting continuous features into discrete features is called data discretization. The process of data discretization is to divide the value range of the continuous feature into several bins, and then use the bins to

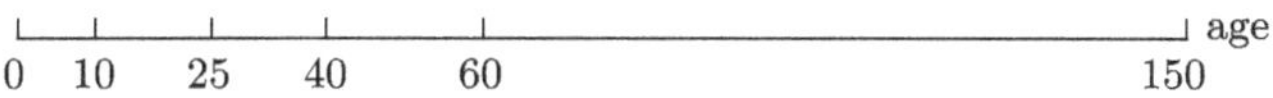

Figure 2.2 Data discretization.

represent the feature values that fall within the bin. The segmentation points between bins are called cut points, and the number of bins segmented by cut points is called arity.

Here is a simple example. Assuming that the "age" continuous feature needs to be divided into k bins, then $(k-1)$ cut points are needed. As shown in Figure 2.2, the "age" feature has values ranging between $[0, 150]$. It is converted into the following five discrete bins by four cut points 10, 25, 40, and 60:

$$0 \leq Age < 10, \qquad 40 \leq Age < 60,$$
$$10 \leq Age < 25, \qquad 60 \leq Age < 150.$$
$$25 \leq Age < 40,$$

The goal of data discretization is to reduce the arity while minimizing the loss of data information. Therefore, how to select the cut points and generate reasonable bins also becomes the key to determining the success or failure of data discretization.

The discretization methods can be divided into two categories: supervised discretization and unsupervised discretization. Unsupervised discretization does not refer to the predicted feature y, and directly proceeds to the data discretization according to the distribution characteristics of the feature. Supervised discretization uses the predicted feature y in the reference dataset to discretize the continuous features. Unsupervised discretization methods include equal-width discretization, equal-frequency discretization, and clustering-based discretization. The supervised discretization methods include information gain discretization, chi-squared discretization, and CAIM discretization.

Data discretization methods generally take the following four steps:

1. **Sort feature:** Sort the values of continuous features in ascending or descending order, this can reduce the computational cost of discretization.

2. **Select cut points:** According to the given evaluation criteria, make a reasonable choice of cut points. Commonly used evaluation criteria are based on information gain or statistics.
3. **Split or merge bins:** Based on the selected cut points, the existing bins are split or merged into new bins. In the process of discretization, the size of the cut points set will change.
4. Repeat steps 1–3 for the D newly generated bins until the termination condition is satisfied. We can set the arity k in advance as a simple termination criterion. One can also set complex decision functions.

After A preliminary understanding of data discretization, we will discretization in detail the discretization methods mentioned earlier.

2.4.1. *Equal-width Discretization*

Equal-width discretization is one of the earliest methods of data discretization.[5] According to the values of the continuous feature, the discretization method divides the feature into k uniform intervals. Then, the values of the feature are divided into corresponding intervals to complete data discretization. We use f to represent the continuous feature that needs to be discretized. We calculate the width of the interval using the feature's maximum value f_{max} and the minimum f_{min}:

$$w = \frac{f_{max} - f_{min}}{k}. \tag{2.11}$$

According to the calculated interval width, and the maximum and minimum values of the feature f, we can find $(k-1)$ cut points, thus completing the discretization of the data.

Equal-width discretization is a simple method of dividing continuous features. For the age feature data distributed over the interval $[0, 90]$, if equal-width discretization is used, we first fix the number of discretization bins to 5, and then obtain the interval width of the bins to $\frac{90-0}{5} = 18$. The four cut points are 18, 36, 54, and 72, respectively. Thus, the discretized feature bins are

$$\begin{aligned} &0 \leq Age < 18, && 54 \leq Age < 72,\\ &18 \leq Age < 36, && 72 \leq Age < 90.\\ &36 \leq Age < 54, && \end{aligned}$$

Table 2.4 Sensibility of equal-width discretization to an outlier.

	Cut points
Normal values	18, 36, 54, 72
Outlier	30, 60, 90, 120

Equal-width discretization demands high quality of input data and is not suitable for features with extreme outliers. For the "age" feature, the normal values should be in the $[0, 90]$ interval. However, if there is an outlier (such as 150), the value of f_{max} could end up being too large, which might cause a severe shift in the cut points, as shown in Table 2.4.

2.4.2. *Equal-frequency Discretization*

When the distribution of feature values is non-uniform (for example, there is an outlier), after the equal-width discretization, the sample size in the bins may be largely uneven. In order to solve this problem, we no longer require the width of the bins to be equal at all times. Instead, we try to equalize the sample size in each interval after discretization. This discretization method is called equal-frequency discretization.[5] According to the total number of continuous feature values n, it is still divided into k bins so that each interval contains $\frac{n}{k}$ values, and then the range of data values in each interval is the corresponding data discretization interval.

Suppose we have 20 age samples, distributed in two ranges of $(0, 10)$ and $(40, 50)$. Now, we apply equal-frequency discretization to this feature. The goal is for each new bin to contain 4 samples, and the result of the discretization is shown in Table 2.5. From Table 2.5, we can see that the age is divided into five bins with different widths and the cut points are 4, 8, 42 and 46. This can cause a problem. That is, samples with close values may be divided into different intervals, for example, 8 and 9, 42 and 43 are respectively assigned to different adjacent intervals.

Equal-width discretization and equal-frequency discretization are the two simplest methods of unsupervised data discretization. Equal-width discretization tends to divide the data unevenly into discrete

Table 2.5 Equal-frequency discretization.

Samples	Interval	Width
1, 2, 3, 4	$[1, 4]$	4
5, 6, 7, 8	$[5, 8]$	4
9, 10, 41, 42	$[9, 42]$	34
43, 44, 45, 46	$[43, 46]$	4
47, 48, 49, 50	$[47, 50]$	4

intervals and is rather sensitive to outliers. Equal-frequency discretization divides the data evenly into discrete intervals. Sometimes the same or close samples are divided into different intervals, making the data in adjacent bins have similar characteristics.

2.4.3. *Clustering Based Discretization*

When discretizing continuous features, if similar samples all fall within the same bin, then such division we can retain the information in the original data. Clustering is a process that aims to divide samples into different classes or clusters.[e] The result of clustering is that samples in the same cluster are very similar, and samples in different clusters are very different. Therefore, we can consider using clustering to discretize continuous features.

The discretization method based on clustering analysis mainly includes the following three steps:

1. For a continuous feature that needs to be discretized, clustering algorithms (such as K-means, EM algorithms, etc.) are used to classify samples into corresponding clusters or classes based on the distribution of the feature.
2. Based on the clustering results and specific strategies, we decide whether to further split or merge clusters. With a top-down strategy, the clustering-based algorithm can continue to divide each cluster into smaller subsets. Using a bottom-up strategy, we can merge similar neighbor clusters to get new clusters.
3. After determining the final clusters, determine the cut points and the number of bins.

[e]We will discuss clustering models in Chapter 6 of this book.

In the entire clustering process, we need to determine the number of clusters in advance and describe the distance Motric between the samples.[f] The number of clusters will affect the effectiveness of the clustering algorithm, which in turn affects data discretization.

2.4.4. *Information Gain Discretization*

The discretization method based on information gain is inspired by the evaluation criterion based on information gain when building the decision tree model.[g] When building a decision tree, the algorithm traverses each feature of the dataset and uses them as candidate splitting nodes by calculating the entropy after the split. Then, the algorithm chooses the feature with the smallest entropy, equivalently the largest information gain, as the formal splitting node. When building the decision tree model, the criterion of splitting continuous features using information gain works well in practice, hence, the discretization method based on information gain can be used for discretization of continuous features.

Discretization based on information gain is divided into the following steps:

1. Sort the continuous feature.
2. Consider each value of the feature as a candidate splitting node (or cut point) and calculate the corresponding entropy. Then, the node with the minimum value of entropy is chosen as the formal cut point, and the original interval is divided into two.
3. Recursively process the two new bins obtained in the second step until the category of the feature in each bin is the same.
4. Merge the adjacent bins with 0 class entropy and same feature category, then recalculate the entropy of the new bins.
5. Repeat step 4 until the termination condition is met. The termination condition may be the depth of the decision tree or the number of leaf nodes.

Among many decision tree algorithms, ID3 and C4.5 are the most commonly used algorithms for feature selection and classification

---

[f]Please refer to Appendix D for commonly used distance calculation method.
[g]We will discuss the decision tree model in Section 4.3 of this book.

based on information gain. The core of applying these two decision tree algorithms to discretize continuous features is to build decision tree models for each individual feature. Then the feature is discretized according to the threshold of splitting node in the decision tree models.

2.4.5. *Chi-squared Discretization*

The discretization method based on information gain adopts a top-down splitting strategy. It first takes the feature values as a whole interval, and then gradually divides the large interval into small intervals according to the classification result of the decision tree, until the discretization process is completed. On the contrary, the chi-squared discretization method takes a bottom-up merge strategy. It considers each value of the feature as an individual interval, and then recursively combines the intervals one by one. The criteria for judging similar merging bins are chi-squared statistics. Below, we introduce the concept of chi-squared test, and then describe the discretization based on chi-squared test.

The chi-squared test is a commonly used hypothesis testing method in statistics. It belongs to the category of nonparametric tests and is a method to compare whether there is significant difference between two given populations.[6,7] The χ^2 statistic of the chi-squared test was first proposed by British statistician Karl Pearson in 1900. Its formula is

$$\chi^2 = \sum_{i=1}^{k} \frac{A_i - E_i}{E_i}, \tag{2.12}$$

where A_i is the number of samples that fall into the ith bin, which is the observation frequency. E_i is the corresponding expected frequency. In the case of a large sample size n, the χ^2 statistic follows approximately the chi-squared distribution with a degree of freedom of $k - 1$.

The most commonly used chi-squared discretization method is the ChiMerge method.[8] The ChiMerge method uses the chi-squared test to determine whether or not the adjacent intervals need to be merged. That is, the category of feature values within the interval is independent of the interval. The ChiMerge discretization process is

1. Treat each value of a continuous feature as a single interval and sort the values.
2. Calculate the chi-squared statistics for each pair of adjacent bins. The adjacent bins, which have the minimum χ^2 value or the χ^2 value lower than the threshold, are merged together. Chi-squared statistics are calculated as
$$\chi^2 = \sum_{i=1}^{k}\sum_{j=1}^{C} \frac{(A_{ij} - E_{ij})^2}{E_{ij}}, \tag{2.13}$$
where A_{ij} is the number of samples of category j in the ith bin, k is the number of bins to compare, C is the number of categories, and $E_{ij} = \sum_{j=1}^{C} A_{ij} \frac{\sum_{i=1}^{k} A_{ij}}{n}$ is the expected frequency of A_{ij}.
3. For the new bins, recursively perform step 1 and 2 until the termination condition is satisfied.

ChiMerge discretization can only merge two intervals ($k = 2$) during each iteration. If the sample size in the dataset is large, the computational cost of the algorithm will be high. The ChiMerge method requires setting a significance level before calculating the chi-squared statistics. The number of bins is controlled by setting the threshold.

2.4.6. *Class-attribute Interdependence Maximization*

Class-attribute interdependence maximization (CAIM) is also an entropy-based data discretization method.[9] Unlike ChiMerge, it takes a top-down approach. By selecting the cut point p, the feature values space is divided into two sub-intervals, $f \leq p$ and $f > p$. The criteria to choose a good cut point is the degree of interdependence between class-attribute.

Suppose there is a continuous feature with n values and C classes. Suppose we want to divide the feature into k sub-intervals. The set of sub-intervals is denoted by $D = \{[d_0, d_1], (d_1, d_2], \ldots, (d_{k-1}, d_k]\}$, where d_0 and d_k are the minimum and maximum value of the feature, respectively. Let $n_{i\cdot}$ denote the number of samples belonging

Table 2.6 Two-dimensional CAIM table.

Category	$[d_0, d_1]$	$(d_1, d_2]$	$\dots$	$(d_{k-1}, d_k]$	Category sample size
1	n_{11}	n_{12}	$\dots$	n_{1k}	$n_{1\cdot}$
2	n_{21}	n_{22}	$\dots$	n_{2k}	$n_{2\cdot}$
$\vdots$	$\vdots$	$\vdots$		$\vdots$	$\vdots$
C	n_{C1}	n_{C2}	$\dots$	n_{Ck}	$n_{C\cdot}$
Interval sample size	$n_{\cdot 1}$	$n_{\cdot 2}$	$\dots$	$n_{\cdot K}$	n

to the category i, $n_{\cdot j}$ denote the number of samples in the interval $(d_{j-1}, d_j]$, and n_{ij} denotes the number of samples in the interval $(d_{j-1}, d_j]$ and belonging to the category i, we can obtain a two-dimensional table composed of class-attribute and discretized feature, as shown in Table 2.6. Based on the two-dimensional CAIM table, we can calculate the following quantity to evaluate the current discretization:

$$\text{CAIM} = \frac{1}{n} \sum_{j=1}^{k} \frac{M_j^2}{n_{\cdot j}}, \tag{2.14}$$

where $M_j = \max\{n_{1j}, n_{2j}, \dots, n_{Cj}\}$. The CAIM value range is $(0, 1]$. The larger the value is, the greater the degree of interdependence between classes and discrete intervals, indicating that the current discretization effect is better.

CAIM discretization is divided into the following steps:

1. Sort the feature to be discretized in ascending order to determine the minimum and maximum values d_0 and d_k for the values range. Initialize the partition strategy to $D = \{[d_0, d_k]\}$.
2. Using each value in the interval as a candidate cut point, calculate the CAIM value after the interval dichotomy, and select the point with the highest CAIM value as the cut point, and update D.
3. For each interval in D, repeat the second step until the termination condition is satisfied.

It is worth noting that CAIM only focuses on the relationship between the class with the largest number of samples in the interval for feature, and ignores the information of other classes in the same interval. The number of discretization intervals that CAIM eventually generates is often close to the number of sample classes.

2.4.7. *Summary*

The essence of discretization is to segment the continuous data, so the outliers in the data will be directly classified into corresponding bins, therefore increasing the robustness of the model. After discretization, the values are converted to the bins numbers with a clear meaning. Compared with the original continuous data, the meaning is clearer, which makes the data more interpretable and the model easier to understand and apply. After discretizing the continuous feature, the values taken by the feature are greatly reduced, which not only reduces the demand for dataset storage space but also reduces the actual amount of computation for model training, thereby improving the computational efficiency of model training.

This section only introduces common discretization methods. More detailed descriptions and comparisons of discretization methods and their applications in supervised learning.[10]

2.5. Outliers

Outliers, also known as abnormal values, are values in a dataset that deviate significantly from other data samples. A definition with statistical significance is: "One observation deviates too much from other observations so that it is doubtful that it is produced by a different mechanism".[11]

Why do outliers exist? Outliers are mainly due to natural variations, errors in data measurement and collection, and errors in manual operations. For these reasons we cannot completely avoid them in real life. Therefore, it is necessary to detect whether there are outliers in the dataset before data analysis.

Outlier detection can be used as a step in data preprocessing to provide high-quality data for data analysis. Outlier detection can also be used directly for many applications such as credit fraud detection, telecommunication fraud detection, disease analysis and computer security diagnosis, etc. In this section, we mainly focus on the former

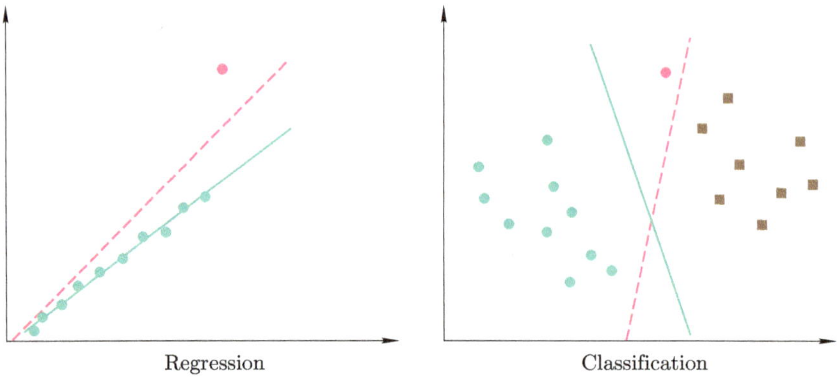

Figure 2.3 Influence of an outlier to data analysis.

one, which discovers noise in the data through discretization and reduces the impact of noise on data analysis.

First, an example is used to illustrate the impact of the presence of outliers on data analysis. As shown in Figure 2.3, the pink circles represent outliers, and the solid green lines and the pink dashed lines represent regression lines and classification decision lines without considering outliers and with consideration of outliers, respectively. The lines in the figure are sensitive to outliers. When there are no outliers, the regression line and the classification decision line can describe the characteristics of the data well. When there are outliers, the slope of the regression line increases, and there is no good fit to the data. The classification decision line also has a large deflection due to the presence of the outlier.

So how to detect outliers in the dataset? The most intuitive idea is to manually determine if there are outliers by visualizing the data (drawing distribution graph, histogram, box plot, etc.). However, this simple approach does not work for high-dimensional data. This section will introduce several commonly used outlier detection methods that can automatically detect outliers from the data.

2.5.1. *Statistics-Based Method*

Statistics-based method assumes that there is only random error in the data. After the corresponding confidence interval is determined based on a given confidence level, the error that exceeds the confidence interval is considered out of the random error range. Then values containing this kind of error are considered outliers. In statistics, the more commonly used detection method is the Pauta criterion. It assumes that the data is normally distributed and the standard deviation is calculated for the data. In the case of a given confidence level, if sample values are 3 times greater than the standard deviation above the mean, we believe that these samples are outliers.

Pauta criterion can only be used when the number of measurements is large. If the number of measurements is very small, the criterion will be invalid. Z-score standardization can be used in outlier detections. The Z-score value represents the distance between the original sample x and the sample mean μ, and this distance is calculated in units of the standard deviation σ:

$$x_{Z\text{-}score} = \frac{x - \mu}{\sigma}. \tag{2.15}$$

After the Z-score values for samples are obtained, a sample satisfying the condition $|x_{Z\text{-}score}| > 3$ is usually considered as an outlier.

2.5.2. *Nearest-neighbors-based Method*

Intuitively, if a sample is an outlier, it will be far away from other samples. We can quantify the outlier degree of a sample. The score is determined by the distance of the sample to its k nearest neighbors, and the value range of the score is $[0, \infty)$. From this we have derived an outlier detection method based on the K-nearest neighbors algorithm[h]:

1. Calculate the distance between each sample and its k-nearest neighbors and place it in set C.
2. Sort all elements in set C in descending order.
3. According to the given distance threshold, select the samples with the distance greater than the given threshold in C as outliers.

Outlier detection based on K-nearest neighbors is simple and effective, and it is commonly used. However, it can only find global

[h]The Section 4.2 in this book discusses K-nearest neighbors algorithm in detail.

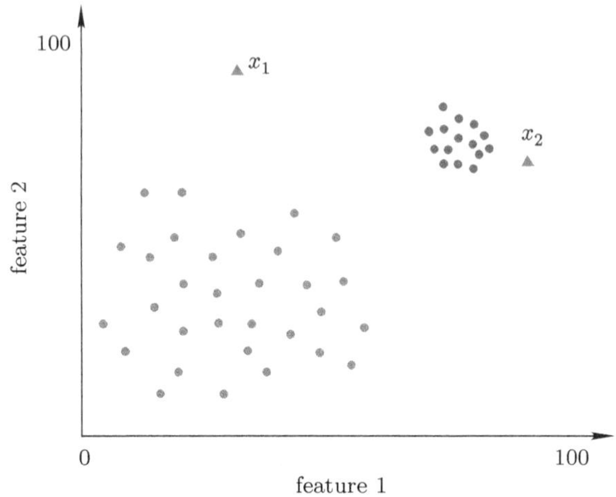

Figure 2.4 Two sample sets with different densities.

outliers and cannot find local outliers. As shown in Figure 2.4, our data mainly include two regions with uneven density distributions. The K-nearest neighbors algorithm can detect the global outlier x_1, but cannot detect the local outlier x_2.

In the following, we introduce an outlier detection method based on the local density, called the local outlier factor (LOF) method.[12] The LOF algorithm calculates a local outlier factor for each sample in the dataset. This factor is calculated from the relative value of the current sample density and its neighbors' density. After calculating the local outlier factor for each sample, we can identify outlier samples directly based on this factor.

First, let's introduce some notations. $d(x_1, x_2)$ represents the distance between sample x_1 and sample x_2. $d_k(x)$ represents the distance between sample x and its k-th nearest neighbor. $N_k(x)$ represents the set of k neighbors for sample x. The reachability distance from sample x_1 to sample x_2 is defined as $rd_k(x_1, x_2) = max\{d_k(x_2), d(x_1, x_2)\}$. Based on the reachability distance, we can define the local reachability density of the sample, which is the inverse of the average reachable distance between the sample and its neighbors:

$$lrd_k(x) = \left(\frac{1}{k} \sum_{y \in N_k(x)} rd_k(x, y)\right)^{-1}, \tag{2.16}$$

Using the local reachability densities of a sample and its neighbors, we can calculate the local outlier factor of the sample:

$$lof_k(x) = \frac{1}{k} \sum_{y \in N_k(x)} \frac{lrd_k(y)}{lrd_k(x)}. \tag{2.17}$$

After calculating the local outlier factor for each sample in the dataset, we can determine whether the sample is an outlier based on the following rule: If $lof_k(x)$ is close to 1, then the local reachability density of sample x and its nearest neighbors are similar, then x can be considered as a normal sample; if $lof_k(x) < 1$, then the local reachability density of sample x is greater than its nearest neighbors, meaning that x is a normal sample; if $lof_k(x) > 1$, then the local reachability density of sample x is less than its nearest neighbors, and x may be an outlier. Figure 2.5 shows an outlier detection application

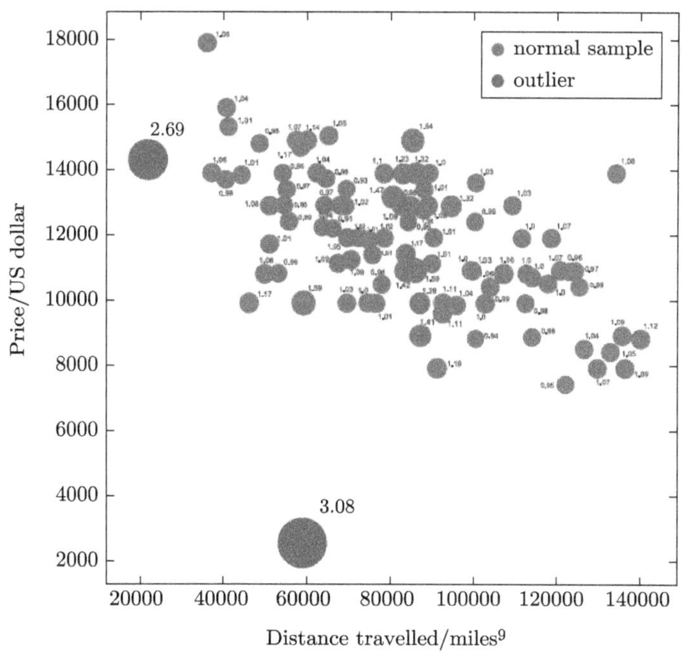

Figure 2.5 One application of LOF.

[i]1 mile = 1.6093 km.

using the LOF algorithm on real data of used cars, where the number next to the sample point indicates its LOF value and the size of the dot is directly proportional to its LOF value.[i]

Both the K-nearest neighbors method and the LOF algorithm need to calculate the distance between samples in the dataset. When the sample size is large and the dimension is large, the computational cost of both methods is high.

2.5.3. *Summary*

When detecting outliers from datasets, we cannot simply think of them as noise. Instead, we must combine specific business requirements to examine the rationality and causes of outliers, so that we can adjust the existing model to improve its explanatory power. If the outliers are filtered during the preprocessing phase, the data analysis task may fail.

In this section, we discuss several simple outlier detection methods for data preprocessing. In fact, outlier detection is an important research topic in the field of data mining. Readers can refer to the existing literature.[13–15]

2.6. Other Preprocessing Methods

Data preprocessing is an important preliminary step of data analysis, and the output directly determines the performance of data analysis. The previous sections introduced some of the most typical preprocessing methods. This section will briefly introduce some other preprocessing methods.

For some problems, we need to center the data first. Centering means to make the mean of features in the dataset equal to 0. We can first calculate the feature mean and then subtract the mean from all the feature values in the dataset. Some of the dimensionality reduction algorithms (such as principal component analysis) cannot run without a data centering process.

Many models assume that the features follow a normal distribution. If the actual features do not follow a normal distribution, we may need to transform them. For example, the distribution of features such as the amount of money and income in real life tends to be right-skewed. If we perform a logistic transformation on those features, we can make them approximately follow a normal distribution.

The above discussion is all about data conversion. In a real data analysis project, it is usually necessary to integrate different types and sources of data before data conversion. In the data integration process, we will encounter data quality issues such as data inconsistency, data redundancy, and data conflicts.[16] Discusses data integration from the perspective of databases systematically.

2.7. Case Studies and Exercises

1. Standardization of concrete strength data

Concrete is an indispensable material for the development of modern civilization. Water-cement ratio (refers to the weight ratio of water content to cement content in concrete) is important to the strength of concrete. Of course, in addition to the water-cement ratio, the strength of concrete is also related to the features of other components, such as blast furnace slag content, superplasticizer content, and coarse aggregate content.

This exercise provides a concrete strength dataset.[17] Each sample in the dataset contains 9 features, the last of which is the concrete strength. Please perform Z-score standardization for each feature in the dataset.

2. Missing value imputation of the dermatology data

The diagnosis of Erythematous-squamous is a thorny issue in the field of dermatology. Because the symptoms and appearances of its different types are very similar. This exercise provides a skin disease dataset.[18] Each sample in the dataset contains 22 histopathological features and 12 clinical features. For some reasons, there are missing values for the "age" feature in the dataset. Please fill in the missing values using the mean imputation method.

3. Youth life survey

Scientific and technological development, and economic growth have caused major changes in the structure of our social population. Today's young people are facing multiple pressures such as employment, housing and responsibility to support parents.

Table 2.7 Feature description of the second-hand car dataset.

Feature	Description
Price	Price
Mileage	Miles traveled
Year	Year of launch
Trim	Trim grade
Engine	Number of engines
Transmission	Manual or automatic

This exercise provides a survey dataset about youth living conditions,[19] which records in detail the music preferences, film preferences, lifestyle habits, consumption habits, and values of young people. Among them, the "Smoking" feature records smoking situation. Please perform the One-Hot encoding method to transform "Smoking" feature.

4. Outlier detection of the car data

Automobile is an indispensable mean of transportation in modern society. With the convenience of automobiles, people can break the distance limit and enjoy life better. A dataset on *Second-hand* car is provided from Vast. There are 417 samples and 6 features in the dataset. The description of the features is shown in Table 2.7.

These samples are manually entered by users and are subject to mistakes. For example, a user may fill in information in the wrong field, or enters "erroneously" values, etc. Please use one method described in this chapter to perform outlier detection on the dataset.

Chapter 3

Regression Model

3.1. Linear Regression

In daily life, we will encounter prediction problems with continuous target features, such as income prediction, sales prediction and commodity inventory prediction. This kind of problem is called regression, which is a typical supervised learning method.

Assume that the input data of the model is a d-dimensional vector $\boldsymbol{x}$, where d represents the dimension of the vector $\boldsymbol{x}$, and the target feature y is continuous. Regression model is equivalent to finding a function f mapping $\boldsymbol{x}$ to y by the equation $y = f(\boldsymbol{x})$. Linear regression and nonlinear regression are the commonly used regression models. Linear regression is the most simple and practical type of regression model and it is also considered as the basis of other regression models. In this section, we will mainly discuss the theory and some practical applications of linear regression.

3.1.1. *Simple Linear Regression*

In simple linear regression, input data contains only one single dimensional variable. Assuming the input feature is x and the target feature is y, the model of simple linear regression is:

$$y = w_1 x + w_0, \tag{3.1}$$

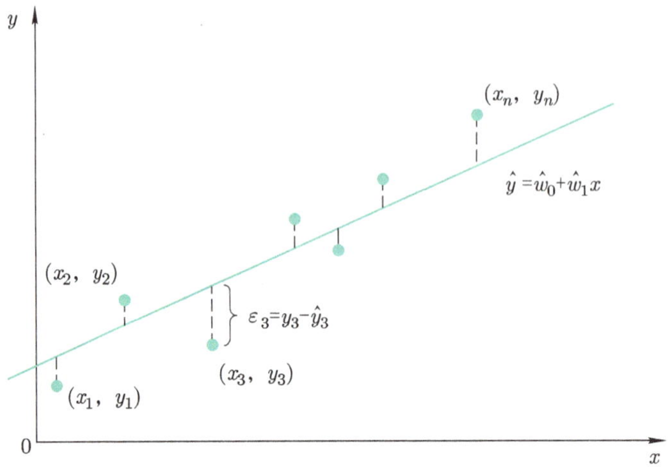

Figure 3.1 Simple linear regression.

where w_1 and w_0 are the parameters. Given a sample (x_i, y_i), the estimated value of the model is evaluated by the equation $\hat{y}_i = w_1 x_i + w_0$. The difference between the true value y_i and the estimated value $\hat{y}_i$ is called the residual of the sample (x_i, y_i), denoted by $\varepsilon_i = y_i - \hat{y}_i$. The residuals are shown in Figure 3.1.

Given the training set $D = \{(x_1, y_1), \ldots, (x_n, y_n)\}$, our goal is to find a line $y = w_1 x + w_0$ with all sample points as close to it as possible. To do this, we can minimize the square sum of residual sum of squares, so as to find the optimal parameters.

It is equivalent to finding the solution of the following formula:

$$\min_{w_1, w_0} RSS(w_1, w_0) = \min_{w_1, w_0} \sum_{i=1}^{n} (y_i - w_1 x_i - w_0)^2. \tag{3.2}$$

Take derivatives of the objective function $RSS(w_1, w_0)$ with respect to w_0 and w_1, respectively, and set them equal to zero.

$$\frac{\partial RSS}{\partial w_0} = \sum_{i=1}^{n} 2(y_i - w_1 x_i - w_0)(-1) = 0, \tag{3.3}$$

$$\frac{\partial RSS}{\partial w_1} = \sum_{i=1}^{n} 2(y_i - w_1 x_i - w_0)(-x_i) = 0. \tag{3.4}$$

The optimal solution is obtained:

$$\hat{w}_1 = \frac{\sum_{i=1}^{n} x_i y_i - n\bar{x}\bar{y}}{\sum_{i=1}^{n} x_i^2 - n(\bar{x})^2}, \quad \hat{w}_0 = \bar{y} - \hat{w}_1\bar{x}, \tag{3.5}$$

where $\bar{x} = \frac{1}{n}\sum_{i=1}^{n} x_i$, $\bar{y} = \frac{1}{n}\sum_{i=1}^{n} y_i$. This method is called the Ordinary Least Square (OLS). As shown in Figure 3.1, the green line is the optimal regression line, which describes the linear relationship between the target feature y and the input feature x.

3.1.2. *Multiple Linear Regression*

In multiple linear regression, the dimension of input data increases from one dimension to d dimension ($d > 1$). Assume the training set is $D = \{(x_1, y_1), \ldots, (x_n, y_n)\}$, and the multiple linear regression can be modeled as:

$$y = \boldsymbol{w}^T \boldsymbol{x}, \tag{3.6}$$

where $\boldsymbol{w} = (w_1, w_2, \ldots, w_d, w_0)^T$ is the model parameter.

Suppose we denote the input features in the training set as the matrix $\boldsymbol{X}$ of $n \times (d+1)$ dimension with the values of all elements of the last column being 1. The target feature of the training set are written in vector form: $\boldsymbol{y} = (y_1, y_2, \ldots, y_n)^T$.

In the multiple linear model, the estimated output of the corresponding input $\boldsymbol{X}$ is:

$$\hat{\boldsymbol{y}} = \boldsymbol{X}\boldsymbol{w}, \tag{3.7}$$

and the residual sum squares is computed by:

$$\mathrm{RSS}(\boldsymbol{w}) = \|\boldsymbol{y} - \hat{\boldsymbol{y}}\|_2^2 = \|\boldsymbol{y} - \boldsymbol{X}\boldsymbol{w}\|_2^2. \tag{3.8}$$

Similar to simple linear regression, we get the derivative of RSS ($\boldsymbol{w}$) and set it equal to $\mathbf{0}$:

$$\frac{\partial RSS}{\partial \boldsymbol{w}} = -2\boldsymbol{X}^T(\boldsymbol{y} - \boldsymbol{X}\boldsymbol{w}) = \mathbf{0}. \tag{3.9}$$

Therefore, the optimal solution is obtained:

$$\hat{\boldsymbol{w}} = (\boldsymbol{X}^T\boldsymbol{X})^{-1}\boldsymbol{X}^T\boldsymbol{y}. \tag{3.10}$$

Note that the solution obtained by Eq. (3.10) will suffer from the overfitting problems, especially when the number of features d is greater than the number of training samples. We can solve this problem by feature selection and regularization. Chapter 9 will discuss feature selection in details. In Section 3.2, we will introduce several classic regularized linear regression models, including the ridge regression and LASSO.

3.1.3. *Summary*

The simple linear regression model is based on the following assumptions:

1. The input features are non-random and uncorrelated to each other;
2. The random errors are of zero-mean, homoscedasticity, and independent from each other;
3. The input feature are uncorrelated with the random errors;
4. The random error follows the normal distribution $N(0, \delta^2)$.

Therefore, after obtaining a linear regression model, it is usually necessary to evaluate the results based on the above assumptions. To illustrate the importance, we will showcase the famous Anscombe's quartet. This dataset consists of four groups of data in total, and each group includes 11 samples denoted by (x, y). The basic statistical characteristics of the four groups of data (including mean, variance, and the linear regression line) are the same, but the scattered diagrams of them are quite different, as shown in Figure 3.2. By applying linear regression on these four groups of datasets, we can get exactly the same model, but quite different results.

After obtaining a regression model, we also need to design some metrics to evaluate its performance. R^2 and RMSE are two Commonly used metrics for the evaluation of regression models. The coefficient R^2 is used to measure the fitness of the model, and the root-mean-square error (RMSE) is used to evaluate the performance of the model on the test set. Refer to Appendix E for more metrics for regression models.

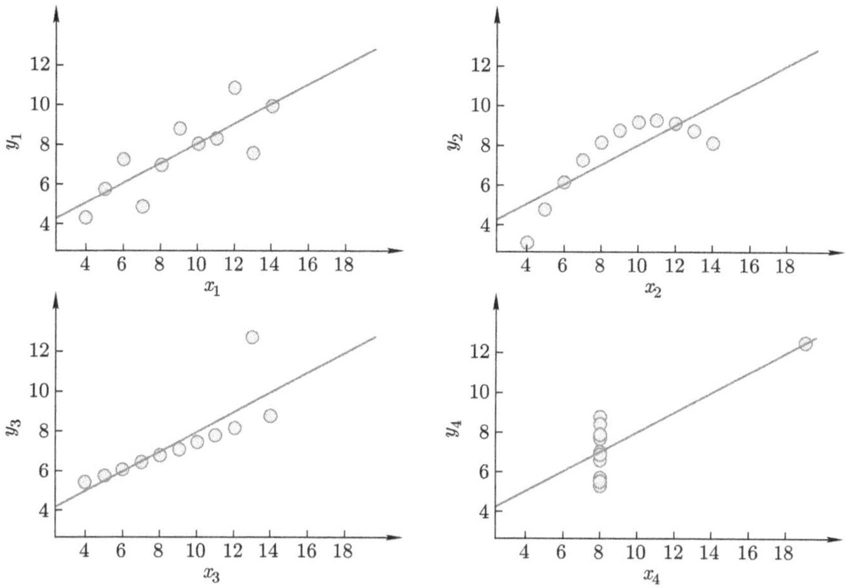

Figure 3.2 Scatter plots of four groups of datasets from the Anscombe's quartet.

3.2. Linear Regression Regularization

The idea of regularization is to add penalty or constraints to the model to control the complexity of the model and to reduce overfitting. Ridge and least absolute shrinkage and selection operator (LASSO) are two popular linear regression methods.

3.2.1. *Ridge and LASSO*

Specifically, ridge and LASSO correspond to L_2 and L_1 regularization, respectively, and the *a priori* assumptions proposed to the coefficient vector $\boldsymbol{w}$ are $\|\boldsymbol{w}\|_2 \leqslant C$ and $\|\boldsymbol{w}\|_1 \leqslant C$, where C is a predetermined constant. In other words, we focus on the following optimization problems with constraints, for ridge and LASSO,

$$\min_{\boldsymbol{w}} \|\boldsymbol{y} - \boldsymbol{X}\boldsymbol{w}\|_2^2, \quad \text{s.t. } \|\boldsymbol{w}\|_2 \leqslant C, \tag{3.11}$$

$$\min_{\boldsymbol{w}} \|\boldsymbol{y} - \boldsymbol{X}\boldsymbol{w}\|_2^2, \quad \text{s.t. } \|\boldsymbol{w}\|_1 \leqslant C. \tag{3.12}$$

By applying the Lagrange Multiplier Method, the above constrained optimization problems are equivalent to the optimization problem of the unconstrained penalty function,

$$\min_{\boldsymbol{w}} \|\boldsymbol{y} - \boldsymbol{X}\boldsymbol{w}\|_2^2 + \lambda\|\boldsymbol{w}\|_2^2, \tag{3.13}$$

as well as

$$\min_{\boldsymbol{w}} \frac{1}{2}\|\boldsymbol{y} - \boldsymbol{X}\boldsymbol{w}\|_2^2 + \lambda\|\boldsymbol{w}\|_1, \tag{3.14}$$

where the regularization coefficient $\lambda > 0$ is a constant that depends on C.

For ridge regression, by directly deriving the objective function (3.13) and making its gradient equal to zero, it can easily be obtained:

$$\begin{aligned} \boldsymbol{w}^{\text{ridge}} &= \underset{\boldsymbol{w}}{\operatorname{argmin}} \left(\|\boldsymbol{y} - \boldsymbol{X}\boldsymbol{w}\|_2^2 + \lambda\|\boldsymbol{w}\|_2^2\right) \\ &= (\boldsymbol{X}^{\mathrm{T}}\boldsymbol{X} + \lambda\boldsymbol{I})^{-1}\boldsymbol{X}^{\mathrm{T}}\boldsymbol{y}. \end{aligned} \tag{3.15}$$

Compared to Eq. (3.10), we can see that the estimated parameters obtained by Ridge Regression have only one more regular term $\lambda\boldsymbol{I}$. The existence of this term makes $(\boldsymbol{X}^{\mathrm{T}}\boldsymbol{X} + \lambda\boldsymbol{I})^{-1}$ more stable in the numerical calculation. Especially when multi-collinearity occurs, $\boldsymbol{X}^{\mathrm{T}}\boldsymbol{X}$ approximates a singular matrix, and ridge regression can still produce stable results. In addition, in ridge regression, it is possible to eliminate multicollinearity features by observing ridge traces.

$\boldsymbol{w}^{LASSO}$, the solution of LASSO, has no analytic expressions, but it has its unique advantages over ridge regression. $\boldsymbol{w}^{LASSO}$ is sparse. In other words, many of its components are close to 0. Therefore, the results obtained by the LASSO model can help us perform feature selection. When correlation between x_i and y is not high or multicollinearity with x_j, it is very likely that w_i^{LASSO} is equal to 0.

The essence of regularization is actually to make a trade-off between the bias and the variance when estimating the parameter $\boldsymbol{w}$. The parameter $\boldsymbol{w}$ obtained by a linear regression model without regularization is already the minimum-variance unbiased estimate. In other words, if you do not abandon unbiasedness, there is no way to improve the model. The ridge regression and LASSO are used as biased estimators to expect a more realistic regression coefficient by abandoning the unbiasedness of the least-squares method.

1. Solution to LASSO

After many years of research and development in the field of statistics and machine learning, there are currently many solutions to LASSO, such as coordinate descent, LARS, and iterative shrinkage-thresholding algorithm (ISTA), which is based on proximal gradients (refer to Section 7.2.1) and FISTA (accelerated version of ISTA). Here, we focus on the ISTA algorithm.

To minimize a smooth function $f(\boldsymbol{w})$, the most commonly used method is the gradient descent. The basic idea is to perform multiple iterations to update the parameters based on the value of current variable $\boldsymbol{w}^{(t)}$:

$$\boldsymbol{w}^{(t+1)} = \boldsymbol{w}^{(t)} - \eta \nabla f(\boldsymbol{w}^{(t)}), \tag{3.16}$$

where η is the learning rate. The above formula can be written as an equivalent "proximal form":

$$\begin{aligned}\boldsymbol{w}^{(t+1)} = \arg\min_{\boldsymbol{w}} & f(\boldsymbol{w}^{(t)}) - \nabla f(\boldsymbol{w}^{(t)})^{\mathrm{T}}(\boldsymbol{w} - \boldsymbol{w}^{(t)}) \\ & + \frac{1}{2\eta}\|\boldsymbol{w} - \boldsymbol{w}^{(t)}\|_2^2.\end{aligned} \tag{3.17}$$

If we minimize the sum of two functions $f+g$, where f is smooth (i.e., it can be approximated by a first-order Taylor expansion) and g is a relatively "simple" function, such as a quadratic function or an absolute sum function. Then we get the following iterative-update method, which is called ISTA:

$$\begin{aligned}\boldsymbol{w}^{(t+1)} &= \arg\min_{\boldsymbol{w}} f(\boldsymbol{w}^{(t)}) - \nabla f(\boldsymbol{w}^{(t)})^{\mathrm{T}}(\boldsymbol{w} - \boldsymbol{w}^{(t)}) + \frac{1}{2\eta}\|\boldsymbol{w} - \boldsymbol{w}^{(t)}\|_2^2 \\ &= \arg\min_{\boldsymbol{w}} g(\boldsymbol{w}) + \frac{1}{2\eta}\|\boldsymbol{w} - (\boldsymbol{w}^{(t)} - \eta\nabla f(\boldsymbol{w}^{(t)}))\|_2^2.\end{aligned} \tag{3.18}$$

The objective functions of LASSO are $f(\boldsymbol{w}) = \frac{1}{2}\|\boldsymbol{y} - \boldsymbol{X}\boldsymbol{w}\|_2^2$ and $g(\boldsymbol{w}) = \lambda\|\boldsymbol{w}\|_1$.

The gradient of f is $\nabla f(\boldsymbol{w}) = \boldsymbol{X}^{\mathrm{T}}(\boldsymbol{X}\boldsymbol{w} - \boldsymbol{y})$. So LASSO can be solved by ISTA with an iterative formula for each step:

$$\boldsymbol{w}^{(t+1)} = \arg\min_{\boldsymbol{w}} \lambda\|\boldsymbol{w}\|_1 + \frac{1}{2\eta}\|\boldsymbol{w} - (\boldsymbol{w}^{(t)} - \eta\nabla f(\boldsymbol{w}^{(t)}))\|_2^2, \tag{3.19}$$

$$\boldsymbol{w}^{(t+1)} = S_{\eta\lambda}(\boldsymbol{w}^{(t)} - \eta\nabla f(\boldsymbol{w}^{(t)})), \tag{3.20}$$

where $S_a(\boldsymbol{v})$ represents a soft thresholding operator, which is defined as

$$S_a(\boldsymbol{v}) = \begin{cases} v_i - a, & \text{if } v_i > a; \\ 0, & \text{if } |v_i| \leqslant a; \\ v_i + a, & \text{if } v_i < -a. \end{cases} \tag{3.21}$$

The ISTA algorithm is a typical approximate gradient method. Interested readers can refer to the related literature.[26]

2. Why LASSO can produce sparse solution

We now analyze why LASSO can produce sparse solutions, while ridge regression cannot. The optimization problems solved by LASSO and Ridge regression are Eq. (3.12) and Eq. (3.11), which are both constrained quadratic programming problems.

We plot the contours of the respective objective function and the feasible region of the solution in Figure 3.3. From the optimization theory we can see that the optimal solution is at the intersection of the contour of the objective function and the feasible region.

Geometrically, we can clearly observe that when we increase the number of contour lines (this corresponds to increasing the value of the objective function) until one of them intersects with the L_1 "ball" (in the two-dimensional case, its shape corresponds to a diamond), the corner is more likely to intersect the contour than the edge. This phenomenon is especially noticeable in the case of high dimensionality, because the high-dimensional corners are more strongly "convex". As the corners are on the axis, sparse solutions appear. Conversely, when we increase the number of contour lines of the objective function of the ridge regression, one of them may intersect with the L_2 "ball" (in the two-dimensional case its shape corresponds to a circle) at any point, so it is not easy to produce a sparse solution.

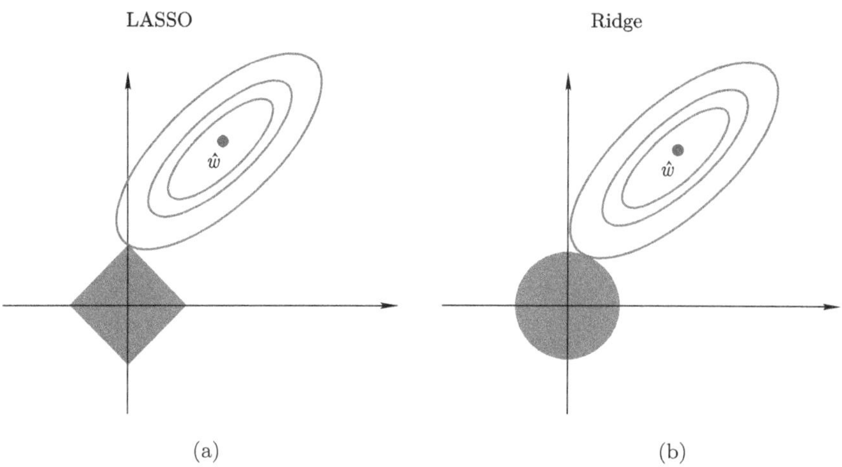

Figure 3.3 Comparison of the ℓ_1 (a) and ℓ_2 (b) regularization in linear regression.

3. Regularization path analysis

In linear regression, the regularization path is a curve on which the estimated value w of the regression coefficient changes as the regularization coefficient increases. It can help us to analyze the correlations between features and how to perform feature selection.

The regularization path of ridge regression is called ridge trace. If the ridge trace fluctuates greatly, it can be concluded that the corresponding feature is related to other features. Suppose we get the regression coefficients under different regularization coefficients λ, as shown in Table 3.1.

According to Table 3.1, ridge traces can be drawn, as shown in Figure 3.4. From the figure, we can see that λ fluctuates greatly in range [0,0.5], while it tends to be stable at $\lambda \geqslant 1$. Therefore we can set $\lambda = 1$. In addition to this qualitative hyper-parameter selection method, we can also use cross validation to select the hyperparameter. More details can be found in Appendix E.

We can also select features based on the ridge traces. Usually, we need to standardize features (such as Z-score standardization). In this way, we can directly compare the regression coefficients. Then we can remove the feature whose regression coefficient is relatively stable and whose absolute value is small. When λ increases, the features

Table 3.1 Ridge trace of two features.

λ	0	0.1	0.15	0.2	0.3	0.4	0.5	1.0	1.5	2.0	3.0
$\widehat{w}_1$	11.31	3.48	2.99	2.71	2.39	2.20	2.06	1.66	1.43	1.27	1.03
$\widehat{w}_2$	–6.59	0.63	1.02	1.21	1.39	1.46	1.49	1.41	1.28	1.17	0.98

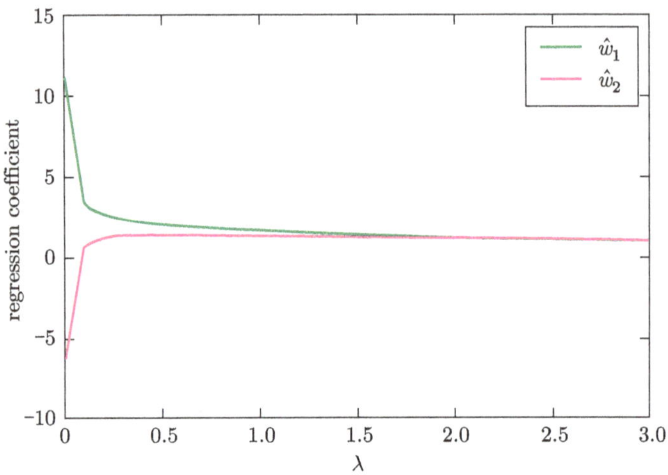

Figure 3.4 Ridge traces.

whose regression coefficients are unstable and tend to zero can also be removed.

Similarly, we can analyze the LASSO regularization path. Take the prostate cancer dataset as an example. This dataset indicates the relationship between prostate antigen (the target feature) and nine other clinical measurements (9-dimensional input). The regularization paths of LASSO and ridge regression with different regularization coefficients are shown in Figure 3.5(a) and (b), respectively.

It is observed that when λ becomes larger, the value of $\sum_j |\boldsymbol{w}_j|$ tends to 0 for both models. Meanwhile, the coefficients of all the features in the model tend to be close to 0. In the ridge regression model, when λ increases gradually, the regression coefficients of almost all the features will reduce to 0 simultaneously. In this case, it is hard to perform feature selection. On the contrary, the advantage of LASSO is that as λ increases, the coefficients of different dimensional features decrease to 0 one by one. Therefore, features can be selected by

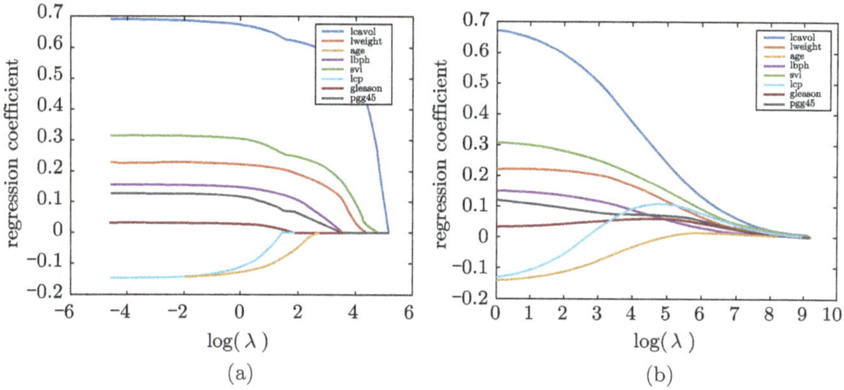

Figure 3.5 Regularization path of (a) LASSO and (b) Ridge regression.

controlling the size of regularization coefficients. This explains why LASSO is one of the most popular linear model for feature selection.

3.2.2. *Other Regularized Linear Regression Models*

Although LASSO is an effective linear model for feature selection, it has some obvious shortcomings in practical applications.[29] First, when there is a group of highly relevant features in the data, LASSO tends to select only one of them, and the choice is quite random. In some applications, it is desirable to select all the features of a certain group. For example, in genetic data, there is a high correlation between multiple genes that express the same protein. We are concerned about the entire group of genes, not only one of them. Second, when the feature dimension d is greater than the number of samples n, LASSO can only select n features at most.

In order to overcome the above shortcomings, Zou *et al.* proposed a regularization method called elastic net in 2005.[29] It combined the regularization of ridge regression and LASSO to minimize the following objective function:

$$J(\boldsymbol{w}) = \|\boldsymbol{y} - \boldsymbol{X}\boldsymbol{w}\|_2^2 + \lambda_1\|\boldsymbol{w}\|_1 + \lambda_2\|\boldsymbol{w}\|_2^2. \tag{3.22}$$

Elastic net has shown the effect of feature grouping, that is, the regression coefficients of highly correlated features tend to be equal (for negative correlated features, the coefficients have a sign difference). It has been successfully applied in the field of microarray classification and gene selection.

In other applications, features do not appear individually, but rather in a pre-defined group (which is often divided based on a specific problem-determining group), such as a group of genes belonging to the same biological process and a group of features encoded by One-Hot. In those cases, we expect to be able to select features by groups. This is the basic idea of Group LASSO.[30] If there are predetermined G groups, Group LASSO minimizes the following objective function for feature selection by groups:

$$J(\boldsymbol{w}) = \|\boldsymbol{y} - \boldsymbol{X}\boldsymbol{w}\|_2^2 + \sum_{g=1}^{G} \lambda_g \|\boldsymbol{w}_g\|_2^2. \tag{3.23}$$

In brief, regularization is an effective method to integrate the data scientists' *a priori* understanding of the problem into model and control the structure of the solution.

3.3. Nonlinear Regression

In practice, the input features and target feature are not linearly correlated. In this case, simple linear regression models may not fit the data well. In this section, we introduce several commonly used nonlinear regression models.

3.3.1. *Spline Regression*

Spline generally refers to a piecewise polynomial curve connected by some control knots. Spline regression refers to a regression model by using a spline curve to fit the function. The simplest spline regression model can be formulated as:

$$y = \beta_0 + \beta_1 x + w_1(x - a_1)_+ + w_2(x - a_2)_+ + \cdots + w_k(x - a_k)_+, \tag{3.24}$$

where k represents the number of control knots holding $a_1 < a_2 < \cdots < a_k$, and the parameter vector is $\boldsymbol{w} = (\beta_0, \beta_1, w_1, \ldots, w_k)^T$. $(u)_+$ outputs u when $u > 0$, otherwise outputs 0.

Assume that the training set is $D = \{(x_1, y_1), \ldots, (x_n, y_n)\}$, the design matrix $\boldsymbol{G} \in R^{n\times(k+2)}$ is:

$$\boldsymbol{G} = \begin{bmatrix} 1 & x_1 & (x_1 - a_1)_+ & \cdots & (x_1 - a_k)_+ \\ 1 & x_2 & (x_2 - a_1)_+ & \cdots & (x_2 - a_k)_+ \\ \vdots & & \vdots & & \vdots \\ 1 & x_n & (x_n - a_1)_+ & \cdots & (x_n - a_k)_+ \end{bmatrix}. \tag{3.25}$$

Then we can get:

$$\boldsymbol{y} = \boldsymbol{G}\boldsymbol{w}. \tag{3.26}$$

The parameter vector $\boldsymbol{w}$ can be obtained by the least squares method. The optimal solution is computed by:

$$\boldsymbol{w} = (\boldsymbol{G}^{\mathrm{T}}\boldsymbol{G})^{-1}\boldsymbol{G}^{T}\boldsymbol{y}. \tag{3.27}$$

The shape of the spline curve can be controlled by the number and position of the control knots. If too many knots are selected, the fitting function will be very complicated, which leads to overfitting. To control the complexity of the model, we often take linear spline regression with penalty,

$$\min_{\boldsymbol{w}} \sum_{i=1}^{n} \left(y_i - (\beta_0 + \beta_1 x_i + \sum_{j=1}^{k} w_j (x_i - a_j)_+) \right)^2 + \lambda \sum_{j=1}^{k} w_j^2, \tag{3.28}$$

where λ is the penalty coefficient. This method is similar to Ridge regression which uses regularization to control model complexity.

In addition to a simple linear spline, we can take higher-order splines, such as a cubic spline,

$$y = \beta_0 + \beta_1 x + \beta_2 x^2 + \beta_3 x^3 + \sum_{k=1}^{k} (x - a_k)_+. \tag{3.29}$$

The B-spline curve is another commonly used spline regression model. Assume the control knots are $a_1 < a_2 < \cdots < a_k$, a k-degree

B-spline curve ($k+1$ order) is defined as follows:

$$B(x) = \sum_{j=0}^{k+m} w_j B_{j,k}(x), \quad x \in [a_0, a_{k+1}], \tag{3.30}$$

where

$$B_{j,0}(x) = \begin{cases} 1, & a_j \leqslant x < a_{j+1}; \\ 0, & \text{otherwise}, \end{cases} \tag{3.31}$$

$$B_{j,k+1}(x) = \alpha_{j,k+1}(x) B_{j,k}(x) + (1 - \alpha_{j+1,k+1}(x)) B_{j+,k}(x), \tag{3.32}$$

where

$$\alpha_{j,k}(x) = \begin{cases} \dfrac{x - t_j}{t_{j+k} - t_j}, & a_{j+k} \neq a_j; \\ 0, & \text{otherwise}. \end{cases} \tag{3.33}$$

We have introduced how to utilize spline regression to solve the fitting problem of one-dimensional nonlinear function. For multidimensional spline regression, please refer to literature.[31]

3.3.2. *Radial Basis Function Network*

Radial basis function (RBF) is a function that depends on the distance between the input $\boldsymbol{x}$ and the center vector $\boldsymbol{c}$, $\phi(\|\boldsymbol{x} - \boldsymbol{c}\|)$, such as Euclidean distance. The function is radially symmetric to the center vector. This is why it is called RBF. Some typical radial basis functions include:

- The Gaussian radial basis function: $\phi(r) = \sqrt{1 + ar^2}$.
- Multi-quadric radial basis function: $\phi(r) = e^{-ar^2}$.
- Inverse quadratic radial basis function: $\phi(r) = \frac{1}{1+ar^2}$.

where a is a hyper-parameter. The radial basis function network (RBF network) approximates the nonlinear function by linearly combining k radial basis.

$$y = \sum_{j=1}^{k} w_j \phi(\|\boldsymbol{x} - \boldsymbol{c}_j\|), \tag{3.34}$$

where k center vectors $\{\boldsymbol{c}_1, \boldsymbol{c}_2, \ldots, \boldsymbol{c}_k\}$ can be obtained by randomly choosing k samples or by clustering the data. $\boldsymbol{w} = (w_1, w_2, \ldots, w_k)^T$

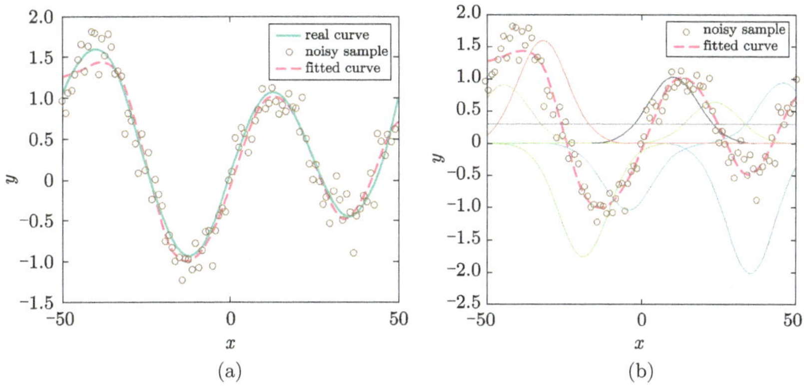

Figure 3.6 Example of data fitting results by using RBF network. (a) Comparison between the fitting result and real function. (b) The position of each radial basis function.

is the model parameter, which can be calculated by using the least squares method. Equation (3.34) can be formulated as by the following vector form:

$$\boldsymbol{y} = \boldsymbol{G}\boldsymbol{w}, \tag{3.35}$$

where $\boldsymbol{y} = (y_1, y_2, \ldots, y_n)^{\mathrm{T}}$, the design matrix $\boldsymbol{G}$ is a $n \times k$ matrix with the element $\phi(\|\boldsymbol{x}_i - \boldsymbol{c}_j\|)$ at row i, column j. Then according to least squares, the parameter $\boldsymbol{w} = (\boldsymbol{G}^{\mathrm{T}}\boldsymbol{G})^{-1}\boldsymbol{G}^{\mathrm{T}}\boldsymbol{y}$ can be estimated similarly to Eq. (3.27).

Figure 3.6 illustrates an example of data fitting results by using RBF network. The training data is generated by $y = \sin(\frac{x}{8}) + (\frac{x}{50})^2 + \varepsilon$, where the Gaussian noise is $N(0, 0.04)$, and the Gaussian radial basis function is used ($k = 8$). The center vector is selected by the clustering method and the hyper-parameter a and σ are defined as $a = 1/(2\sigma^2)$, $\sigma = 8$. Figure 3.6(a) shows the comparison between the fitting result and the real function. The colored line in Figure 3.6(b) shows the position of each radial basis function. For more detailed discussion on the radial basis network, refer to literature.[32]

From the above example, we can find that Spline regression and RBF regression are both based on linear combination of the generalized basis functions. Therefore, we should select appropriate basis functions for different problems. In addition to the above nonlinear regression models, more nonlinear regression methods include support vector regression (SVR),[33] Kernel Ridge Regression, Gaussian process,[34] and multi-layer perceptron (introduced in Chapter 14 of this book).

3.4. Case Studies and Exercises

1. Predicting the area of forest fire

As a "regulator" of nature, forests have various functions such as conserving water sources and purifying air. As an integral part of the primary industry, they have important economic value. However, forest fires frequently occur and consequently lead to a great loss of resources. So, we need to make predictions to prevent forest from fires.

The dataset consists of 517 samples in total, and each sample contains 12 features and the target feature is the conflagration area.[35] Due to the correlation between some features, we should select the features. In addition, the dataset contains a considerable number of samples with a conflagration area of zero, so how to transform the target feature should also be taken into consideration.

Please use the above dataset to build a regression model to predict the conflagration area of forest fires. Please use appropriate metrics to evaluate the performance of the regression model (see Appendix E).

2. Predicting of public bicycle rental

Bicycle sharing systems have great advantages over the traditional bike rentals. Many cumbersome and complex processes have been circumvented, e.g., registration, renting, and returning. Users can complete all these procedures online. Meanwhile, some useful information such as bicycle travel time and routes can be fully recorded.

Given a public bicycle rental dataset,[36] please predict the number of bicycles rented based on the environmental information including the city's weather, the temperature, the humidity, and the wind speed.

3. Prediction of music origin

Audio tracks are parallel lines in sequencer software. In this exercise, our problem aims to make the machine predict the origin of a piece of music by given some features of that music.

Table 3.2 Features of the medical insurance cost dataset.

Feature	Description
Age	Age of primary beneficiary
Sex	Primary beneficiary's gender
BMI	Body mass index (providing an understanding of the body, weights that are relatively high or low relative to height)
Children	Number of children covered by health insurance
Smoke	Smoking (yes, no)
Region	Beneficiary's residential area in the US (northeast, southeast, southwest, northwest)
Charges	Individual medical costs billed by health insurance

The dataset contains 1,059 samples with 68 audio features and 2 target features (representing the latitude and longitude of the place where the music originated).[37] Build a regression model based on the features of the audio track to predict the longitude of the origin of the music.

4. Medical insurance cost prediction

Insurance companies usually raise more annual premiums than the cost of the beneficiary's medical services. Therefore, precise prediction of the medical cost is valuable for the insurers.

The dataset is acquired from the demographic data of the U.S. Census Bureau. The dataset consists of 1,338 samples with 7 features as shown in Table 3.2.

Please use "charges" as the target feature to build a regression model to predict the average medical cost of a primary beneficiary.

Chapter 4

Classification Model

Classification is another type of supervised learning. In the credit risk assessment, it is necessary to predict whether the customer will default in the future according to the customer's historical repayment record and the age of credit account; in medical diagnosis, a typical problem is to classify tumor cells as positive or negative according to their characteristics; in marketing, one often needs to predict the customer's preference for new products according to her historical purchase records and behaviors in our daily email service, classifying email into normal email or spam email is also a typical classification problem.

A classification task can mainly divided into two stages: (1) training a classifier; (2) perform prediction using the trained classifier. Usually when we get a dataset, we divide the data into two parts: the training set and the test set. The training set is used to train the classifier, and the test set is used to evaluate the performance of the classifier. Each sample in the training set has a class label. After the classifier is trained, it is possible to predict for samples without class labels. A typical classification model building process is shown in Figure 4.1.

Suppose the input data of the model is a d-dimensional vector $\boldsymbol{x}$, and the output y is a finite set of categories. The setting of classification model is equivalent to finding a function f to establish a mapping from $\boldsymbol{x}$ to y, i.e., $y = f(\boldsymbol{x})$. Many classical algorithms can be applied to building classification model and the differences are mainly their assumptions about the form of f, such as commonly used logistic regression, K-nearest neighbor model, decision

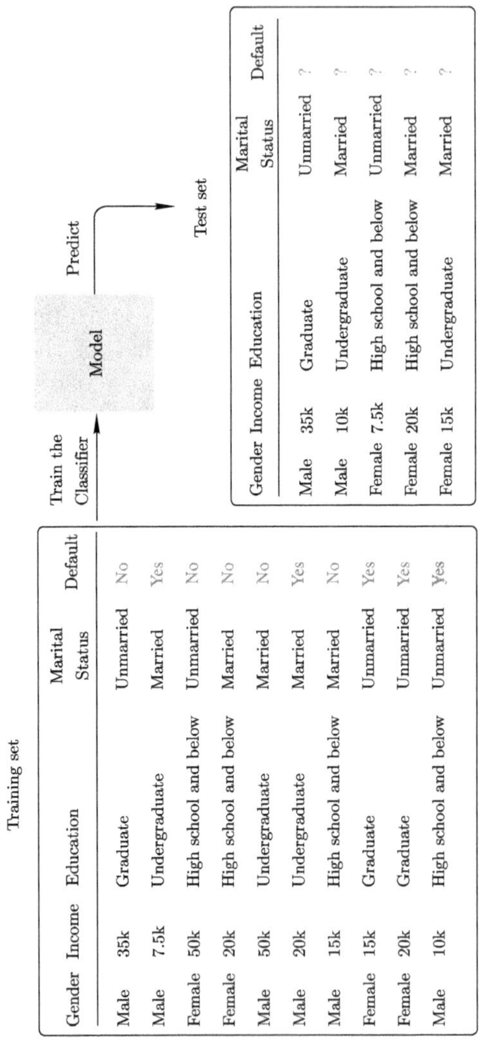

Training set

Gender	Income	Education	Marital Status	Default
Male	35k	Graduate	Unmarried	No
Male	7.5k	Undergraduate	Married	Yes
Female	50k	High school and below	Unmarried	No
Female	20k	High school and below	Married	No
Male	50k	Undergraduate	Married	No
Male	20k	Undergraduate	Married	Yes
Male	15k	High school and below	Married	No
Female	15k	Graduate	Unmarried	Yes
Female	20k	Graduate	Unmarried	Yes
Male	10k	High school and below	Unmarried	Yes

Test set

Gender	Income	Education	Marital Status	Default
Male	35k	Graduate	Unmarried	?
Male	10k	Undergraduate	Married	?
Female	7.5k	High school and below	Unmarried	?
Female	20k	High school and below	Married	?
Female	15k	Undergraduate	Married	?

Figure 4.1 Classification model construction procedure.

tree model, naive Bayesian and support vector machines. The metrics for evaluating classification models include accuracy, F-measure, precision and recall, etc., see Appendix E for details.

In this chapter we will introduce the five commonly used classification models mentioned above in detail. In Chapter 5, we will discuss two more complex ensemble classification methods: Random Forests and AdaBoost.

4.1. Logistic Regression

There are many algorithms for solving classification problem in machine learning, in which logistic regression is one of the most widely used ones. Logistic regression is the most classic and commonly used model in risk assessment. In addition, it has been widely used in Internet ad click prediction.

Logistic regression is called "regression" because it takes the idea of regression analysis. However it is a model used to solve classification problems, especially binary classification problem. In this section, we first introduce the relationship between logistic regression and linear regression; then we discuss how to estimate the parameters finally, we discuss how to extend logistic regression to solve multiclass or multinomial classification problem.

4.1.1. *From Linear Regression to Logistic Regression*

The target feature y is continuous in linear regression and can be written as a linear weighted form of each feature of the sample $\boldsymbol{x}$:

$$y = w_1x_1 + \cdots + w_dx_d + w_0 = \boldsymbol{w}^{\mathrm{T}}\boldsymbol{x}, \tag{4.1}$$

where $\boldsymbol{w}$ is the parameter vector and d is the feature dimension. Now, let's assume that we are solving the binary classification problem, that is, y takes values in the range of $\{1, -1\}$. Obviously, linear regression cannot directly solve this problem, because y takes a continuous real number rather than a discrete value. In order to be able to solve the binary classification problem using the regression method, we need to change the output of the linear regression.

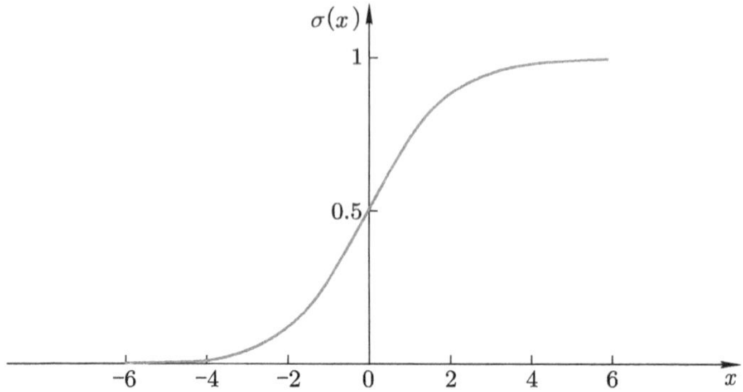

Figure 4.2 Logistic function.

The logistic regression method introduces a Logistic function that maps the continuous output to $(0,1)$. The Logistic function is defined as follows:

$$\sigma(x) = \frac{1}{1+\mathrm{e}^{-x}}, \tag{4.2}$$

where the corresponding function curve is shown in Figure 4.2. When the input x is large or small, the output is close to 0 or 1, and $\sigma(0) = 0.5$. The derivative of the Logistic function can be expressed function value, $\sigma'(x) = \sigma(x)(1-\sigma(x))$.

With Logistic function $\sigma(\cdot)$, we can map any real number to the interval $(0,1)$. The output can be interpreted as the probability that a sample belongs to the positive class $(y=1)$. We denote $f(x) = \boldsymbol{w}^{\mathrm{T}}\boldsymbol{x}$, in the logistic regression, the probability that sample $\boldsymbol{x}_i$ belongs to positive class is

$$p(y_i = 1|\boldsymbol{x}_i) = \sigma(f(\boldsymbol{x}_i)) = \frac{1}{1+\mathrm{e}^{-\boldsymbol{w}^{\mathrm{T}}\boldsymbol{x}_i}}, \tag{4.3}$$

the probability that sample $\boldsymbol{x}_i$ belongs to negative class $(y=-1)$ is

$$p(y_i = -1|\boldsymbol{x}_i) = 1 - p(y_i = 1|\boldsymbol{x}_i) = 1 - \sigma(f(\boldsymbol{x}_i)) = \frac{1}{1+\mathrm{e}^{\boldsymbol{w}^{\mathrm{T}}\boldsymbol{x}_i}}. \tag{4.4}$$

Comparing Eq. (4.3) and (4.4), considering that $y_i \in \{1,-1\}$, we can write them in a unified way

$$p(y_i|\boldsymbol{x}_i) = \sigma(f(\boldsymbol{x}_i)) = \frac{1}{1+\mathrm{e}^{-y_i\boldsymbol{w}^{\mathrm{T}}\boldsymbol{x}_i}}. \tag{4.5}$$

Eq. (4.5) defines the basic form of logistic regression. Based on linear regression, a nonlinear function is used to establish the relationship between the binary target feature and the original inputs.

4.1.2. *Parameter Estimation*

Suppose the training set is $D = \{(\boldsymbol{x}_1, y_1), (\boldsymbol{x}_2, y_2), \ldots, (\boldsymbol{x}_n, y_n)\}$. We can use the maximum likelihood method to estimate the parameter $\boldsymbol{w}$ of logistic regression. Using Eq. (4.5), we can write the likelihood function of the training set as

$$L(\boldsymbol{w}) = \prod_{i=1}^{n} p(y_i|\boldsymbol{x}_i). \tag{4.6}$$

Taking the logarithm of $L(\boldsymbol{w})$, we get the negative log likelihood of the training set

$$NLL(\boldsymbol{w}) = -\ln \prod_{i=1}^{n} p(y_i|\boldsymbol{x}_i) = -\sum_{i=1}^{n} \ln p(y_i|\boldsymbol{x}_i). \tag{4.7}$$

Now, our goal becomes: given a training set D, find the parameter $\boldsymbol{w}$ which makes the negative loglikelihood function (4.7) minimum, that is, solve the following optimization problem

$$\min_{\boldsymbol{w}} \quad NLL(\boldsymbol{w}). \tag{4.8}$$

A commonly used method for finding the minimum value of the objective function is gradient descent.[a] After initializing the parameter, the gradient descent method uses the following iteration formula to update the parameter:

$$\boldsymbol{w} \leftarrow \boldsymbol{w} - \eta \nabla NLL(\boldsymbol{w}), \tag{4.9}$$

where η is the learning rate and $\nabla NLL(\boldsymbol{w})$ is the gradient of the objective function for the parameter $\boldsymbol{w}$.

It can be seen that as long as we can calculate $\nabla NLL(\boldsymbol{w})$, we can use Eq. (4.9) to find the parameter of logistic regression. Now let's see how to calculate $\nabla NLL(\boldsymbol{w})$. Denote $f_i(\boldsymbol{w}) = y_i \boldsymbol{w}^{\mathrm{T}} \boldsymbol{x}_i$, and apply

[a]The introduction of the gradient descent method can be found in Appendix C.

the chain rule to compute the derivative of the composite function, which is:

$$\begin{aligned}\nabla NLL(\boldsymbol{w}) &= -\sum_{i=1}^{n} \frac{\partial \ln p(y_i|\boldsymbol{x}_i)}{\partial \boldsymbol{w}} \\ &= -\sum_{i=1}^{n} \frac{1}{p(y_i|\boldsymbol{x}_i)} \frac{\partial p(y_i|\boldsymbol{x}_i)}{\partial \boldsymbol{w}} \\ &= -\sum_{i=1}^{n} \frac{1}{p(y_i|\boldsymbol{x}_i)} \frac{\partial p(y_i|\boldsymbol{x}_i)}{\partial f_i(\boldsymbol{w})} \frac{\partial f_i(\boldsymbol{w})}{\partial \boldsymbol{w}}.\end{aligned} \tag{4.10}$$

Observing Eq. (4.5), we can see that $p(y_i|\boldsymbol{x}_i) = \sigma(f_i(\boldsymbol{w}))$, thus we have

$$\frac{\partial p(y_i|\boldsymbol{x}_i)}{\partial f_i(\boldsymbol{w})} = p(y_i|\boldsymbol{x}_i)(1 - p(y_i|\boldsymbol{x}_i)). \tag{4.11}$$

Since $f_i(\boldsymbol{w}) = y_i \boldsymbol{w}^{\mathrm{T}} \boldsymbol{x}_i$, it is easy to obtain

$$\frac{\partial f_i(\boldsymbol{w})}{\partial \boldsymbol{w}} = \frac{\partial y_i \boldsymbol{w}^{\mathrm{T}} \boldsymbol{x}_i}{\partial \boldsymbol{w}} = y_i \boldsymbol{x}_i. \tag{4.12}$$

Substituting Eq. (4.11) and (4.12) into Eq. (4.10), we get

$$\nabla NLL(\boldsymbol{w}) = -\sum_{i=1}^{n} \left(1 - \sigma(y_i \boldsymbol{w}^{\mathrm{T}} \boldsymbol{x}_i)\right) y_i \boldsymbol{x}_i. \tag{4.13}$$

Now substituting Eq. (4.13) into Eq. (4.9), we can use the following iteration formula to estimate $\boldsymbol{w}$:

$$\boldsymbol{w} \leftarrow \boldsymbol{w} - \eta \sum_{i=1}^{n} \left(1 - \sigma(y_i \boldsymbol{w}^{\mathrm{T}} \boldsymbol{x}_i)\right) y_i \boldsymbol{x}_i. \tag{4.14}$$

4.1.3. *Summary*

Logistic regression can be used for binary classification problems, such as classifying e-mails into normal email and spam email, or predicting loans as default and non-default. In real applications, the prediction outcome may belong to multiple classes. For example, a product review can be classified as positive, negative or neutral. In the optical character recognition task, a handwritten number is classified as 0-9. It is a 10 class classification problem. Logistic regression cannot be directly used to solve multiclass classification problems, we need to extend it. To simplify, we only need to replace the logistic function with the softmax function (4.2). Assuming that the input $\mathbf{z}$ is a C-dimensional vector, the softmax function is defined as

$$\sigma(\mathbf{z})_j = \frac{\mathrm{e}^{z_j}}{\sum_{c=1}^{C} e^{z_c}}, \quad j \in \{1, 2, \ldots, C\}. \tag{4.15}$$

Now, suppose the training set is $D = \{(\boldsymbol{x}_1, y_1), (\boldsymbol{x}_2, y_2), \ldots, (\boldsymbol{x}_n, y_n)\}$, where each sample may belong to one of C categories, i.e., $y_i \in 1, 2, \ldots, C$. In logistic regression, we only have a d-dimensional parameter vector $\boldsymbol{w}$. In the multi-class case, we need one parameter vector $\boldsymbol{w}_c$ for each category c. Our parameter is $\mathbf{W} = \{\boldsymbol{w}_1, \boldsymbol{w}_2, \ldots, \boldsymbol{w}_C\}$. For a particular training sample $(\boldsymbol{x}_i, y_i)$, the conditional probability $p(y_i|\boldsymbol{x}_i)$ can be written as

$$p(y_i|\boldsymbol{x}_i) = \frac{e^{\boldsymbol{w}_{y_i}^{\mathrm{T}} \boldsymbol{x}_i}}{\sum_{c=1}^{C} e^{\boldsymbol{w}_i^{\mathrm{T}} \boldsymbol{x}_i}}, \tag{4.16}$$

the estimation of parameter $\mathbf{W}$ can also adopt the maximum likelihood method, please refer to the previous section.

4.2. K-Nearest Neighbor

K-nearest neighbor (KNN) is one of the most classic and simple supervised learning methods. KNN is a good choice when there is little or no prior knowledge of the data distribution. KNN can be used to solve both classification and regression problems. The intuition is simple: when classifying a test sample, the training set is first scanned, k training samples most similar to the test sample are found, and the categories of the test sample are determined according to the categories of the k samples. A weighted vote can also be made by the degree of similarity between the k samples and the test sample. If you want to output the probability that the test sample belongs to each class, you can estimate it by the empirical distribution of the sample belonging to each class.

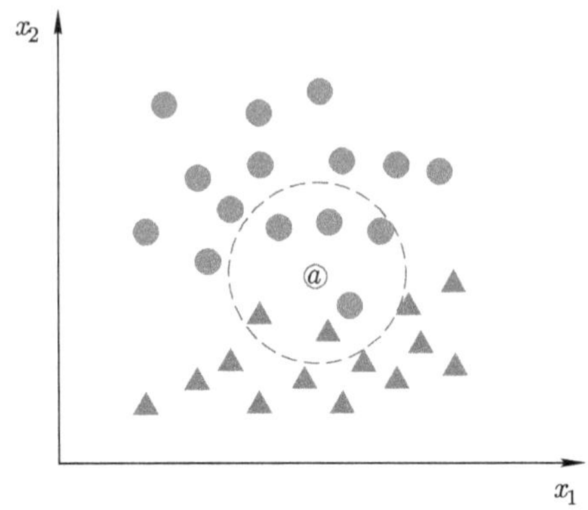

Figure 4.3 Illustration of the prediction of the KNN algorithm.

We first use an example to illustrate the idea of K-nearest neighbor. Suppose our samples are divided into positive and negative, which are represented by solid circles and solid triangles respectively in Figure 4.3. Now, given a test sample a, as shown by the open circle in the figure, we need to predict its label. We first look for the 6 samples closest to a, and find that 4 of them are positive and 2 are negative. Among the neighbor samples of a, there are more positive samples than negative samples, so we predict a as a positive class.

Specifically, the detailed process of the K-nearest neighbor algorithm is as follows:

1. Determine the size k and the method to calculate distance;
2. Obtain k samples that are most similar to the test sample from the training samples;
3. Determine the class label of the test sample by voting according to the class of the k most similar training samples.

For K-nearest neighbor algorithm, there are three key problems:

1. How to find the "nearest neighbors" of the test sample, that is, how to calculate the distance or similarity between samples;
2. How to choose the size k to achieve the best prediction performance;
3. How to make predictions faster when the number of samples in the training set is large or the dimension is very large.

An important problem in the K-nearest neighbor algorithm is to calculate the distance between samples to determine which samples in the training set are closer to the test sample. The distance metric is very important, which often affects the final predictions. For example, Euclidean distance is a good choice when the sample data can be represented as a multidimensional continuous value.

In practical applications, we need to choose the distance calculation method according to the characteristics of the application and the data itself. Commonly used distance metrics include Euclidean distance, Manhattan distance, Mahalanobis distance, Hamming distance, cosine similarity, and Jaccard distance (see Appendix D for details). When the existing distance metrics cannot meet the actual application requirements, it is also necessary to propose a *suitable* distance metric.

4.2.1. *Choosing the Value of k*

In the K-nearest neighbor algorithm, different choices for the value of k have an impact on the final results. For example, in Figure 4.4, when $k = 3$ and $k = 6$, the same test sample will be classified into different categories.

In general, starting from $k = 1$, as k increases, the classification performance will gradually increase. After increasing k to a certain

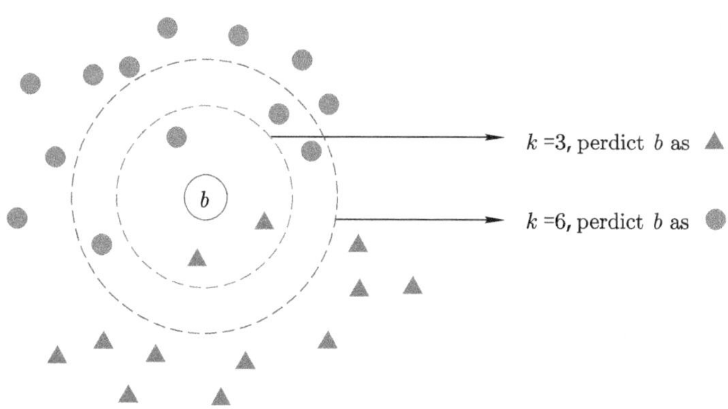

Figure 4.4 The influence of different values of k on the prediction results.

value, the classification performance will gradually decrease as k further increases. When k increases to the value such that the sample size of training set is the same as the whole dataset, the K-nearest neighbor algorithm will have the same prediction result for any test sample. Determining the optimal k is difficult for a specific application problem. It is often necessary to evaluate the performance of the model under different values through cross-validation or other methods.

4.2.2. *Improving Prediction Speed*

For a single test sample, the distance between the test sample and each training sample needs to be calculated. The time complexity is $O(n)$,[b] where n is the number of samples in the training set. This complexity is not particularly high in itself. Unfortunately, we need $O(n)$ time for each test sample. The time complexity is $O(mn)$ (m is the number of test set samples). From this perspective, the time cost of the prediction of the K-nearest neighbor algorithm is very high. Therefore, we need to use some indexing technology to shorten the prediction time and improve the prediction speed. One method is to design a data structure for the training data for indexing and hierarchically partition the search space. With this data structure we are able to search for the k training samples that are most similar to any test sample at higher speed.

The most commonly used data structure is the k-d tree (k-dimensional tree), which is the extension of the binary search tree in the multi-dimensional space. Each internal node in the k-d tree holds two kinds of information: (1) a hypercube in a multidimensional space; and (2) a hyperplane perpendicular to a certain dimension. A node is split into two child nodes by the hyperplane of each node. When the number of samples that falls in a hypercube on a node is less than a given threshold, the node no longer splits further. In the K-nearest neighbor algorithm, the role of the k-d tree

[b]O is a representation of the complexity of the algorithm. Suppose $f(n)$ and $g(n)$ are functions defined on a set of positive integers. If there is a positive integer n_0 and a positive constant c, such that $f(n) \leqslant cg(n)$ when $n \geqslant n_0$, then this is denoted as $f(n) = O(g(n))$.

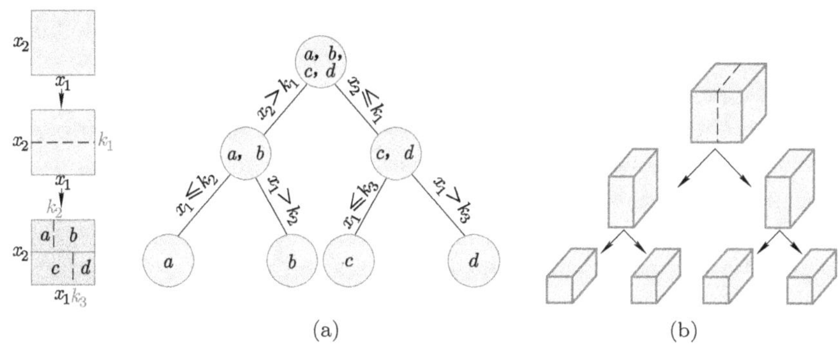

Figure 4.5 k-d trees. (a) k-d tree in two-dimensional space. (b) k-d tree in three-dimensional space.

is to build an index on the training dataset, so that when predicting, it is possible to quickly find a sample that similar to the test sample. Figure 4.5(a) and Figure 4.5(b) show the k-d tree in two-dimensional and three-dimensional space, respectively.

As shown in Figure 4.5(a), in two-dimensional space, compare with k_1, from x_2 dimension, we can divide the space into two regions, and then on the x_1 dimension, compare respectively with k_2 and k_3, then we can divide the space into four regions. In this way, one only needs to compare twice, then a test sample can be classified into one of a, b, c, and d.

4.2.3. *Summary*

Although the K-nearest neighbor algorithm is simple, it has a solid theoretical foundation. Researchers have shown that for the binary classification problem, if the training set is infinite, the upper bound of the generalization error is twice the *Bayes error rate*.[38]

In addition, the K-nearest neighbor algorithm has the advantage of insensitive to outliers. When the training set is sufficiently large, the K-nearest neighbor algorithm can obtain good classification results. However, it is not computationally efficient because each test sample requires a large number of calculations to obtain its neighbors, especially when the feature dimension is large. When the training set is small, the K-nearest neighbor algorithm tends to cause overfitting.

It should be noted that as a general idea, the K-nearest neighbor algorithm can be used not only to solve classification problem, but also to solve regression problem. For a test sample $\boldsymbol{x}$, we can first calculate the distance between the test sample and each sample in the training set using Euclidean distance or other distance metrics, assuming $d(\boldsymbol{x}_i, \boldsymbol{x}_j)$ represents the distance between sample $\boldsymbol{x}_i$ and sample $\boldsymbol{x}_j$. Then sort the training samples by ascending order according to the distance, select the first k training samples, and denote them as $\{(\boldsymbol{x}_1, y_1), \ldots, (\boldsymbol{x}_k, y_k)\}$. Finally, use the following formula to calculate the predicted value $\hat{y}$ of the sample $\boldsymbol{x}$:

$$\hat{y} = \sum_{i=1}^{k} \frac{y_i}{d(\boldsymbol{x}, \boldsymbol{x}_i)}. \tag{4.17}$$

4.3. Decision Tree

There are many scenarios in real life that require us to make decisions through effectively asking questions. Let us take the flu diagnosis as an example. The doctor asks about the patient's various symptoms (headache, fever, etc.) to get the basic symptoms of the patient, and then according to his own rules of experience, he diagnoses whether the patient has a cold or a flu. When approving a loan the bank also needs to make decision about whether or not to lend money based on the borrower's basic information such as income, education level, and marital status.

The rules used by banks may be as follows:

Rule 1: If the borrower's income is high, the borrower will not default;
Rule 2: If the borrower has a medium income and a bachelor's or postgraduate degree, the borrower will not default;
Rule 3: If the borrower has a medium income and a high school degree or below, the borrower will default;
Rule 4: If the borrower's income is low, the borrower will default.

The above decision rules of the bank can be represented as a tree structure, as shown in Figure 4.6. Each non-leaf node in the tree represents a question on a feature. There are two levels of questions:

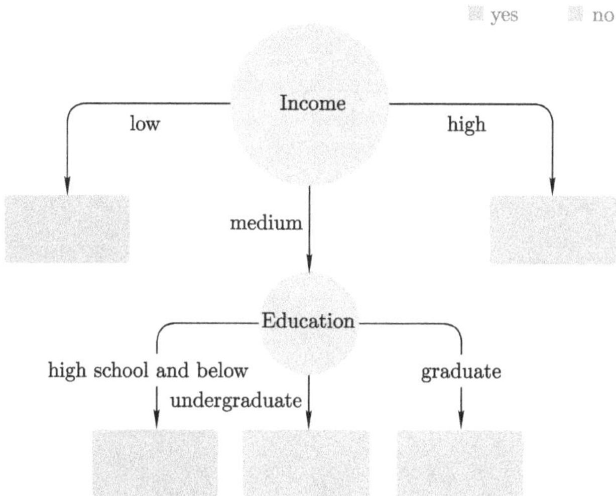

Figure 4.6 The decision tree for bank loans.

The first is on which feature to ask questions about, such as on education level. The second is what questions to ask with regards to the feature, such as "graduate?" "undergraduate?" or "high school and below?". The leaf nodes in the tree represent the decision result, which is determined according to a series of questions on the path from the root node of the tree to the leaf node.

In data science, the model represented by the above form is called decision tree. Suppose we have a training set as shown in Table 4.1. The core problem of the decision tree model is how to automatically generate the decision tree shown in Figure 4.6 based on the training set.

4.3.1. *Decision Tree Generation*

The decision tree is generated from the root node and selects the corresponding feature; then, the splitting point of the node feature is selected, and the node is split according to the splitting point. For discrete features, the node is split according to the feature values. In Figure 4.6, the root node feature is "Income", according to its values "high", "medium" and "low", the root node is split into three child nodes. For continuous features, the root node is split to child

Table 4.1 The bank loan dataset.

No.	Gender	Income	Education	Martial status	Default or not
1	M	high	graduate	single	no
2	M	low	undergraduate	married	yes
3	F	high	high school and below	single	no
4	F	medium	high school and below	married	no
5	M	high	undergraduate	married	no
6	M	medium	undergraduate	married	yes
7	M	medium	high school and below	married	no
8	F	medium	graduate	single	yes
9	F	medium	graduate	single	yes
10	M	low	high school and below	single	yes

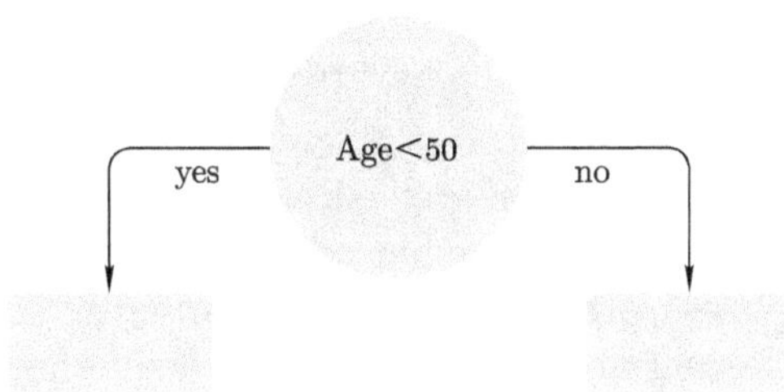

Figure 4.7 Example for decision tree node splitting.

nodes according to the specific feature value (splitting point). If the dataset of bank loans includes the "Age" feature, the splitting point can be selected as "50", and the corresponding node splitting diagram is shown in Figure 4.7. In general, when a node belongs only to a certain class (classification problem) or the variance is small (regression problem), the node is no longer split further.

The key question of decision tree generation is: how to choose node feature and feature splitting points? To answer the above question, we first introduce the concept of impurity. Impurity is used to indicate the degree of equilibrium of the class distribution of the samples that fall on the current node. The goal of the splitting node is to make the class distribution of the samples more uneven after the node splitting, that is, the impurity needs to be reduced. Therefore,

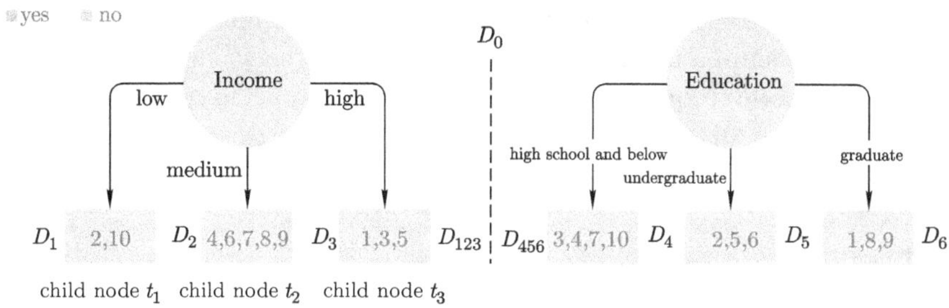

Figure 4.8 Choose node feature and splitting point according to the decrease in impurity.

each time we split a node, we choose the feature and splitting point pair that maximizes the decrease in impurity after the node splits.

As shown in Figure 4.8, assume the dataset is $D_0 = \{1, 2, 3, 4, 5, 6, 7, 8, 9, 10\}$, in which samples are represented by numbers. For the bank loan dataset, we take the features "Income" and "Education" to split the node. The first choice is to select feature "Income". After the node splitting, the dataset is divided into three parts: D_1, D_2 and D_3, which is denoted as D_{123}. The second choice is to select feature "Education level", after the node is split, the dataset is divided into three parts: D_4, D_5 and D_6, which is denoted as D_{456}. Suppose $Imp(\cdot)$ is the impurity of the node. Then, before and after splitting the node according to the first method, the decrease in the impurity is $Imp(D_0) - Imp(D_{123})$. Before and after splitting the node according to the second method, the decrease in the impurity is $Imp(D_0) - Imp(D_{456})$. Comparing the magnitude of the decrease in the impurity under the two split methods, the split method with larger decrease value is the better split method.

So, how to measure the impurity of nodes? Below we introduce three commonly used *metrics*: Gini index, entropy and misclassification error.

1. Gini index

Gini index is an indicator used by Italian researcher Gini in the 20th century to judge the fairness of social income distribution. It can reflect the equilibrium of the population distribution of various income levels in society.[39] In decision tree we can use Gini index to measure the impurity of the distribution of different categories of samples falling in a node. Assuming that the dataset has C categories, and the relative frequency of the category c samples in node t is $p(c|t)$, the Gini index of node t is

$$Gini(t) = 1 - \sum_{c=1}^{C} [p(c|t)]^2. \tag{4.18}$$

When the number of samples of each *category* in a node is evenly distributed, Gini index *takes* maximum value of $(1 - \frac{1}{C})$ when the samples all belong to one category, Gini index takes minimum value of. In the bank loan dataset, $C = 2$, as shown in Figure 4.6. If the root node is split according to "Income", the Gini indices for the three child nodes are as follows:

$$\begin{aligned} Gini(t_1) &= 1 - \left(\left(\frac{2}{2}\right)^2 + \left(\frac{0}{2}\right)^2\right) = 0; \\ Gini(t_2) &= 1 - \left(\left(\frac{3}{5}\right)^2 + \left(\frac{2}{5}\right)^2\right) = 0.480; \\ Gini(t_3) &= 1 - \left(\left(\frac{0}{3}\right)^2 + \left(\frac{3}{3}\right)^2\right) = 0. \end{aligned} \tag{4.19}$$

We have already introduced the calculation of Gini index by the number of samples of different classes in a node t. Assume the number of samples in node t is n, node t is split into K child nodes, wherein the number of samples in the k-th child node t_k is n_k. For each child node t_k, we can calculate its Gini index $Gini(t_k)$ using Eq. (4.18). The weighted average of Gini indices of K child nodes represents the Gini index after splitting:

$$Gini_{split} = \sum_{k=1}^{K} \frac{n_k}{n} Gini(t_k). \tag{4.20}$$

We choose the splitting which maximizes the decrease of the Gini index, $Gini(t_0) - Gini_{split}$,where t_0 is the root node. It is obvious that $Gini(t_0) = 1 - ((\frac{5}{10})^2 + (\frac{5}{10})^2) = 0.500$. From Eq. (4.19), the Gini index after splitting according to the feature "Income" can be obtained as

$$Gini_{split} = \frac{2}{10} \times 0 + \frac{5}{10} \times 0.480 + \frac{3}{10} \times 0 = 0.240. \tag{4.21}$$

Then the Gini index decrease is $Gini(t_0) - Gini_{split} = 0.260$. Similarly, the Gini index decrease is 0.173, 0.020 and 0 respectively according "Education", "Marital status" and "Gender". Since the Gini index decrease of feature "Income" is largest we choose feature "Income" to split.

After splitting the root node, we find that node t_2 contains both default samples and non-default samples. So we need to split the node t_2. The strategy for selecting features is similar to that when splitting the root node. The complete decision tree is shown in Figure 4.9. The leaf nodes contain only samples of the same class.

2. Information entropy

Information entropy is a concept used to measure information uncertainty. Researchers use information entropy to measure node impurity. Suppose the dataset has C classes, $p(c|t)$ is the relative frequency of c-th class of node t, then the information entropy of note t is

$$Entropy(t) = -\sum_{c=1}^{C} p(c|t) \log_2 p(c|t). \tag{4.22}$$

When the samples in the node are evenly distributed in each category, information entropy takes maximum value of $\log_2 C$. This indicates that the node has the highest impurity. When all the samples belong to a certain class, the entropy takes minimum value of. For the three example nodes in Table 4.2, the

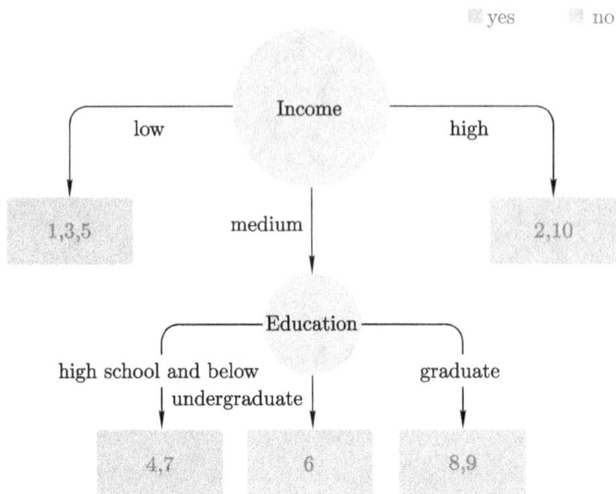

Figure 4.9 Complete decision tree based on Gini index.

Table 4.2 Number of samples of the two classes in different nodes.

No. of node	Default samples size	Non-default samples size
t_1	2	0
t_2	3	2
t_3	0	3

entropy is calculated as follows:

$$\begin{aligned} Entropy(t_1) &= -\frac{2}{2}\log_2\frac{2}{2} - \frac{0}{2}\log_2\frac{0}{2} = 0; \\ Entropy(t_2) &= -\frac{3}{5}\log_2\frac{3}{5} - \frac{2}{5}\log_2\frac{2}{5} = 0.971; \\ Entropy(t_3) &= -\frac{0}{3}\log_2\frac{0}{3} - \frac{3}{3}\log_2\frac{3}{3} = 0. \end{aligned} \tag{4.23}$$

In the above calculation we defined $0\log_2 0 = 0$. Based on node information entropy, we can calculate information entropy decrease (also called information gain):

$$InfoGain = Entropy(t_0) - \sum_{k=1}^{K}\frac{n_k}{n}Entropy(t_k). \tag{4.24}$$

For the current node, we choose the split with largest information gain. In practice, information gain tends to split the nodes of the decision tree into many leaf nodes (the number of samples of each node is small), which is easy to cause overfitting problems.

One way to overcome the above disadvantage is to adjust information gain using the sample size of each child node. Suppose node t_0 contains n samples and is split into k child nodes. The sample sizes of child nodes are $\{n_1, n_2, \ldots, n_K\}$, then the splitting information is

$$SplitInfo = -\sum_{k=1}^{K} \frac{n_k}{n} \log_2 \left(\frac{n_k}{n}\right). \tag{4.25}$$

Then, use the information gain ratio instead of the information gain as a criterion for evaluating the split:

$$\begin{aligned} &InfoGainRatio \\ &= \frac{InfoGain}{SplitInfo} = \frac{Entropy(t_0) - \sum_{k=1}^{K} \frac{n_k}{n} Entropy(t_k)}{-\sum_{k=1}^{K} \frac{n_k}{n} \log_2 \left(\frac{n_k}{n}\right)}. \end{aligned} \tag{4.26}$$

Information gain *ratio* adjusts information gain by splitting information, which can prevent a node from splitting into a large number of leaf nodes.

3. Misclassification error

Misclassification error is another method to measure node impurity. Assuming that the dataset has C categories, and the relative frequency of c-th class at node t is $p(c|t)$, the misclassification *error* of node t is

$$Error(t) = 1 - \max\left(p(1|t), p(2|t), \ldots, p(C|t)\right). \tag{4.27}$$

Misclassification error is the proportion of misclassified samples when the current node predicts majority category. When samples evenly distributed in each category, misclassification error reaches the maximum value of $\left(1 - \frac{1}{C}\right)$, indicating that impurity is the largest.

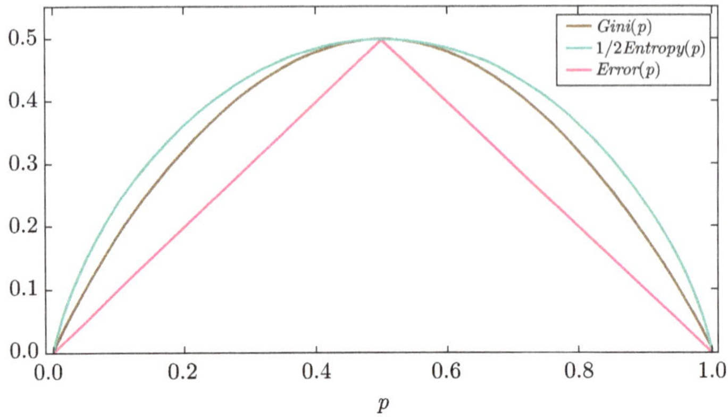

Figure 4.10 Impurity function curves for binary classification.

When all samples belong to one category, misclassification error gets the minimum value of. For the three nodes in Table 4.2, their misclassification errors are calculated as follows:

$$\begin{aligned} Error(t_1) &= 1 - \max\left(\frac{2}{2}, \frac{0}{2}\right) = 0; \\ Error(t_2) &= 1 - \max\left(\frac{3}{5}, \frac{2}{5}\right) = 0.400; \\ Error(t_3) &= 1 - \max\left(\frac{0}{3}, \frac{3}{3}\right) = 0. \end{aligned} \tag{4.28}$$

For binary classification problem, assuming that relative frequency of positive class is p, then relative frequency of negative class is $1 - p$. The above three impurity *metrics* are respectively

$$\begin{aligned} Gini(p) &= 2p(1-p); \\ Entropy(p) &= -p\log_2 p - (1-p)\log_2(1-p); \\ Error(p) &= 1 - \max(p, 1-p). \end{aligned} \tag{4.29}$$

The corresponding function curves are shown in Figure 4.10.

For the three impurity metrics, when the relative frequency is 0 or 1, the classification performance is the best, and the impurity is 0. When the relative frequency is around 0.5, the classification performance is the worst, and the three methods have the largest value. When p fixed, the impurity measured by information entropy is the largest, which also means the largest penalty for poor splitting.

4.3.2. *Decision Tree Algorithms*

Based on different impurity metrics, features types and target type, researchers propose several decision tree algorithms. In this section we introduce three of them: ID3, C4.5 and CART.

ID3 (Iterative Dichotomiser 3) is a classic decision tree algorithm proposed by Australian computer scientist Quinlan in 1986. It was the first decision tree algorithm.[40] ID3 uses information entropy as the impurity metric at a node, and chooses information gain as the node split evaluation criterion. Its detailed procedure is shown in Algorithm 1. ID3 tends to select features with more values and cannot handle continuous

---

Algorithm 1 ID3

Input: Training set D, target feature Y, feature set F.
Output: Decision tree T.
1: Create a root node.
2: **if** the Samples' target feature are all of positive class: **then**
3: Return a decision tree T with a single node predicting only positive class;
4: **if** the Samples' target are all of negative class: **then**
5: Return a decision tree T with a single node predicting only negative class;
6: **if** feature set F is empty: **then**
7: Return a decision tree T with a single node predicting the majority class;
8: Using (4.24) to compute information gain of features in F, denote A as the feature with maximum information gain;
9: Set A as the root node of the decision tree T;
10: Denote $D(i)$ the subset of samples whose feature A takes the value i;
11: **for** each value i of the feature A **do**
12: **if** $D(i)$ is empty: **then**
13: Set a new leaf node in this sub-branch, and predicting the majority class
14: **else**: recursively invoke the ID3 algorithm in this sub-branch to build a new sub-tree.

---

features. Furthermore, ID3 may cause the sample size of the leaf node to be small, leading to the problem of overfitting.

Quinlan proposed C4.5 to overcome the above shortcomings.[41] First, for a continuous feature, the node is no longer split by enumerating all of its possible values, but by a splitting threshold. For example, for age feature, if the splitting threshold is 50, then two child nodes will be generated, respectively containing samples with ages less than or equal to 50 and greater than 50. Secondly, to avoid generating too many child nodes, C4.5 exploits information gain ratio instead of information gain as the node split evaluation metric. Since information gain ratio considers the node splitting information, it is no longer biased to select those discrete features with more values.

ID3 and C4.5 are mainly used to solve classification problem. Breiman, Friedman, *et al.* proposed CART algorithm (Classification And Regression Tree algorithm) to solve both classification and regression problem.[42] For classification CART uses Gini index to measure node impurity, and uses the decrease in Gini index value as the node split evaluation metric. For regression, variance of the target feature of the node is used as a measure of impurity, and variance decrease value is used as an evaluation metric for node splitting.

A major difference between CART and ID3 or C4.5 is that each node is only split into two child nodes. For continuous feature, by selecting splitting threshold, a node can be split into two child nodes. For discrete feature, CART no longer generates one child node for each feature value, but selects a specific feature value each time. The node is split based on whether a sample takes that feature value. For example, for discrete feature "Education", assume its possible values are {high school or below, undergraduate, graduate}. Assume value "high school and below" is currently considered when splitting a node, then the node is split into two child nodes according to whether a sample takes the value "high school or below" or not.

Table 4.3 compares the above three algorithms from the dimensions of feature type, impurity and number of splitting child nodes.

4.3.3. *Pruning of Decision Trees*

The decision tree algorithm may generate too complex tree (for example, too many layers, too many nodes in the tree, etc.). The learned

Table 4.3 Comparison of three decision tree algorithms.

Algorithm	Input feature type	Target feature type	Impurity	Number of child nodes
ID3	discrete	discrete	Information gain	$K \geqslant 2$
C4.5	discrete and continuous	discrete	Information gain ratio	$K \geqslant 2$
CART	discrete and continuous	discrete and continuous	Gini index, variance	$K = 2$

decision tree model may not perform well on the test set, leading to the over-fitting problem. One reason may be that the training set contains outliers. As shown in Figure 4.11, the impurity of node t_j has already met the requirements. Splitting node t_j into two leaf nodes t_m and t_{m+1} may improve the classification accuracy on training set, while leads to poor performance on test set.

We can control the complexity of tree through pruning strategy. According to whether the pruning is performed during the decision tree generation process or after the decision tree is generated, it can be divided into prepruning and postpruning. Prepruning refers to setting a threshold for the decrease value in impurity during the generation of the decision tree. If the decrease in impurity is less than

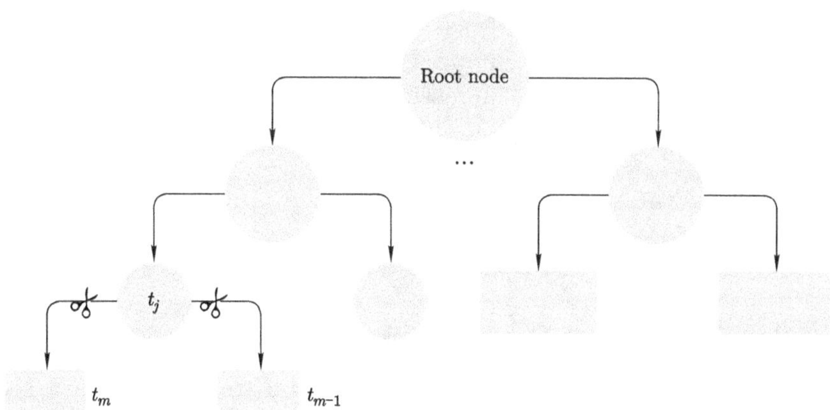

Figure 4.11 Overly complex splits may lead to poor performance.

a given threshold, then the splitting of the node is stopped and the node is marked as a leaf node. Postpruning refers to the procedure in which first a decision tree model is completely built using the training set, and after that, the decision tree is pruned. Prepruning is simple and straightforward, but postpruning has proven to be more successful in practice.

Let's take postpruning as an example to discuss how to pruning a decision tree. Before this, we first need a metric to measure the performance of the whole decision tree. Through this metric, we can compare the decision tree before and after pruning to determine whether a specific pruning operation is to be performed. Assuming that there are C classes in the data, a tree T includes $|T|$ nodes, and the sample size in node t is n_t. Then, the performance metric decision tree (here we call the overall loss function) can be defined as

$$Cost_\alpha(T) = \sum_{t=1}^{|T|} n_t Imp(t) + \alpha|T|, \tag{4.30}$$

where $Imp(t)$ is impurity for node t, such as the Gini index, information entropy or misclassification error we introduced in Section 4.3.1. The first term of Eq. (4.30) is the degree of fitting of the decision tree to the training set, and the second term is the complexity of the decision tree model. The decrease in the first term will cause the second item to increase, and vice versa. By minimizing Eq. (4.30), the fitting and complexity can be weighed, and the complexity control parameter α represents a trade-off we set in advance. A larger α value means a higher penalty for complex decision tree models.

4.3.4. *Analysis of Decision Trees*

After a decision tree is generated, it can be converted into a set of IF-THEN rules. This kind of rules coincide with human habits of decision-making behaviors, which makes the decision tree model highly interpretable. At the same time, through a certain visualization method, the learned decision tree model can be easily visualized. Strong interpretability and easy visualization make decision tree models widely used, such as in finance, health care, industrial production and manufacturing, astronomy and molecular biology.

We can also understand decision tree from the perspective of spatial division. The node splitting process of the decision tree is actually equivalent to the division of space. Assuming that a decision tree divides the space into M disjoint regions, the prediction function of the decision tree can be expressed as the form

$$f(\boldsymbol{x}) = \sum_{m=1}^{M} c_m I(\boldsymbol{x} \in R_m), \tag{4.31}$$

where $I(\cdot)$ is the indicator function. c_m refers to some average of the output values of the samples in the subspace R_m: in regression problem, it returns the average of the predicted values of the samples in the subspace; and in classification problems, it returns the majority class of the samples in the subspace.

Decision tree is simple in principle, easy to understand, has high precision, and is highly interpretable. It can handle missing features well. For example, missing value can be treated as new values. Decision tree uses the idea of greedy algorithm. It selects the feature with maximum information gain to split at each time, and is easy to fall into a local optimal solution. When used to determine the classification boundary, the decision boundary is a straight line parallel to the coordinate axis. The limited ability to express classification boundary may lead to poor performance in practical problem.

4.4. Naive Bayes

Naive Bayes is a classification method based on Bayes' theorem and feature conditional independence hypothesis. Feature conditional independence means that the value of any feature is not related to other features in a given sample class. Although this assumption does not hold in reality for many scenarios, Naive Bayes can achieve good performance. It has been widely used in text analysis, such as text classification, spam filtering, and sentiment analysis. In addition, Naive Bayes is used in medical diagnostics and recommendation systems.

4.4.1. *Bayes Theorem*

Consider the games between two basketball teams A and B. Suppose that 75% of games A wins and the remaining games B wins. In the winning games of A, only 20% is played in team B's home ground; and in the winning games of B, 60% are home wins. If the next game is played at team B's home ground, which team is more likely to win? This problem can be solved by the famous Bayes theorem. Suppose X, Y is a pair of random variables, and their joint probability $p(X = x, Y = y)$ refers to the probability that X takes x and Y takes y. The conditional probability $p(X = x|Y = y)$ refers to the probability that the variable X takes x when Y takes y. The joint probability and conditional probability of X and Y satisfy:

$$p(X, Y) = p(Y|X)p(X) = p(X|Y)p(Y), \tag{4.32}$$

then the Bayes theorem is

$$p(Y|X) = \frac{p(X|Y)p(Y)}{p(X)}. \tag{4.33}$$

Suppose X stands for the sample and Y stands for the class label of the sample. Bayes theorem is a statistical principle that combines prior knowledge with evidence obtained from samples. In Eq. (4.33), $p(Y)$ is called a prior distribution, $p(X|Y)$ is called a likelihood function, $p(X)$ is called evidence, $p(Y|X)$ is called the posterior distribution.

Now let's answer the above question. For ease of presentation, the random variable X represents the host and the random variable Y represents the winner of the game. X and Y can take value in the set

$\{\text{A}, \text{B}\}$. Based on the known information, the following results can be obtained:

- The probability of A winning the game is $p(Y = \text{A}) = 0.75$;
- The probability of B winning the game is $p(Y = \text{B}) = 1 - p(Y = \text{A}) = 0.25$;
- The probability of B being the host when B wins is $p(X = \text{B}|Y = \text{B}) = 0.60$;
- The probability of B being the host when A wins is $p(X = \text{B}|Y = \text{A}) = 0.20$.

Our goal is to calculate $p(Y = \text{B}|X = \text{B})$. Using Bayes theorem, we obtain

$$
\begin{aligned}
p(Y &= \text{B}|X = \text{B}) \\
&= \frac{p(X = \text{B}|Y = \text{B})p(Y = \text{B})}{p(X = \text{B})} \\
&= \frac{p(X = \text{B}|Y = \text{B})p(Y = \text{B})}{p(X = \text{B}, Y = \text{B}) + p(X = \text{B}, Y = \text{A})} \\
&= \frac{p(X = \text{B}|Y = \text{B})p(Y = \text{B})}{p(X = \text{B}|Y = \text{B})p(Y = \text{B}) + p(X = \text{B}|Y = \text{A})p(Y = \text{A})} \\
&= \frac{0.60 \times 0.25}{0.60 \times 0.25 + 0.20 \times 0.75} \\
&= 0.5. \qquad (4.34)
\end{aligned}
$$

It can be seen that if the next game is played at B's home ground, the probability of B winning is only 50%, and the probability of A winning is also 50%. A and B have the same probability to win the game.

4.4.2. *Naive Bayes Model*

Suppose the training set $D = \{(\boldsymbol{x}_1, y_1), (\boldsymbol{x}_2, y_2), \cdots, (\boldsymbol{x}_n, y_n)\}$ contains n samples. The sample $\boldsymbol{x} = \{x_1, x_2, \cdots, x_d\}$ is a d-dimensional vector and the class label y takes value in $\{1, 2, \cdots, C\}$. Suppose that $\mathbf{X} = \{X_1, X_2, \cdots, X_d\}$ means input d-dimensional random variables and Y means categorical variables. Using the Bayes *theorem*, we can find a strategy to classify test samples $\boldsymbol{x}$:

$$p(Y=c|X=\boldsymbol{x})=\frac{p(X=\boldsymbol{x}|Y=c)p(Y=c)}{p(X=\boldsymbol{x})}. \tag{4.35}$$

For a given test sample $\boldsymbol{x}$, $p(X=\boldsymbol{x})$ is independent of the class label. Therefore, the above formula can be further simplified to

$$p(Y=c|X=\boldsymbol{x}) \propto p(X=\boldsymbol{x}|Y=c)p(Y=c). \tag{4.36}$$

The prior distribution $p(Y=c)$ can be estimated by counting the samples in the training set. The key now is to calculate the conditional probability distribution $p(X=\boldsymbol{x}|Y=c)$:

$$p(X=\boldsymbol{x}|Y=c)=p(X_1=x_1,X_2=x_2,\ldots,X_d=x_d|Y=c). \tag{4.37}$$

Naive Bayes makes the conditional independence of the above conditional probability distributions, that is, conditioning on a given class label, the features are independent of each other. This is a strong condition and the reason why the naive Bayes is called "naive". Now, $p(X=\boldsymbol{x}|Y=c)$ can be written in a multiplicative form:

$$p(X=\boldsymbol{x}|Y=c)=\prod_{i=1}^{d}p(X_i=x_i|Y=c). \tag{4.38}$$

Combining (4.36) and (4.38), we can get the naive Bayes model. It uses the following formula to predict sample's class label $\hat{y}$:

$$\hat{y}=\underset{c\in\{1,2,\cdots,C\}}{\operatorname{argmax}}\ p(Y=c)\prod_{i=1}^{d}p(X_i=x_i|Y=c). \tag{4.39}$$

4.4.3. *Parameter Estimation*

The parameters of naive Bayes algorithm are the priori parameter $p(Y=c)$ and the conditional parameter $p(X_i=x_i|Y=c)$. Given the training set $D=\{(\boldsymbol{x}_1,y_1),(\boldsymbol{x}_2,y_2),\ldots,(\boldsymbol{x}_n,y_n)\}$, the maximum likelihood estimate for $p(Y = c)$ is

$$p(Y = c) = \frac{\sum_{i=1}^{n} I(y_i = c)}{n}, \tag{4.40}$$

where $I(\cdot)$ is the indication function, its value is 1 when the input is true, otherwise 0. $\sum_{i=1}^{n} I(y_i = c)$ is equivalent to the number of samples in the training set belong to class c. It can be seen that the estimate of the prior probability $p(Y = c)$ can be completed by simply counting the number of samples in each category of the training set.

Now let's see how to estimate the conditional parameter $p(X_i = x_i|Y = c)$. We need to distinguish whether feature X_i is a discrete or continuous. If X_i is discrete, we can use a simple counting method to estimate the conditional distribution. Suppose X_i can take values of $\{v_1, v_2, \ldots, v_K\}$, the maximum likelihood estimate of the conditional probability $p(X_i = v_k|Y = c)$ is

$$p(X_i = v_k|Y = c) = \frac{\sum_{i=1}^{n} I(x_i = v_k, y_i = c_k)}{\sum_{i=1}^{n} I(y_i = c)}. \tag{4.41}$$

If X_i is continuous, there are usually two ways to deal with it. The first method is to discretize the continuous feature, as discussed in Section 2.4, and then use Eq. (4.41) to estimate the conditional parameters. The second method assumes that the continuous feature obeys a certain probability distribution, and then uses the training set to estimate the parameters of the distribution. Given a certain class c, assume that X_i obeys a normal distribution $N(\mu, \sigma^2)$ of mean μ and variance of σ^2, and $p(X_i = x|Y = c)$ can be written as follows:

$$p(X_i = x|Y = c) = \frac{1}{\sqrt{2\pi}\sigma} \mathrm{e}^{-\frac{(x-\mu)^2}{2\sigma^2}}. \tag{4.42}$$

By using the maximum likelihood estimation, it can be seen that the mean and variance of class c in the training set on feature X_i can be used as an estimate of the mean μ and the variance σ^2.

4.4.4. *Algorithm Analysis*

The most common application scenario for Naive Bayes is spam filtering. The basic idea is to calculate the posterior probability distribution of a given text by identifying words that are often found in some spam messages, and to determine whether the text is spam, which is a typical binary classification problem.

The Naive Bayes classifier is stable because outliers are averaged when estimating conditional probabilities from the training set. Naive Bayes classifier can also handle missing values by ignoring certain samples. For data with irrelevant features, Naive Bayes classifier is robust. If X_i is an irrelevant feature, then $p(X_i|Y)$ is likely to a uniform distribution whose conditional probability does not affect the total posterior probability. However, if the conditional independence assumption is not met, the performance may be degraded.

4.5. Support Vector Machine (SVM)

Support vector machine (SVM) achieve good generalization by maximizing the classification margin to determine the optimal decision hyperplane. SVM maps low-dimensional data to high-dimensional space by means of kernel function, and makes data in high-dimensional space linearly separable, so that it can deal with the non-linearly separable problems in low-dimensional space. SVM is mainly used in text recognition, text classification, face recognition in pattern recognition, and also applied to many engineering techniques and information filtering.

4.5.1. *Margin and Support Vector*

When we have a dataset with class information, how do we classify new sample using the SVM model? Suppose we have a dataset that describes if a patient has cancer:

$$D = \{(\boldsymbol{x}_1, y_1), (\boldsymbol{x}_2, y_2), \dots, (\boldsymbol{x}_n, y_n)\}, \quad y_i \in \{-1, +1\},$$

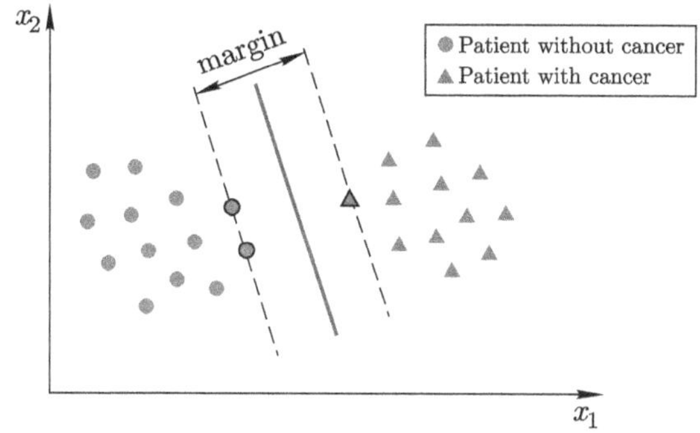

Figure 4.12 The meaning of margin.

where n is the number of samples, the feature dimension is d, and the sample is divided into positive class ($y_i = +1$, indicating patient with cancer) and negative class ($y_i = -1$, indicating patient without cancer). Our goal is to find a suitable line[c] which makes the margin between the samples of positive and negative classes the largest, as shown in Figure 4.12, where the triangle represents positive class and the circle represents negative class.

First we need to define margin. For a given training set and the separating hyperplane $\boldsymbol{w}^{\mathrm{T}}\boldsymbol{x} + b = 0$, the functional margin for the separating hyperplane with respect to the sample point $(\boldsymbol{x}_i, y_i)$ is

$$\widehat{\gamma}_i = y_i(\boldsymbol{w}^{\mathrm{T}}\boldsymbol{x}_i + b). \tag{4.43}$$

The function margin represents the correctness and confidence of the classification prediction. If the function margin is greater than zero, the classification is correct; if the function margin is less than zero, the classification is wrong, and the larger the value, the greater the confidence of the classification. When $\boldsymbol{w}$ and b are scaled by the same factor, the function margin will change, but the separating hyperplane $\boldsymbol{w}^{\mathrm{T}}\boldsymbol{x} + b = 0$ will not change at this time, so the function

[c]In the case of high dimensions, this line is called the separating hyperplane, also known as the decision hyperplane.

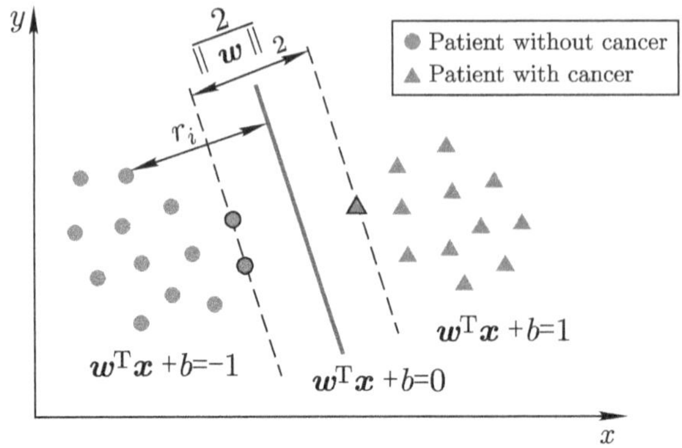

Figure 4.13 Margin and support vectors.

margin is not conducive to finding the separating hyperplane with the maximum margin. We define the geometric margin as

$$r_i = \frac{\widehat{\gamma}_i}{\|\boldsymbol{w}\|_2} = \frac{y_i(\boldsymbol{w}^{\mathrm{T}}\boldsymbol{x}_i + b)}{\|\boldsymbol{w}\|_2}. \tag{4.44}$$

The geometric margin is numerically equal to the distance from the sample point to the separating hyperplane. The geometric margin does not change as $\boldsymbol{w}$ and b are scaled by the same factor at the same time, so we use geometric margin to find the optimal separating hyperplane. As shown in Figure 4.13, the distance of the positive sample $(\boldsymbol{x}_i, y_i = +1)$ to the hyperplane is $\frac{\boldsymbol{w}^{\mathrm{T}}\boldsymbol{x}_i + b}{\|\boldsymbol{w}\|_2}$, and the distance of the negative sample $(\boldsymbol{x}_i, y_i = -1)$ to the hyperplane is $\frac{-(\boldsymbol{w}^{\mathrm{T}}\boldsymbol{x}_i + b)}{\|\boldsymbol{w}\|_2}$, where $\boldsymbol{w}$ is the normal vector of the separating hyperplane, b is the intercept. Therefore, the distance from any sample $(\boldsymbol{x}_i, y_i)$ to the hyperplane can be uniformly expressed as $\frac{y_i(\boldsymbol{w}^{\mathrm{T}}\boldsymbol{x}_i + b)}{\|\boldsymbol{w}\|_2}$. Margin means the minimum geometric margin from the training samples to the separating hyperplane.

Support vectors refer to points closest to the separating hyperplane. The optimal separating hyperplane is mainly determined by these support vectors, which is why this method is called "support vector machine", as shown by the circle and triangle data points with boxes in Figure 4.13.

The learning goal of SVM is to find a separating hyperplane, so that the minimum distance of training set to the hyperplane is maximized. The mathematical formula is expressed as follows:

$$\max_{\boldsymbol{w},b} \quad \min_{\boldsymbol{x}_i} \quad \frac{y_i(\boldsymbol{w}^{\mathrm{T}}\boldsymbol{x}_i + b)}{\|\boldsymbol{w}\|_2}, \tag{4.45}$$

$\min_{\boldsymbol{x}_i} \frac{y_i(\boldsymbol{w}^{\mathrm{T}}\boldsymbol{x}_i+b)}{\|\boldsymbol{w}\|_2}$ will not change when $\boldsymbol{w}$ and b are scaled by the same factor. For the convenience of subsequent analysis, assume that $\min_{\boldsymbol{x}_i} y_i(\boldsymbol{w}^{\mathrm{T}}\boldsymbol{x}_i + b) = 1$, that is, let the minimum function margin be 1. When the minimum function margin is 1, the function margin of other samples should not be less than 1, and a linear inequality constraint $y_i(\boldsymbol{w}^{\mathrm{T}}\boldsymbol{x}_i + b) \geqslant 1$ is introduced for each sample. Now, our optimization goal is:

$$\max_{\boldsymbol{w},b} \quad \frac{1}{\|\boldsymbol{w}\|_2}, \quad \text{s.t.} \quad y_i(\boldsymbol{w}^{\mathrm{T}}\boldsymbol{x}_i + b) \geqslant 1, \quad i = 1, 2, \ldots, n. \tag{4.46}$$

The above optimization problem is equivalent to

$$\min_{\boldsymbol{w},b} \quad \frac{1}{2}\|\boldsymbol{w}\|_2^2, \quad \text{s.t.} \quad y_i(\boldsymbol{w}^{\mathrm{T}}\boldsymbol{x}_i + b) \geqslant 1, \quad i = 1, 2, \ldots, n. \tag{4.47}$$

Equation (4.47) is the primal problem of the linear support vector machine. It is a quadratic programming problem with linear inequality constraints. It can be solved by existing optimization calculation package, using the Newton method, quasi-Newton method and other numerical optimization algorithms. The complexity is $O(d^3)$, where d is the feature dimension. However, when the feature dimension is large, the optimization algorithm needs to invoke optimization calculation package multiple times, leading to low computation efficiency.

4.5.2. *Dual Problem and SMO Algorithm*

Now we adopt the Lagrangian duality theory to obtain the dual problem of the linear support vector machine. The advantages include: the dual problem is often easier to solve; it is natural to introduce the kernel function to generalize the problem to nonlinear case.

Considering the original problem (4.47), we convert it into an unconstrained optimization problem using the Lagrangian method, and the corresponding Lagrangian function $L(\boldsymbol{w}, b, \boldsymbol{\alpha})$ can be written as

$$L(\boldsymbol{w}, b, \boldsymbol{\alpha}) = \frac{1}{2}\|\boldsymbol{w}\|_2^2 - \sum_i \alpha_i \left(y_i(\boldsymbol{w}^{\mathrm{T}}\boldsymbol{x}_i + b) - 1\right). \tag{4.48}$$

Now our problem is

$$\min_{\boldsymbol{w},b} \quad \max_{\alpha_i \geqslant 0} \quad L(\boldsymbol{w}, b, \boldsymbol{\alpha}), \tag{4.49}$$

Since this is a convex quadratic programming problem that satisfies the Slater condition, the above problem is equivalent to

$$\max_{\alpha_i \geqslant 0} \quad \min_{\boldsymbol{w},b} \quad L(\boldsymbol{w}, b, \boldsymbol{\alpha}). \tag{4.50}$$

When the Slater condition is satisfied, the solution of the original problem is equivalent to the solution of the dual problem. We can solve the dual problem using KKT conditions (see Eq. (C.12)). To solve the internal minimization problem, we compute the partial derivatives of the Lagrangian function with respect to $\boldsymbol{w}$ and b, and make the partial derivatives equal to zero.

$$\frac{\partial L(\boldsymbol{w}, b, \boldsymbol{\alpha})}{\partial \boldsymbol{w}} = \boldsymbol{w} - \sum_i \alpha_i y_i \boldsymbol{x}_i = \boldsymbol{0}, \tag{4.51}$$

$$\frac{\partial L(\boldsymbol{w}, b, \boldsymbol{\alpha})}{\partial b} = -\sum_i \alpha_i y_i = 0, \tag{4.52}$$

thus,

$$\boldsymbol{w} = \sum_i \alpha_i y_i \boldsymbol{x}_i, \tag{4.53}$$

$$\sum_i \alpha_i y_i = 0. \tag{4.54}$$

Substituting $\boldsymbol{w} = \sum_i \alpha_i y_i \boldsymbol{x}_i$ into Lagrangian function (4.48), our objective function is equivalent to

$$L(\boldsymbol{w}, b, \boldsymbol{\alpha}) = \sum_i \alpha_i - \frac{1}{2}\sum_i \sum_j \alpha_i \alpha_j y_i y_j (\boldsymbol{x}_i^{\mathrm{T}} \boldsymbol{x}_j). \tag{4.55}$$

Now, our optimization problem is equivalent to

$$
\begin{aligned}
\max_{\alpha_i \geqslant 0} \quad & \sum_i^n \alpha_i - \frac{1}{2}\sum_i \sum_j \alpha_i \alpha_j y_i y_j (\boldsymbol{x}_i^{\mathrm{T}} \boldsymbol{x}_j), \\
\text{s.t.} \quad & \sum_i^n \alpha_i y_i = 0, \quad \alpha_i \geqslant 0.
\end{aligned} \tag{4.56}
$$

The above problem (4.56) is a dual problem equivalent to the original problem and it is a quadratic programming problem. The SMO algorithm[44] does not directly solve the optimization problem of Eq. (4.47), but solving its equivalent dual problem.

The SMO algorithm was proposed by John C. Platt of Microsoft Research Institute in 1998. The basic idea is that if the solutions of all variables satisfy the KKT condition of the optimization problem, then the optimization problem is solved. Otherwise, select two variables, fix other variables, and construct a quadratic programming problem for the two variables. The solution of the quadratic programming problem is closer to the solution of the original quadratic programming problem. Specifically, after initializing all variables $\alpha_i, i = 1, 2, \ldots, n$, the SMO algorithm continuously performs the following three steps until convergence:

(1) Selects two variables α_i and α_j according to a certain heuristic strategy;
(2) Treat other variables as constants and solve the optimization problem (4.56) with only two variables, α_i and α_j;
(3) Iteratively update α_i and α_j, and stop if the variables converges. Otherwise, repeat steps 1 and 2.

How to get the solution of $\boldsymbol{w}$ and b based on the solution to the dual problem $\boldsymbol{\alpha}$? We get $\boldsymbol{w}$ by formula (4.53)

$$
\boldsymbol{w} = \sum_i \alpha_i y_i \boldsymbol{x}_i. \tag{4.57}
$$

So how to get b? According to the KKT condition, the optimal solution of the original problem $\boldsymbol{w}^*$ satisfies the following conditions:

$$\begin{cases} \alpha_i^* \geqslant 0, \\ y_i(\boldsymbol{w}^{*\mathrm{T}}\boldsymbol{x}_i + b^*) \geqslant 1, \\ \alpha_i^* \left(y_i(\boldsymbol{w}^{*\mathrm{T}}\boldsymbol{x}_i + b^*) - 1\right) = 0. \end{cases} \tag{4.58}$$

For support vector $(\boldsymbol{x}_s, y_s) \in S(\alpha_i > 0)$, where S is the set of all support vectors, it satisfies $y_s(\boldsymbol{w}^{\mathrm{T}}\boldsymbol{x}_s + b) = 1$. Considering $y_s \in \{-1, 1\}$, it satisfies for any support vector

$$b = y_s - \sum_{i \in S} \alpha_i y_i \boldsymbol{x}_i^{\mathrm{T}} \boldsymbol{x}_s. \tag{4.59}$$

It can be seen that as long as any one of the support vectors is selected, b can be directly calculated by the above formula. In practice, in order to ensure the stability of the solution, b is often calculated for each support vector, and then the average is taken as an estimate of b:

$$b = \frac{1}{|S|} \sum_{s \in S} \left(y_s - \sum_{i \in S} \alpha_i y_i \boldsymbol{x}_i^{\mathrm{T}} \boldsymbol{x}_s \right). \tag{4.60}$$

The complexity of the dual problem is $O(n^3)$. In general, when the number of samples n is small, the SMO algorithm is computationally efficient and one should choose the dual problem to solve. However, when the number of samples is large, the complexity of the dual problem depends on n, and the computational efficiency decreases. At this point, one should choose to solve the original problem.

4.5.3. *Soft Margin SVM*

For data that is approximately linear in boundary, a soft margin method can be used, allowing samples to cross the separating hyperplane (allowing for misclassification), but penalizing samples crossing the separating hyperplane. For data of the complex separating hyperplane, low-dimensional data is mapped to high-dimensional or even infinite-dimensional space by means of kernel functions, so that non-linearly separable data in low-dimensional space can be handled.

A sample is allowed to cross the separating hyperplane with penalty ξ_i. The optimization problem is expressed as

$$\begin{aligned} \min_{\boldsymbol{w},b,\boldsymbol{\xi}} \quad & \frac{1}{2}\|\boldsymbol{w}\|_2^2 + C\sum_i \xi_i, \\ \text{s.t.} \quad & y_i(\boldsymbol{w}^{\mathrm{T}}\boldsymbol{x}_i + b) \geqslant 1 - \xi_i, \qquad i = 1, 2, \ldots, n, \\ & \xi_i \geqslant 0, \qquad i = 1, 2, \ldots, n. \end{aligned} \tag{4.61}$$

Similarly, we can use the Lagrangian multiplier method to solve the problem. The corresponding Lagrangian function is

$$\begin{aligned} L(\boldsymbol{w}, b, \boldsymbol{\alpha}, \boldsymbol{\xi}, \boldsymbol{\mu}) = \frac{1}{2}\|\boldsymbol{w}\|_2^2 + C\sum_i \xi_i - \sum_i \alpha_i \\ \times \left(y_i(\boldsymbol{w}^{\mathrm{T}}\boldsymbol{x}_i + b) - 1 + \xi_i\right) \sum_i \mu_i \xi_i. \end{aligned} \tag{4.62}$$

Taking the partial derivatives of the *Lagrangian function* with respect to $\boldsymbol{w}$, b, $\boldsymbol{\xi}$, and making the corresponding partial derivatives equal to zero, we get the corresponding KKT conditions

$$\begin{cases} \boldsymbol{w} = \sum_i \alpha_i y_i \boldsymbol{x}_i, \\ \sum_i \alpha_i y_i = 0, \\ \alpha_i = C - \mu_i, \\ \alpha_i \geqslant 0, \quad \mu_i \geqslant 0, \\ y_i(\boldsymbol{w}^{\mathrm{T}}\boldsymbol{x}_i + b) - 1 + \xi_i \geqslant 0, \\ \alpha_i \left(y_i(\boldsymbol{w}^{\mathrm{T}}\boldsymbol{x}_i + b) - 1 + \xi_i\right) = 0, \\ \mu_i \xi_i = 0, \quad \xi_i \geqslant 0. \end{cases} \tag{4.63}$$

Substituting the above results into the Lagrangian function (4.62), our original problem is equivalent to the following dual problem:

$$\begin{aligned} \max_{\alpha_i \geqslant 0} \quad & \sum_i \alpha_i - \frac{1}{2}\sum_i \sum_j \alpha_i \alpha_j y_i y_j (\boldsymbol{x}_i^{\mathrm{T}} \boldsymbol{x}_j), \\ \text{s.t.} \quad & \sum_i \alpha_i y_i = 0, \quad 0 \leqslant \alpha_i \leqslant C, \quad i = 1, 2, \ldots, n. \end{aligned} \tag{4.64}$$

This problem can also be solved efficiently using the SMO algorithm. After $\boldsymbol{\alpha}$ is obtained, $\boldsymbol{w}$ can be derived via $\boldsymbol{w} = \sum_i \alpha_i y_i \boldsymbol{x}_i$. According to the support vectors, one can calculate b:

$$b = \frac{1}{|S|}\sum_{s\in S}\left(y_s - \sum_{i\in S}\alpha_i y_i \boldsymbol{x}_i^{\mathrm{T}}\boldsymbol{x}_s\right). \tag{4.65}$$

Unlike a simple linear support vector machine, after the soft margin is introduced, the support vectors are samples that satisfies the condition $\alpha_i \in (0, C]$.

4.5.4. *Kernel Functions and Kernel Methods*

A unique feature of the support vector machine model is the ability to handle non-linearly separable data through the kernel method. This is also an advantage of converting the original problem of the support vector machine into a dual problem. In reality, most datasets are not linearly separable, as shown in Figure 4.14. The basic idea of kernel method is to map the non-linearly separable data in the original space to another space using a kernel function, and try to make the data linearly separable in the new space.

To use a kernel function in support vector machine, one only needs to replace the inner product term of the objective function in the dual problem with the kernel function. Commonly used kernel functions include polynomial kernel function, Gaussian kernel function,[d] Laplacian kernel function and Fisher kernel function,[e] as shown in Table 4.4.

---

[d]Also known as Gaussian Radial Basis Function (Gaussian RBF).
[e]Fisher kernel function is also known as sigmoid kernel function.

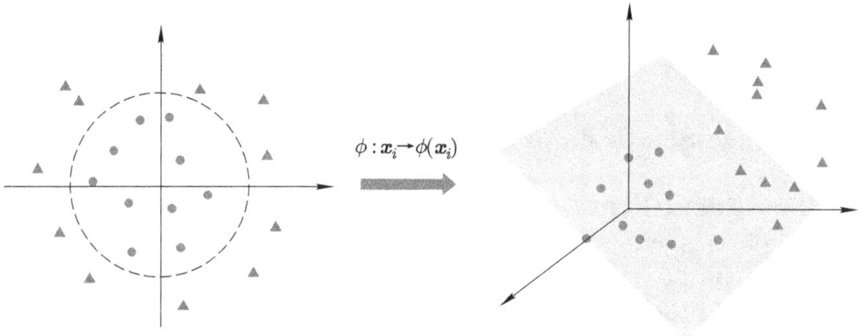

Figure 4.14 Mapping data with kernel method.

Table 4.4 Commonly used kernel functions.

Kernel function	Definition	Description
Polynomial kernel function	$(\boldsymbol{x}_1^{\mathrm{T}}\boldsymbol{x}_2 + 1)^d$	d is an integer
Gaussian kernel function	$\mathrm{e}^{-\frac{\Vert\boldsymbol{x}_1-\boldsymbol{x}_2\Vert_2^2}{2\delta^2}}$	$\delta > 0$
Laplacian kernel function	$\mathrm{e}^{-\frac{\Vert\boldsymbol{x}_1-\boldsymbol{x}_2\Vert_2}{\delta}}$	$\delta > 0$
Fisher kernel function	$\tanh(\beta\boldsymbol{x}_1^{\mathrm{T}}\boldsymbol{x}_2 + \theta)$	$\beta > 0, \theta < 0$

4.5.5. *Advantages and Disadvantages of Support Vector Machine*

Support vector machine has a history of more than 20 years, it has shown great power in text classification and has become the mainstream technology of machine learning from the end of the 20th century to the beginning of the 21st century. It uses a kernel function to map data to a high-dimensional space to handle non-linearly separable data. The decision boundary is mainly determined by a small number of support vectors. Therefore, SVM usually can achieve good generalization ability.

When feature dimension is much larger than sample size, the prediction performance of the support vector machine will not be so good. When sample size is large, the use of the nonlinear kernel function will result in lower computational efficiency. As SVM is originally designed for binary classification task, extensions are needed to solve multiclass classification tasks. Support vector machine cannot directly predict probabilistic outputs.

4.6. Case Studies and Exercises

1. Glass identification
Glass debris is one of the common things in crime scenes. If the glass category can be accurately identified based on the debris, it will be of great help for case investigation. This question uses a glass types dataset.[45] The dataset has a total of 214 samples, each sample containing 9 features, with the type feature indicating the glass category.

Please randomly extract 70% samples as training set, and the remaining 30% samples as test set. Then choose a classification algorithm studied in this chapter to train a glass category prediction model using the training set. Finally, select appropriate matrics to evaluate the classification performance on test set (refer to Appendix E for the evaluation matrics of classification model).

2. Optical character recognition
Optical character recognition (OCR) refers to a process in which an electronic device, such as a scanner, determines a character shape by detecting dark and bright modes, and uses a character recognition method to translate a shape into a computer character.

The dataset contains 20,000 samples of the 26 English capital letters.[46] Each sample represents a rectangular area in the optical image. This area contains only a single character. Each sample contains 16 input features and a target feature "letter", and the "letter" feature indicates which letter the current sample is.

Use the classification methods in this chapter to build a classifier that recognizes optical characters.

3. Personal credit risk assessment
In many countries, government agencies closely monitor loan operations. The bank needs to clearly explain why the applicant's loan application was rejected or approved. This interpretability is also important for loan applicants. When the loan application is rejected by the bank, the applicant needs to know why his credit rating does not meet the bank's requirements.

By building an automated credit rating model, instant online credit approvals can save the bank a lot of labor costs.

The dataset[47] contains 1000 loan samples. Each sample contains 20 input features and a target feature "default" indicating whether the loan is in default. Logistic regression and decision tree are suitable for this problem for their good interpretablity.

Use logistic regression or decision tree to build a personal credit risk assessment model.

4. Breast cancer diagnosis

Early breast cancer testing primarily examines abnormal lumps of breast tissue. If machine learning algorithm can be used, automatic diagnosis through the detection data of the breast lumps will bring great benefits to the medical system: not only can the detection efficiency be greatly improved, but also the risk of misjudgment can be reduced.

The dataset contains 569 samples, each containing 30 features of the breast cell.[48] These 30 features include the mean, standard deviation and maximum value of 10 basic features (including radius, texture, perimeter, area, symmetry, etc.) of the digitalized nuclei. Use the classification methods in this chapter to build a classifier for breast cancer diagnosis.

Chapter 5

Ensemble Method

5.1. Overview of Ensemble Method

There is a Chinese folk saying: "Two heads are better than one". This folk saying tells us the importance of considering more opinions when making decisions. An individual always has his own shortcomings, and if he can learn widely from others and consider the problem from different perspectives, he will often make better decisions.

This chapter introduces ensemble method, which is inspired from the above simple idea. By integrating multiple base models, ensemble method usually achieves better predictive performance than single base model. It is assumed that each base model should outperform random guess, and there exist diversity between different models. Ensemble method is usually applied to supervised learning. Taking classification as an example, the basic framework of ensemble method is shown in Figure 5.1.

Why can ensemble method achieve better performance than a single model? We can understand it from the following two aspects:

Firstly, ensemble method enhances the model expressive ability. As shown in Figure 5.2, when using a single perceptron, it is difficult to distinguish two classes of samples correctly due to the limited

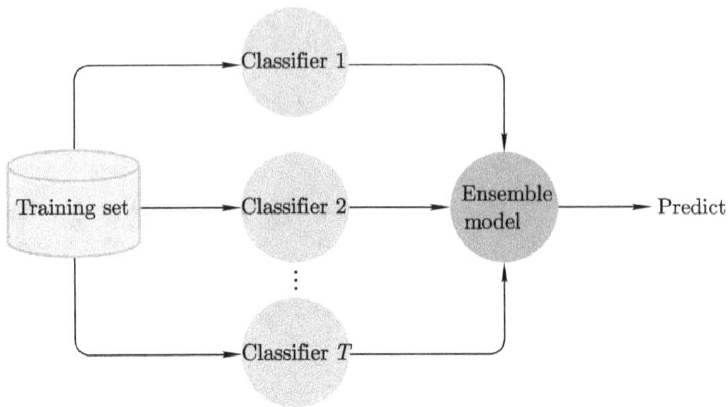

Figure 5.1 Basic framework of ensemble method.

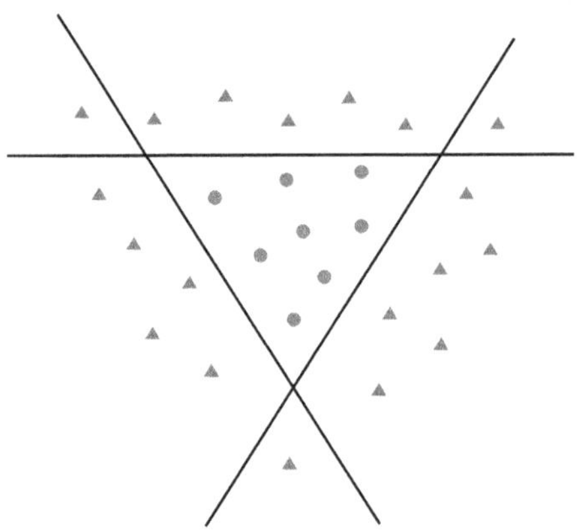

Figure 5.2 Multiple perceptrons strengthen the model expressive ability.

expressive ability. If we integrate three perceptron models, the two classes of samples can be easily distinguished.

Secondly, ensemble method tends to reduce errors compared with a single model. Suppose there are T independent base models, and the error rate for any single base model is p. Assume an ensemble model H is the majority voting of the T base models, then the error rate of H is

$$\text{Error}_H = \sum_{k \leqslant \frac{T}{2}} \mathrm{C}_T^k p^{T-k}(1-p)^k. \tag{5.1}$$

When $T = 5$, and $p = 0.1$, the error rate of H is lower than 0.01. When the base models are independent from each other, the ensemble method can drastically improve the model performance.

Building an ensemble model is divided into two steps: building several base models and integrating these base models according to a certain strategy. The base models can be classification models discussed in Chapter 4, such as decision trees and support vector machines. Note that the base models can be the same type or different types. There are many strategies to integrate base models. Simple ones include majority voting, mean averaging, and weighted averaging. More complex approaches take the outputs of base models as input, and take the actual label as prediction target to train a new model. Stacking is one of these most typical approaches. We introduce three classic ensemble methods in this chapter: Bagging, Boosting, and Stacking.

5.1.1. *Bagging*

Bagging as known as Bootstrap Aggregating, is an ensemble method proposed by Breiman in 1994 to improve the performance of multiple classifiers.[49] As the name implies, it includes two steps: bootstrapping[a] and aggregating. Bootstrap is utilized to obtain samples from the training set to train base model. Then the prediction results of base models are aggregated to as the ensemble model. The basic idea of bagging is shown in Figure 5.3.

Suppose there is a training set containing n samples $D = \{(\boldsymbol{x}_1, y_1), \ldots, (\boldsymbol{x}_n, y_n)\}$. The procedure of building a bagging model with T base models is shown in Algorithm 2. Bootstrap adopts sampling with replacement to obtain the same sample size as the original training set. In average, 36.8% of the samples in D do not appear in D_t. In this way, we can obtain multiple training sets with diversity to build base models.

[a]See Appendix E for the detailed introduction of bootstrapping.

Algorithm 2 Bagging

Input: Training set $D = \{(\boldsymbol{x}_1, y_1), \ldots, (\boldsymbol{x}_n, y_n)\}$, learning algorithm h.
Output: Ensemble model $H(\boldsymbol{x})$.
1: **for** $t = 1$ to T **do**
2: Use bootstrapping to obtain a sample set D_t from the training set D;
3: Based on D_t, use the learning algorithm to train a base model $h_t(\boldsymbol{x})$;
4: Output the final classifier: $H(\boldsymbol{x}) = \text{sign}\left(\sum_{t=1}^{T} h_t(\boldsymbol{x})\right)$.

Bagging uses a simple integration strategy to obtain an ensemble model: majority voting is used for the classification and mean averaging is used for the regression. Mean averaging can drastically reduce the variance. Therefore, Bagging is particularly suitable for improving the performance of base models with high variance and low bias. High variance means that any slight change in the training set will significantly affect the performance, such as decision trees and neural networks. We will introduce random forest, a typical bagging algorithm using decision tree as base model, in Section 5.2.

In Figure 5.3, we can find there is no dependency between the acquisition of multiple samples and the training of the base models, so it is convenient to perform parallel computation. Especially when

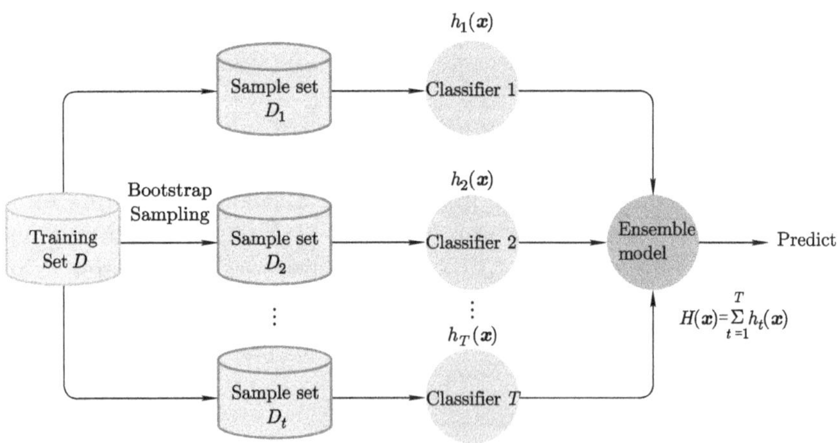

Figure 5.3 The basic idea of bagging.

the size of the training set or the number of base models is large, the ability of parallel computing will be more obvious.

5.1.2. *Boosting*

Boosting is another ensemble method for integrating multiple base models to improve performance. Taking classification as an example, boosting assumes that each base model is a weak classifier, that is, a classifier slightly better than random guess. The aim of boosting is to get a comprehensive strong classifier based on these weak classifiers. Unlike bagging that can train multiple base classifiers in parallel, the weak classifiers in boosting must be trained sequentially. The basic idea is that a weak classifier is firstly trained from the training set, and the next weak classifier is trained based on the previous one by focusing on the misclassified samples. This idea is similar to the process of human learning. Take the process of exam review as an example. After finishing a test paper, we will examine the learning effect by referring to the answer, and focus on reviewing the previous wrong answers. In this way, we can enhance the grasp of knowledge.

So how to ensure the subsequent classifiers focus on the missclassified samples more? As shown in Figure 5.4, boosting implements this by adjusting weights of samples. The weights of the misclassified samples are increased. The subsequent classifier will put more emphasis on predicting these samples right. Bagging uses majority voting or mean averaging to integrate base models, which is equivalent to assigning equal weight to each base model. However, boosting assigns weights to base models according to their performances. Taking classification as an example, the higher error rate of a base classifier, the lower the weight of the base classifier is. Note that when calculating the error rate of each base classifier, the sample weights should be taken into consideration.

Boosting needs to know the error rate upper bound of base models in advance, which is difficult in practice. In 1995, Freund and Schapire proposed AdaBoost algorithm. AdaBoost has been widely applied since we do not need to know the error rate upper bound inadvance. AdaBoost is a classic boosting algorithm, which we will introduce in details in Section 5.3.

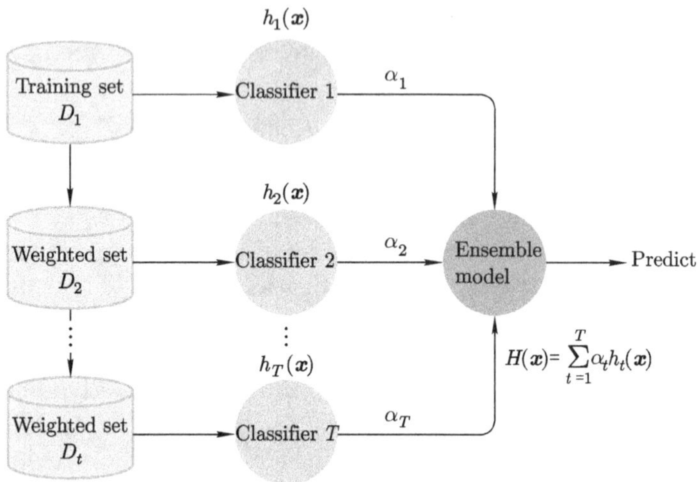

Figure 5.4 The framework of boosting.

5.1.3. *Stacking*

Bagging and boosting usually consists of members having a single-type base model. The final prediction results are obtained by directly integrating the prediction results of base models. Another method of integrating multiple models is stacking.[50] Stacking is usually used to integrate heterogeneous base models, such as supporting vector machine, logistic regression, decision tree, etc. Stacking regards the prediction results of each base model as input features. The actual label of each sample is used as the prediction target, then a high-level model is then trained. The framework of stacking is shown in Figure 5.5.

Stacking algorithm consists of the following steps:

1. Divide the training set D into two disjoint subsets D_1 and D_2;
2. Train different base models with training set D_1;
3. Apply base models to get the prediction results on D_2;
4. Take the prediction results of the base models as input, and the sample labels of D_2 as the learning objective to train a high-level model.

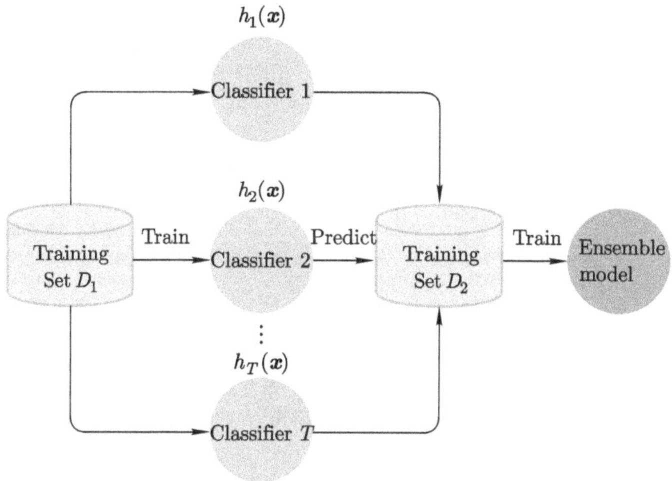

Figure 5.5 The framework of stacking.

5.2. Random Forest

Random forest is a classic bagging algorithm, which has received significant attention in recent years.[51] It has been widely used in many practical problems, such as marketing, stock analysis, financial fraud detection, genomic data analysis and disease risk prediction. A "forest" is constructed by randomly sampling samples and features from the dataset to train different decision trees with diversity (we have already introduced the principle of decision trees in Section 4.3). The core of random forest algorithm is "random", and "forest" means that it uses decision trees as base models.

Each decision tree in the "forest" makes independent prediction on test set. The prediction results of all trees will be averaged (majority voting for classification and mean averaging for regression). Since each tree is different and the prediction results are independent, it can reduce error rate. Bagging trains decision trees and reduce their correlations by bootstrap sampling. In addition to bootstrap sampling, random forest introduces randomness of feature and feature threshold to reduce correlation between decision trees.

5.2.1. *Random Forest Algorithm*

Assume that the training set consists of n samples $D = \{(\boldsymbol{x}_1, y_1), \ldots, (\boldsymbol{x}_n, y_n)\}$, and each sample has d features. A random forest with T decision trees is trained as shown in Algorithm 3.

Algorithm 3 Random forest

Input: Training set D, feature dimension d, the number of randomly selected features m, decision tree learning algorithm h.
Output: Ensemble model $H(\boldsymbol{x})$.
1: **for** $t = 1$ to T **do**
2: Use bootstrapping to obtain a training set D_t of size n from training set D;
3: Select m features from d features randomly, based on the m features of D_t, use learning algorithm to train a decision tree $h_t(\boldsymbol{x})$;
4: For regression output ensemble model: $H(\boldsymbol{x}) = \frac{1}{T}(\sum_{t=1}^{T} h_t(\boldsymbol{x}))$;
5: For classification, output ensemble model: $H(\boldsymbol{x}) = \text{majority\_vote}(\{h_t(\boldsymbol{x})\}_{t=1}^{T})$.

The training of a decision tree starts with bootstrapping sampling. Bootstrapping ensures that the data characteristics learnt by each individual tree is different, which enhances the independence between decision trees. For each tree in the forest, m $(m < d)$ features are randomly selected. How to determine the size of m? A larger m may improve performance of a single decision tree, but the correlation between different decision trees may also be increased. To the contrary, a smaller m will reduce the correlation between different decision trees, but the performance of each tree may decrease due to insufficient features. Therefore, to determine the value of m, one typical method is to use cross validation to select the optimal value. The other one is set m as the empirical value $\log_2 d + 1$.

5.2.2. *Performance Evaluation and Feature Evaluation*

Taking classification as an example, there are mainly two metrics to evaluate the performance of a random forest. Classification *Margin*: the classification margin of a random forest on a sample refers to the difference between the proportion of the decision trees that correctly classify the sample and the proportion of the decision trees misclassifying the sample. Assuming that 75% of the trees correctly classifying sample A, then the classification margin is $75\% - 25\% = 50\%$. The margin of a random forest on test set is obtained by averaging the classification margin of each sample. In practice, we expect that the larger margin is, the better performance it achieves. Since large margin indicates that the classifier has a strong generalization ability. For each tree, some samples are not used during training. They are called *out-of-bag samples*. The prediction error rate of a random forest for out-of-bag samples is called the out-of-bag error(OOBE), which can be computed as follows:

1. For each sample S, calculate the classification results of the trees in which S is an out-of-bag sample;
2. A majority voting rule is used as the classification results of the samples;
3. The out-of-bag error of a random forest is computed as the ratio of the number of misclassified samples over the total number of samples.

In random forest, it is convenient to measure feature importance to help with feature selection and to explain the results. There are the following two main methods:

The first method is to calculate the average information gain of a feature. When training the decision tree, one can calculate how much information gain of a feature obtained in each tree. For a random forest, the importance of a feature can be cumulated by its information gain in each individual tree.

Another method is to evaluate the effect of each feature on the accuracy of the model. A new sample set is generated by disrupting the value order of a feature. Then evaluate the accuracy of the random forest model on the new sample set. The importance of a feature can be measured according to the changes of the accuracy between the new sample set and the original sample set. For those important features, the accuracy of the model will be greatly decreased, while little impact will be generated on the accuracy by reordering those unimportant features.

Both methods have their own pros and cons. The first method will be more beneficial for those features with more values. For those features with large divergence in different types, it cannot explain the importance of the features. Besides, when there exists a strong correlation between features, the first method will enlarge the weight of the first selected feature. In other words, any one of the features can be used as an important feature. Once the first feature is selected, the importance of other features will drops rapidly. In this case, the first selected feature will be mistakenly considered to be much more important than the others. Although such a problem will be alleviated by the random selection method, it has not been completely solved. Moreover, the second method will be greatly influenced by the size of samples.

5.2.3. *Algorithm Analysis*

In this subsection, we discuss random forest algorithm from the perspective of bias and variance. The bias of random forest is consistent with the bias of a single decision tree. The main characteristic of random forest is to reduce the variance.

Suppose the variance of a single decision tree is $\sigma^2(\boldsymbol{x})$, and the correlation between any two decision trees is $\rho(\boldsymbol{x})$. A random forest is composed of T decision trees, then the variance of the random forest $\mathrm{Var}(\boldsymbol{x})$ can be formulated as:

$$\mathrm{Var}(\boldsymbol{x}) = \rho(\boldsymbol{x})\sigma^2(\boldsymbol{x}) + \frac{1-\rho(\boldsymbol{x})}{T}\sigma^2(\boldsymbol{x}). \tag{5.2}$$

It can be seen that as T increases, the second term of the variance will decrease. This demonstrates that random forest can reduce the total error. In the case study of classification of red wine quality, a random forest model was trained by using different number of decision trees, and the *out-of-bag* errors are shown in Figure 5.6.

From Figure 5.6, we can see that as T increases, the *out-of-bag* error drops significantly. When $T \to \infty$, the variance of random forest is only related to the correlation between the decision trees and the variance of a single decision tree. In the extreme case, the variance of the random forest is 0 if the decision trees are completely uncorrelated. If there is a strong correlation between the decision trees, the variance of the random forest is equal to the variance of the single decision tree $\sigma^2(\boldsymbol{x})$.

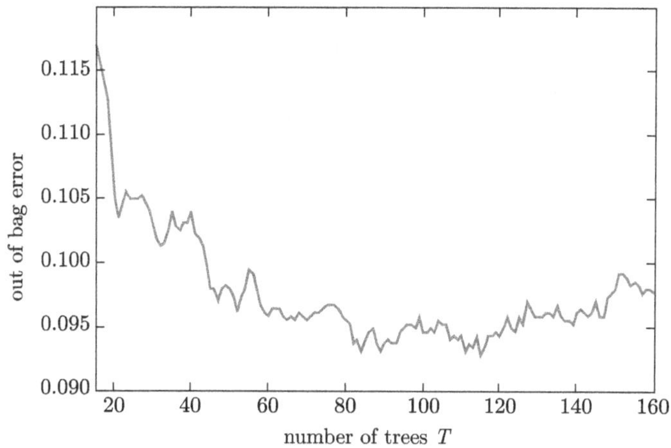

Figure 5.6 The OOB error of the random forest with different number of trees.

Sample sampling and feature sampling in random forest can actually reduce the correlation $\rho(\boldsymbol{x})$ between trees by means of randomness. That is to reduce the variance by lowering the first term $\rho(\boldsymbol{x})\sigma^2(\boldsymbol{x})$ in Eq. (5.2). The strategy of integrating models of random forests, that is, the majority voting or averaging method, is actually to reduce the variance by lowering the second term in Eq. (5.2).

Through the above analysis, we can see that random forest can greatly reduce model variance and improve prediction performance by sample sampling, feature sampling and averaging strategy. Caruana *et al.* conducted the evaluations on more than 20 different datasets and found that random forest achieved the best performance on most dataset.[52] The random feature sampling will further make random forest insensitive to the multicollinearity between features, and capable of handling missing data and imbalanced data. Due to the sampling of both samples and features, and the parallel training of base models, random forest can process large data set, and easy to be parallelized. Random forest can use out-of-bag errors to achieve the effect of cross validation evaluate the model.

Through the combination of multiple decision trees, random forest provides a different way of computing the similarity between the samples. A similarity matrix is generated by recording the amount of times the two samples co-occurs in the same leaf node. For a specific sample, the amount of times that its co-occurrence can be normalized by the number of trees using the sample as training sample.

However, like other ensemble methods, random forest combines many models (decision trees) into a single model, so the interpretability is strongly weakened. In addition, random forest is likely to cause over-fitting problems when dealing with small data set or low-dimensional data with high noise, since bootstrapping may increase the impact of noise.

5.3. AdaBoost

AdaBoost, short for Adaptive Boosting, is a typical boosting algorithm, which was proposed by Freund and Schapire in 1994.[53]

The key difference between AdaBoost and the general boosting algorithm introduced in Section 5.1.2 is that it no longer needs to predetermine error upper bound of weak classifier, which is the meaning of the word "adaptive" in its name. This advantage of AdaBoost makes it widely applied in practice. For example, it has been widely used in the field of face recognition. In 2003, Freund and Schapire won the Gödel Prize for their work of AdaBoost, which is the highest award in the field of theoretical computer community. In the IEEE International Data Mining Conference in 2006, AdaBoost was awarded as one of the top ten data mining algorithms.

The basic idea of AdaBoost is to train a set of weak classifiers $\{h_1(\boldsymbol{x}), h_2(\boldsymbol{x}), \ldots, h_T(\boldsymbol{x})\}$ by using same training set with different sample weights, and these weak classifiers are then linearly integrated with weights into a strong classifier.

$$H(\boldsymbol{x}) = \sum_{t=1}^{T} \alpha_t h_t(\boldsymbol{x}), \tag{5.3}$$

where α_t is the weight of the t-th weak classifier. During the training process of each weak classifier, the samples in the training set will be assigned weights $\boldsymbol{w} = (w_1, w_2, \ldots, w_n)$. For simplicity, the initial weight of the sample can be set as the same value as $1/n$. When training the next weak classifier, the weights of the samples that are misclassified by the current weak classifier will increase, while the weights of the samples that are correctly classified by the current weak classifier will decrease.

5.3.1. *The Process of AdaBoost Algorithm*

Let us take binary classification problem as an example to introduce the process of AdaBoost algorithm. Suppose the training set is $D = \{(\boldsymbol{x}_1, y_1), \ldots, (\boldsymbol{x}_n, y_n)\}$, $y_i \in \{1, -1\}$. AdaBoost firstly trains a set of weak classifiers sequentially and then combines them into the final classifier in a weighted manner. Note that each weak classifier is trained based on a weighted sample set from the original training set. The sample weights are determined on whether the samples are correctly classified by the previous weak classifier. The core idea is to increase the weights of misclassified samples.

The process of AdaBoost is shown in Algorithm 4.

Algorithm 4 AdaBoost

Input: Training set $D = \{(\boldsymbol{x}_1, y_1), \ldots, (\boldsymbol{x}_n, y_n)\}$, weak learning algorithm h.
Output: Ensemble classifier $H(\boldsymbol{x})$.

1: Initialize sample weights $\boldsymbol{w}^1$: $w_i^1 = \frac{1}{n}, i = 1, 2, \ldots, n$;
▷ In initialization the sample has the same weight
2: **for** $t = 1$ to T **do**
3: Train a classifier $h_t(\boldsymbol{x})$ using training set D and weight $\boldsymbol{w}^t$;
4: Calculate the error of the current weak classifier:

$$\varepsilon_t = \frac{\sum_{i=1}^{n} w_i^t I(y_i \neq h_t(\boldsymbol{x}_i))}{\sum_{i=1}^{n} w_i^t};$$

5: Calculate the classifier weight based on its error:

$$\alpha_t = \frac{1}{2} \ln \left(\frac{1 - \varepsilon_t}{\varepsilon_t} \right);$$

6: Update sample weights $\boldsymbol{w}^{t+1}$ as follows: ▷ Increase the weights of samples misclassified by $h_t(\boldsymbol{x})$

$$w_i^{t+1} \leftarrow \frac{w_i^t \cdot \mathrm{e}^{-\alpha_t y_i h_t(\boldsymbol{x}_i)}}{Z_t},$$

where Z_t is a normalization factor;
7: Output the final classifier: $H(\boldsymbol{x}) = \text{sign}\left(\sum_{t=1}^{T} \alpha_t h_t(\boldsymbol{x}) \right)$.

It is obvious that there are two types of weights need to compute for each weak classifier. The first is the weight of the current weak classifier, which represents the importance of the current weak classifier in the final strong classifier. Intuitively, the weight of the weak classifier should be determined by its performance. The better performance the weak classifier achieves, the higher weight should be assigned. In AdaBoost, the weight of each weak classifier α_t is calculated according to the performance on the current training samples. Specifically, the performance of the weak classifier can be evaluated by the weighted error rate ε_t (see step 4 of the algorithm). It is easy to find that ε_t is equal to the ratio between the sum of weights of misclassified samples and sum of weights of all samples in the training set.

The second weight is the sample weight. In the initial stage, samples are assigned same weight as $\frac{1}{n}$. Then sample weights are updated by step 6 of the algorithm. Specifically, if a sample is correctly classified by a weak classifier, $y_i h_t(\boldsymbol{x}_i) = 1$, and its weight is multiplied by a factor $\mathrm{e}^{-\alpha_t}$, which is less than 1. That means the sample weight is decreased by some proportion. When the weak classifier misclassifies a sample, $y_i h_t(\boldsymbol{x}_i) = -1$, then the sample weight is multiplied by a factor e^{α_t}, which is greater than 1. It is equivalent to increasing the weight of a misclassified sample. In fact, during the process of training each weak classifier, AdaBoost adjusts half of the total weight to correctly classified samples and the other half is assigned to misclassified samples.

In AdaBoost, sample distribution is changed by weighing the samples so that each weak classifier puts emphasis on misclassified samples by the previous weak classifier. This requires each weak classifier can be learned on a weighted training set.

Some learning algorithms can directly handle weighted training set by modifying the loss function, that is, to multiply the sample error term with its weight. For other learning algorithms, sampling method can be adopted. The probability of drawing a sample is proportional to its weight. The weak classifier is learned from a training set which is constructed according to the distribution of sample weights.

We demonstrate the process of AdaBoost algorithm by the following example. In this example, decision stump[b] is chosen as the

[b]A decision stump is a decision tree with only one non-leaf node.

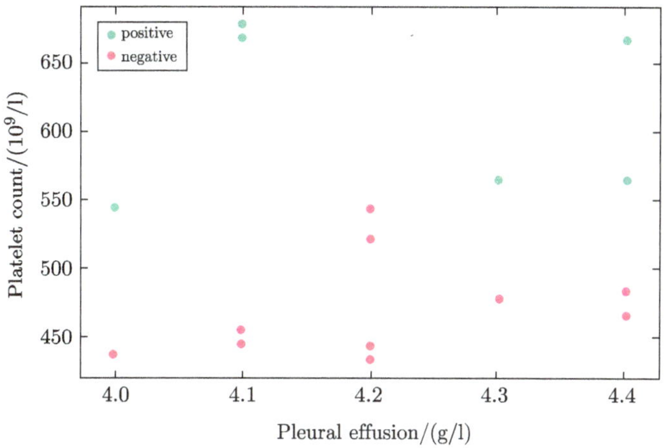

Figure 5.7 16 samples of two features from the mesothelioma's disease dataset.

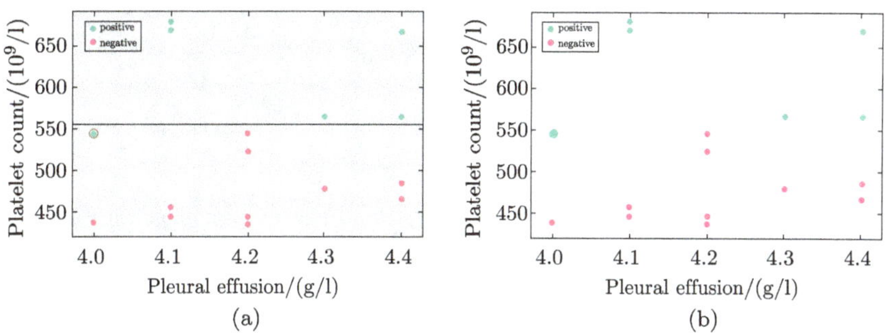

Figure 5.8 The first base classifier $h_1(x)$.

weak classifier. The mesothelioma's disease dataset is used.[54] In this dataset, there are 324 samples and 34 features. For ease of description, we choose two features, "pleural effusion" and "platelet count", and randomly select 16 samples, including 6 positive samples and 10 negative samples. Initially, all samples are assigned the same weight, as shown in Figure 5.7. The diameter of the circle indicates the sample weight.

In the first iteration, base classifier $h_1(\boldsymbol{x})$ is learned, as shown in Figure 5.8(a). The error rate of the classifier is $\varepsilon_1 = 0.11$, and the weight of the base classifier is $\alpha_1 = 2.08$. Then, the sample weights are adjusted according to ε_1, as shown in Figure 5.8(b).

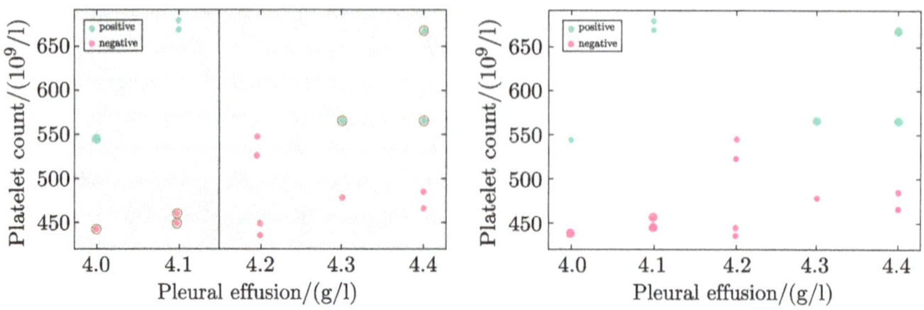

(a) (b)

Figure 5.9 The second base classifier $h_2(\boldsymbol{x})$.

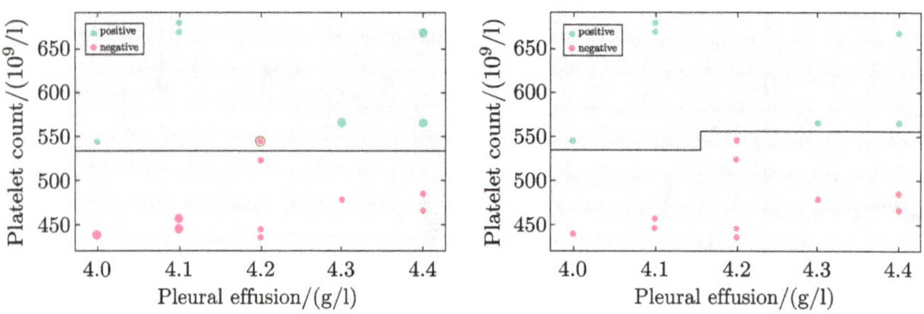

(a) (b)

Figure 5.10 The third base classifier $h_3(\boldsymbol{x})$ and the ensemble classifier $H(\boldsymbol{x})$.

In the second iteration, base classifier $h_2(\boldsymbol{x})$ is learned, as shown in Figure 5.9(a). The error rate of $h_2(\boldsymbol{x})$ is $\varepsilon_2 = 0.19$ and the weight of $h_2(\boldsymbol{x})$ is $\alpha_2 = 1.47$. The sample weights are also adjusted, as shown in Figure 5.9(b).

In the third iteration, base classifier $h_3(\boldsymbol{x})$ is learned, as shown in Figure 5.10(a). The error rate of $h_3(\boldsymbol{x})$ is $\varepsilon_3 = 0.17$ and the weight of $h_3(\boldsymbol{x})$ is $\alpha_3 = 1.56$. Since the overall error rate reaches 0, the learning is stopped. The final classifier is: $H(\boldsymbol{x}) = \text{sign}(2.08 \cdot h_1(\boldsymbol{x}) + 1.47 \cdot h_2(\boldsymbol{x}) + 1.56 \cdot h_3(\boldsymbol{x}))$.

5.3.2. *Error Analysis of AdaBoost*

There are mainly two metrics to analyze the error of AdaBoost, training error and generalization error. The training error is used to evaluate the error of the model on the training set. The training error is

$$\text{Error}_{\text{train}} = \frac{1}{n}\sum_{i=1}^{n} L(y_i, H(\boldsymbol{x}_i)), \tag{5.4}$$

where n is the number of samples in the training set and $L(\cdot)$ denotes the loss function. The most important theoretical property of AdaBoost is to reduce the training error. Suppose the error ε_t of the t-th weak classifier is denoted by $\frac{1}{2} - \gamma_t$. In a binary classification problem, if the samples are distributed evenly, the error of random guess is $\frac{1}{2}$, so γ_t measures the performance of weak classifier. Freund and Schapire have proved that the upper bound of the training error of AdaBoost's final classifier $H(\boldsymbol{x})$ is

$$\prod_{i}^{T} 2\sqrt{\varepsilon_t(1-\varepsilon_t)} = \prod_{i}^{T} \sqrt{1-\gamma_t^2} \leqslant \mathrm{e}^{-2\sum_t \gamma_t^2}. \tag{5.5}$$

If each weak classifier is better than random guess, and there exists $\gamma > 0$ making $\gamma_t \leqslant \gamma$ for all classifiers, then the training error of AdaBoost decreases exponentially with the increase of the number of weak classifiers T. In practice, γ is not required to be predetermined, which is the main difference from other boosting algorithms. Note that it is very difficult to predetermine the value of γ.

The generalization error is used to evaluate the error of the model on the test set. The generalized error can be computed as

$$\text{Error}_{\text{test}} = E_{\boldsymbol{x},y}(L(y_i, H(\boldsymbol{x}_i))), \tag{5.6}$$

where $(\boldsymbol{x}, y)$ denotes any sample point in the entire sample space (including the test set).

Theoretically, as the number of weak classifiers T increases, the upper bound of the generalization error will increase, which may cause overfitting. However, empirical studies have shown that AdaBoost is quite resistant to overfitting even with thousand of weak classifiers. The test error can continue to decrease, even if the training error drops to zero, as shown in Figure 5.11.

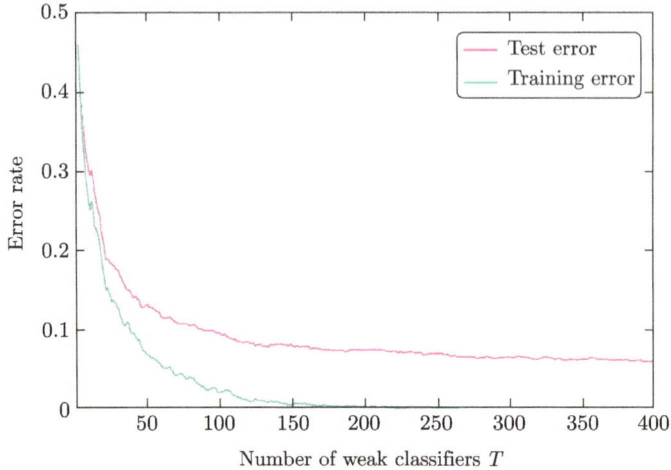

Figure 5.11 The training and test error of AdaBoost with different number of weak classifiers.

5.3.3. *Objective Function of AdaBoost*

In Figure 5.11, after several rounds of iteration, the training error has dropped to zero. As the rounds of iteration increase, the test error continues to decrease. Why can the test error continue to decrease when the training error has been reduced to zero? The reason is that the error function of AdaBoost is the exponential loss function:

$$L(y, f(\boldsymbol{x})) = \mathrm{e}^{-yf(\boldsymbol{x})}, \tag{5.7}$$

where $f(\boldsymbol{x}) = \sum_{t=1}^{T} \alpha_t h_t(\boldsymbol{x})$ is called the margin of $\boldsymbol{x}$. Based on the above exponential loss function, the objective function of AdaBoost is formulated as:

$$\sum_{i=1}^{n} \mathrm{e}^{-y_i f(\boldsymbol{x}_i)}. \tag{5.8}$$

In AdaBoost, the ensemble classifier H is the weighted averaging of weak classifiers. Actually, its prediction depends on the sign of $f(\boldsymbol{x})$. That is, when $f(\boldsymbol{x}) > 0$, the prediction of H is 1; when $f(\boldsymbol{x}) < 0$, the prediction of H is –1. In Figure 5.11, the training error and test error only measure the accuracy of the prediction, while it do not measure the confidence of the predictions. $|f(\boldsymbol{x})|$ is equal to the margin $yf(\boldsymbol{x})$ (because $y \in \{1, -1\}$), so it can be used to represent the confidence of the prediction. Since the objective function of AdaBoost is the exponential loss function, the margin can be further increased even if the classification error is reduced to zero. As a result, the confidence of the prediction is continuously improved so that the performance on test set can be further improved.

5.3.4. *Summary*

Many machine learning algorithms suffer from the overfitting problem, while AdaBoost algorithm is resistant to it, which is one of its key advantages. Besides, AdaBoost can not only reduce the variance, but also reduce the bias. Therefore, compared with individual weak classifier, AdaBoost is usually able to dramatically improve the performance of classification. Meanwhile, AdaBoost provides a learning framework for different types of weak classifiers, e.g., decision tree or other algorithms. In addition, AdaBoost is simple to be applied, as there are almost no additional parameters to be tuned except for setting the number of weak classifiers T (parameters of each weak classifier are not included).

However, AdaBoost algorithm also has some drawbacks. Since the final classifier is combined by multiple weak classifiers, AdaBoost is always lack of interpretability. Especially in some applications that require high interpretability, such as financial risk controls. In addition, AdaBoost is sensitive to outliers. Since some outliers are likely to be misclassified by a weak classifier, it will cause the following weak classifiers to focus on how to correctly classifying these outliers, and result in a decrease of performance.

5.4. Case Study: Personal Credit Risk Assessment

In the 1980s, FICO credit scoring system was built based on the logistic regression algorithm, which was then becoming the giant of the US credit scoring market. In the financial field, traditional analysis mostly uses logistic regression algorithms. However, with the rapid development of big data modeling technology, many algorithms and technologies are emerging. The Big Data Scoring Algorithm Research

Group[c] selected five machine learning algorithms, including support vector machine (SVM), decision tree, random forest, AdaBoost, and Gradient Boosted Decision Tree (GBDT) to build models to assess personal risk based on a large-scale samples from the People's Bank of China. To evaluate the performance of these models, we conducted a lot experiments based on the real dataset from three aspects. The accuracy and interpretability on the personal credit risk assessment were firstly comprehensively compared among different models. Then the stability of each algorithm was also evaluated based on the external samples of time points.

5.4.1. *Background*

Previously, a digital reading system was developed by the credit center of the People's Bank of China. This system refers to the FICO credit scoring system. FICO is a US company on the personal consumption credit evaluation. The core algorithm of the FICO credit scoring system adopts logistic regression.

With the advancement of statistical analysis and big data technology, there are many machine learning methods including decision tree, random forest, neural network, and AdaBoost, and so on. The performance of these algorithms on personal credit, e.g., accuracy, stability and interpretability have not been evaluated in the dataset of credit center.

To this end, we select five popular machine learning algorithms, including support vector machine, decision tree, random forest, AdaBoost and GBDT, to comprehensively evaluate and compare the performance of them on the personal credit risk assessment, including, accuracy, stability, and interpretability. The result can further facilitate understanding the advantages and shortcomings of the algorithms in personal risk scoring.

5.4.2. *Model Building*

In this study, credit center provides the personal credit data of 12.65 million persons, including their loan records, credit card records,

[c]Member of the big data scoring algorithm research group include Gaoyan Ou, Ranran Wang, ChuWang, Liang Yang, Suo Cao Huiying Zhang.

quasi-credit card records, special transactions records, and inquiry records during the period from 2010.7.31 to 2012.7.31. The record of personal default is defined as a binary predictive variable, where 0 denotes not overdue and 1 denotes overdue over 90 days. The process of model building is as follows:

1. Explorative analysis is conducted on the raw data in order to study the meaning of each attribute and assess the data quality;
2. Based on the above analysis, seven categories of statistical features are determined to characterize personal credit, including historical repayment information, account type and quantity, account in use and closed, credit duration, new account information, inquiry information and special transaction information;
3. Single-variable decision tree is adopted for discretization of continuous features;
4. Build the models for personal credit risk assessment based on machine learning algorithms;
5. Evaluate and analyze is conducted on the performance of the models built by each algorithm.

5.4.3. *Evaluation*

The models are evaluated from three aspects: accuracy, stability and interpretability.

The metrics of ROC_AUC, Kolmogorov–Smirnov (KS) index are used to assess the accuracy, while population shift index (PSI) is used to assess the stability. The interpretability can be assessed by feature importance, which is used to measure the impact of each feature on the prediction results. The KS index and PSI are defined as follows:

1. KS is used to measure the maximum distance between the normal samples and the default samples. Firstly, sort the samples according to the credit scores or predicted default rates in an ascending order, and then calculate the cumulative proportion of default/normal samples for each score or default rate. Then KS can be computed by the maximum value of the cumulative differences between the normal samples and the default samples. The illustration of KS is shown in Figure 5.12. The model is empirically considered to be poor when KS less than 0.2, while good when KS is greater than 0.75.

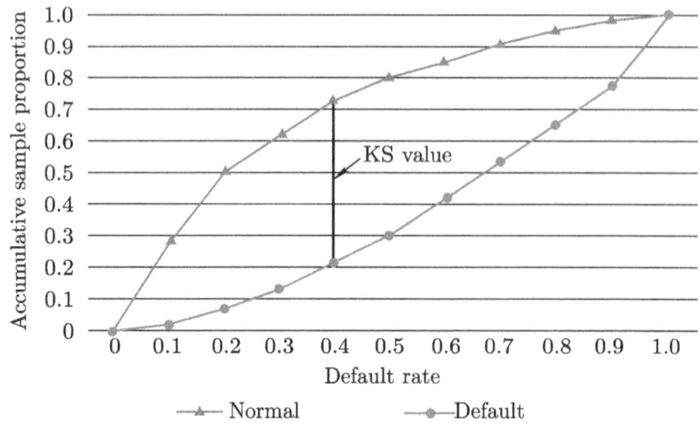

Figure 5.12 Illustration of KS.

2. Generally speaking, a model is developed based on a dataset before a specific time point, PSI is used to measure if the model is still effective for other external samples at another time point. In other words, PSI indicates the stability of a model, which is the most popular used index in stability evaluation.
The formula for calculating the PSI is

$$\text{PSI} = \sum(\text{Actual\%} - \text{Expected\%}) \ln\left(\frac{\text{Actual\%}}{\text{Expected\%}}\right), \qquad (5.9)$$

where Expected% and Actual% represent the corresponding ratio of each score or default rate of the population on training set and test set, respectively. In general, the model is considered to have a high stability if PSI is less than 0.1, but has a poor stability if PSI is greater than 0.25, and has a moderate stability when between them.

1. Evaluating on the 2010 dataset

We constructed our experiments on the 2010 dataset. The 2010 dataset include 10 million persons whose credit was labeled as 0 or 1. The 2010 dataset is randomly divided into training set and test set with the ratio as 7:3. The ROC_AUC and KS of different models are shown in Figures 5.13 and 5.14.

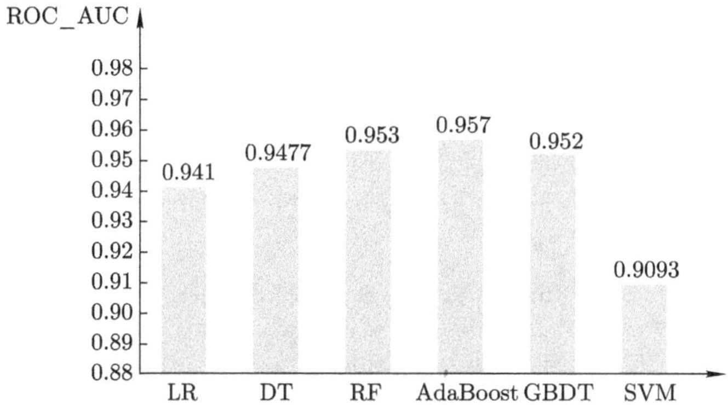

Figure 5.13 ROC_AUC of models on 2010 test set.

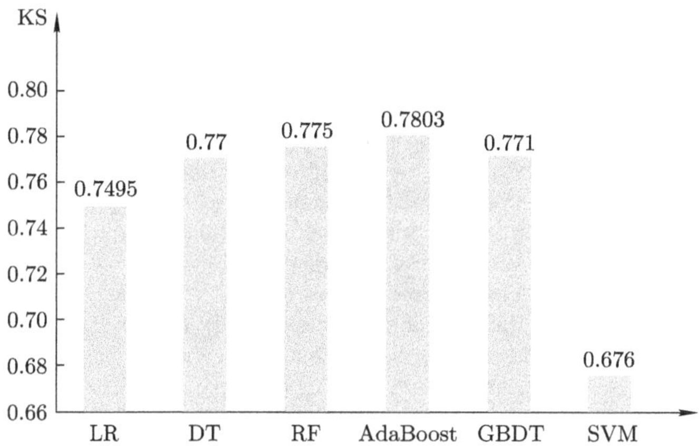

Figure 5.14 KS index of models on 2010 test set.

From the results, we can find that the models that adopt ensemble algorithms (random forest and GBDT) perform better, and the ROC_AUC can reach 0.95 or even higher. Support vector machine gets the worst performance, because the training set is large and the kernel functions cannot be used in this study. The decision tree performs relative well with the ROC_AUC of 0.9477. Besides, AdaBoost performs best in terms of KS, which reaches 0.7803. The performance of other models has the following order: RF > GBDT > DT > SVM. As to the interpretability, feature importance is measured to help interpret the results. The comprehensive ranking

Table 5.1 Ranking of feature importance.

Comprehensive rank of importance	Feature
1	Current overdue amount of credit card
2	Percentage of normal repayments of credit card in last two years
3	Total overdue periods
4	Total amount of credit loans in use
5	Total amount of current overdue loan
6	Percentage of normal repayments of currently in-use credit cards in last one year
7	Number of state "23" for credit account in last two years
8	Percentage of normal repayments of currently in-use credit accounts
9	Total periods of current overdue loan
10	Average ratio of currently in-use quasi-credit accounts' overdue amount to credit limit
11	Percentage of normal repayments of currently in-use credit cards in last one year
12	Number of state "1" of credit accounts in last 12 months
13	Average loan credit limit
14	Average credit years of household loan
15	Average credit years of credit card
16	Average credit years of loan
17	Maximal credit years of credit card
18	Maximal credit years of other loan
19	Maximal credit years of loan
20	Number of credit accounts in use

is obtained based on the importance of the features, as shown in Table 5.1.

We can see that features with greatest impact on personal credit are: overdue situation, normal repayment ratio, unsecured loan, credit limit usage rate, credit period, etc. The ranking results indicate that the features that affect the prediction of machine learning algorithms are with high interpretability. Compared with the traditional logistic regression model, these machine learning algorithms do not require too many feature selections and feature evaluations, and can automatically select important features to build the model.

2. External sample test

In order to evaluate the performance of the models on the external samples of another time point, we evaluate the machine learning models based on the 2011 dataset. The 2011 dataset contains personal credit information of 13 million samples. That means all the models are trained based on the 2010 dataset and tested on 2011 dataset, so as to evaluate the accuracy and the stability on the external data. The results are shown in Figure 5.15 and Table 5.2, where 2010 represents the performance of the models on the 2010 test set, and 2011 represents the performance of the models on the 2011 dataset.

According to the comparison of prediction results on both datasets, decision tree, random forest and AdaBoost decreases against 2010 in terms of accuracy, while GBDT increases. The KS of decision tree and AdaBoost decrease, while that of logistic regression, random forest, GBDT and support vector machine increase.

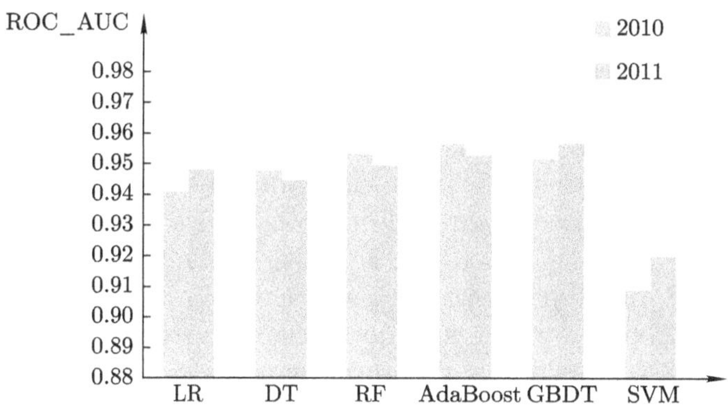

Figure 5.15 Comparison of models in terms of ROC_AUC in the 2010 dataset and 2011 dataset.

Table 5.2 KS and PSI of different models in the 2011 dataset.

Evaluation index	Logistic regression	Decision tree	Random forest	AdaBoost	GBDT	Support vector machine
KS	0.7616	0.7629	0.7772	0.7741	0.7844	0.7009
PSI	0.0450	0.2243	0.6050	0.0627	0.1899	0.0476

In summary, there was no significant difference between the 2010 dataset and 2010 data set in accuracy. As for the PSI, AdaBoost and SVM achieve good stability (PSI < 0.1), while decision tree and GBDT are less stability (0.1 < PSI < 0.25). The stability of random forest is quite low (PSI > 0.25).

5.4.4. *Summary*

To evaluate the performance of machine learning algorithms in personal credit risk assessment, we make comprehensive comparison among several machine learning algorithms from different aspects, including accuracy, stability and interpretability. Five machine learning algorithms were selected, including decision tree, random forest, AdaBoost, GBDT, and support vector machine, to build risk assessment models on large-scale datasets of 10 million persons.

Results on the 2010 dataset show that three ensemble algorithms of AdaBoost, GBDT and random forest achieve better performance, while the support vector machine was the worst in terms of accuracy. Results on external dataset show that AdaBoost and SVM have high stability, decision tree and GBDT are moderately stable, while random forest has relatively low stability. As for interpretability, machine learning algorithms can evaluate the importance of features with comprehensive explanation. Some machine learning algorithms (such as AdaBoost) perform well in both accuracy and stability.

Machine learning is a tool for model development. To obtain a good model, we need to know the characteristics of both algorithms and data. Domain expert and data scientist are required to cooperate on the understanding of data and model selection. In addition, data scientist needs to deeply understand the core principles of machine learning algorithms, and be capable of quickly implementing algorithms and processing large-scale data, so as to fully utilize machine learning to develop a high-performance model for credit risk assessment.

5.5. Case Studies and Exercises

1. Diagnosis of Mesothelioma's disease

Mesotheliomas' disease is a kind of aggressive pleural tumors. The main causes of the disease are: asbestos exposure time, sputum virus

40 infection, genetic predisposition and smoking habits. In this exercise, a Mesothelioma's disease dataset was provided.[54] From the 34 features we can diagnose if a person is healthy or ill. Use the ensemble algorithms in this chapter to build a classifier for diagnosis of Mesothelioma's disease.

2. Kobe's shot prediction
Kobe Bryant is one of the greatest stars in the NBA history. He was drafted by NBA at the age of 17 and won five championship rings. He officially retired on April 14, 2016. Now we are given the shot data during Kobe's 20 year NBA career from 1996 to 2016, including the shooting position, the match against the two sides, whether home or away, predict whether he can hit the shot based on the information of the shooting position and the way he shot. Use the ensemble algorithms in this chapter to build a classifier to predict whether a shot will hit.

3. Classification of red wine quality
Recently, red wine is becoming more and more popular. However, the quality of red wine is mainly determined by sommeliers, which cannot meet the needs of the market. Therefore, some researchers have tried to use machine learning algorithms to automatically identify the quality of red wine, which has greatly improved the speed and accuracy.

For the classification of red wine quality, the physical and chemical indicators of red wine (such as alcohol concentration, PH value, sugar content, non-volatile acid content, volatile acid content, twisting acid content, etc.) can be used as features.

In this exercise, we use a wine dataset consisting of 1599 samples and 11 features.[55] 80% of the samples are extracted from each category as the training set, and the remaining samples are used as test set. Please use the classification algorithm in this chapter to classify the quality of red wine.

Chapter 6

Clustering Model

Clustering is a typical unsupervised learning method. It can group similar samples into clusters for an unlabeled dataset. A cluster corresponds to a potential category of samples.

Clustering can simplify data and help to find intrinsic structure of data. It is an important tool for unlabeled data exploration. For example, the world is divided into different climate regions based on the observed climate patterns. Clustering can also be used as a preprocessing step for other tasks. For example, clustering is extremely critical in information retrieval (such as search engines). Search engine usually clusters the existing web pages in advance. When user submits a query, the search engine returns web pages in the same cluster that are most relevant to the query.

We generally classify the clustering methods into two categories based on input data: similarity-based clustering method and feature-based clustering method. $n \times n$ similarity matrix or distance matrix $\boldsymbol{D}$ is used as input for the former, while feature data $\boldsymbol{X} \in n \times d$ is used as input for the latter. The advantage of similarity-based clustering method is that it can fuse the measure of data similarity in different domains, and even consider adding a kernel function to measure the similarity. Feature-based clustering method can directly consider the original data, avoiding information loss caused by measuring distance. In this chapter, we focus on similarity-based clustering method.

In addition to input types, the output results of clustering methods may be different. Flat clustering, also known as partition clustering, directly divides samples into multiple non-intersect subsets. Hierarchical clustering, divides samples into different clusters by constructing a hierarchial tree structure.

6.1. K-means Clustering

K-means, which originated from signal processing, is a widely used clustering method. The goal of K-means is to divide n samples into K clusters. Each sample is categorized into its nearest cluster. Figure 6.1 is an example of K-means in image processing.

The figure shows the original image and the segmentation results obtained by K-means with different K values. Each pixel of the image is represented by three values of RGB (red, green and blue). The clustering process uses each pixel in the image as the sample, where the closer pixels are divided into the same cluster. After being partitioned, the centroid of the cluster is used to replace the value of each pixel in the cluster. As shown in Figure 6.1, when $K = 2$, the result image contains only two colors. It can be seen that K-means clustering can perform image information compression. Below we specifically introduce the K-means clustering method.

6.1.1. *K-means Clustering Model*

Given a dataset $D = \{\boldsymbol{x}_1, \boldsymbol{x}_2, \ldots, \boldsymbol{x}_n\}$, where $\boldsymbol{x}_i$ is a d-dimensional real-value vector, a set of d-dimensional vectors $\boldsymbol{c}_k$ $(k = 1, 2, \ldots, K)$ is introduced, where $\boldsymbol{c}_k$ is the centroid of the k-th cluster. Our goal is to find a cluster for each sample to minimize the sum of the squares of its distance from its nearest centroid $\boldsymbol{c}_k$. For each sample $\boldsymbol{x}_i$, we introduce a corresponding set of binary indicator variables $r_{ik} \in \{0, 1\}$, where k is the k-th cluster. If $\boldsymbol{x}_i$ is divided into cluster k, then $r_{ik} = 1$, and for $j \neq k$, $r_{ik} = 0$, i.e., $\sum_{j=1}^{K} r_{ij} = 1$. We define the objective

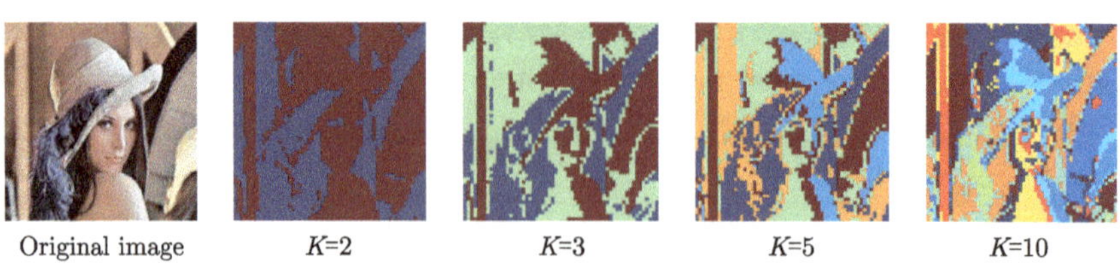

Figure 6.1 K-means clustering for image segmentation.

function as

$$J = \sum_{i=1}^{n} \sum_{k=1}^{K} r_{ik} \|\boldsymbol{x}_i - \boldsymbol{c}_k\|^2. \tag{6.1}$$

J is also called the distortion measure, which represents the sum of the squares of the distance between each sample and the centroid $\boldsymbol{c}_k$ of the cluster in which it belongs to. Our goal is to find the values of $\{r_{ik}\}$ and $\{\boldsymbol{c}_k\}$, and minimize the value of J. After initializing $\boldsymbol{c}_k$, we can keep $\boldsymbol{c}_k$ fixed and minimize J by adjusting r_{ik}. Then we can keep r_{ik} fixed and adjust $\boldsymbol{c}_k$ to minimize J. Repeat these two steps until convergence. We can get the analytical solution because J is a linear function of r_{ik}. Since the items related to different i are independent, we can optimize each i separately. For cluster k with minimal $\|\boldsymbol{x}_i - \boldsymbol{c}_k\|^2$, we make r_{ik} equal to 1, otherwise r_{ik} is equal to 0.

Now we consider the optimization of the $\boldsymbol{c}_k$ value when r_{ik} is fixed. The objective function J is a quadratic function of $\boldsymbol{c}_k$. Set the derivative of J with respect to $\boldsymbol{c}_k$ equal to zero:

$$\frac{\partial J}{\partial \boldsymbol{c}_k} = 2\sum_{i=1}^{n} r_{ik}(\boldsymbol{x}_i - \boldsymbol{c}_k) = 0, \tag{6.2}$$

then J reaches minimum when

$$\boldsymbol{c}_k = \frac{\sum_{i=1}^{n} r_{ik}\boldsymbol{x}_i}{\sum_{i=1}^{n} r_{ik}}, \tag{6.3}$$

where $\boldsymbol{c}_k$ is the mean of all samples in the k-th cluster.

The above process is repeated until the samples' clusters no longer change (or the number of iterations exceeds the threshold). Since the value of objective function J decreases in each iteration, it will eventually converge. However, it may converge to a local optimum. Formally, K-means works as follows:

1. Randomly select K samples as the initial cluster centroids;
2. Repeat the following steps until convergence:
   (a) Assign each sample to its nearest centroid to form K clusters;
   (b) Update the cluster centroids using Eq. (6.3).

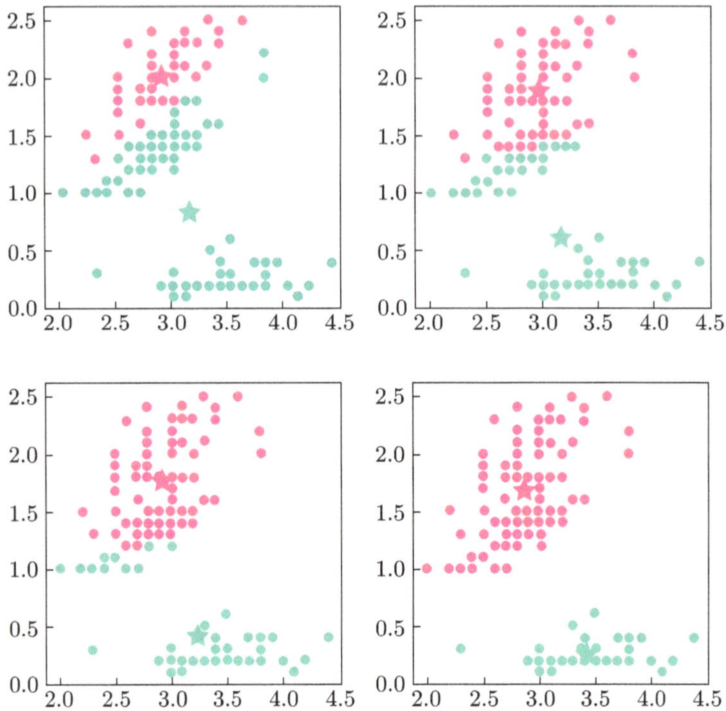

Figure 6.2 The process of K-means clustering.

Figure 6.2 shows how K-means works on a toy dataset. Each subfigure represents an iteration of the clustering algorithm, where the pink and green pentagrams represent the centroids of the two clusters respectively.

6.1.2. *Choice of K*

The choice of K value is closely related to K-means clustering. Essentially, the number of selected clusters is a model selection problem. We will introduce several common selection methods below.

1. Choose K by maximizing a model selection criterion called Bayesian Information Criterion (BIC):

$$\mathrm{BIC}(\mathcal{M}|\boldsymbol{X}) = \mathrm{LL}(\mathcal{M}|\boldsymbol{X}) - \frac{p}{2}\ln(n), \tag{6.4}$$

where $\mathrm{LL}(\mathcal{M}|\boldsymbol{X})$ is the log likelihood function of the dataset $\boldsymbol{X}$ on model $\mathcal{M}$, p is the number of parameters in the model $\mathcal{M}$, and n is the number of samples in the dataset $\boldsymbol{X}$;[56]

2. Initialize K with a large value and gradually decrease until the minimum description length (MDL) no longer decreases. The MDL principle selects the model which can generate the shortest encoding of the data;[57]
3. Starting from $K = 1$, a cluster is divided into smaller clusters until each cluster obeys the Gaussian distribution.[58]

6.1.3. *Choice of the Centroids*

Randomly initializing the centroids is a common method, but we are easily get trapped in local optimal solution. To avoid this scenario, one strategy is to perform multiple runs with randomly initialized centroids, and then select the clusters with the smallest J value. A more effective strategy is using hierarchical clustering (refer to Section 6.2) to cluster the samples, then extract K clusters from the clustering result before selecting the centers of these clusters as the initial centroids. This strategy produces effective results and does not require large number of samples.

We summarize the K-means algorithm as follows. K-means is intuitive and easy to implement, and it supports a variety of distance metrics (usually using Euclidean distance; for more distance metrics refer to Appendix D). The K value should be set and K centroids should be initialized. The clustering result depends on the selection of initial K centroids. K-means is easy to fall into local optimum. K-means cannot handle non-clustered data well and is susceptible to outliers.

6.1.4. *Variants of K-means*

Many variants of K-means have been proposed. Next, we introduce several common variant algorithms.

1. **Bisecting K-means:** The bisecting K-means algorithm selects the cluster with the largest J value from the existing clusters, and then divides it into two clusters, and repeats the process until K clusters are contained in the cluster set. We can also use

other cluster selection strategies, such as choosing the cluster with the largest number of samples to divide. Bisecting K-means selects the centroid multiple times in each division process and selects the optimal division. Even so, bisecting K-means cannot guarantee the global optimal solution.
2. **K-medians clustering:** Use median instead of mean as the cluster centroid.
3. **K-medoids clustering:** Use the sample at the center of the cluster as the centroid, instead of the mean of each cluster.
4. **K-means++ clustering**[59]**:** In the process of initializing cluster centroids, it selects samples as far as possible as the initial centroids.
5. **K-means clustering based on rough set**[60]**:** In each iteration of the original K-means algorithm, each sample can only be assigned to one cluster. In K-means based on rough set, a sample can be assigned to multiple clusters.

6.2. Hierarchical Clustering

Hierarchical clustering clusters samples at different levels and a tree structure is gradually formed. There are two basic methods: bottom-up agglomerative clustering and top-down divisive clustering. Distances between samples (distance matrix) are used as input in both methods. Agglomerative clustering merges similar clusters at each level from bottom to up; conversely, divisive clustering splits a cluster into smaller ones according to some criteria from top to bottom. Both methods are heuristic strategies and do not optimize a clear objective function, so it is difficult to strictly evaluate the performance of clustering.

6.2.1. *Agglomerative Clustering*

Agglomerative clustering starts with each sample as a cluster and then merges the most similar (closest) two clusters in each iteration until all clusters are merged into one. The process of agglomerative clustering is as follows:

1. Treat each sample as a cluster: $C_i \leftarrow \{i\}, i = 1, \ldots, n$;
2. Initialize clusters to be merged as $S \leftarrow \{1, \ldots, n\}$;
3. Repeat the following iterations until S is empty:
   (a) Select the two most similar clusters to merge: $(j, k) \leftarrow \operatorname{argmin}_{j,k \in S} d_{j,k}$;
   (b) Create a new cluster $C_l \leftarrow C_j \cup C_k$;
   (c) Remove cluster j and k from $S \leftarrow S \backslash \{j, k\}$;
   (d) If $C_l \neq \{1, \ldots, n\}$ then add cluster l to $S \leftarrow S \cup \{l\}$;
   (e) For each $i \in S$, update the inter-cluster distance matrix $d(i, l)$.

The time complexity of selecting the two most similar clusters is $O(n^2)$. Since it takes n steps in the algorithm, the total algorithm complexity is $O(n^3)$. We can use the priority list method to reduce the complexity to $O(n^2 \ln(n))$.[61] A large dataset can be first clustered by K-means, and then hierarchical clustering is performed on the centroids of the clusters.

We can visualize the agglomerative clustering using a binary tree, as shown in Figure 6.3. The initial samples are all on the leaf nodes, and each step selects two clusters for merging. The height of the branches represents the distance between the merged two clusters. The root node of the binary tree represents the cluster containing all the samples. Based on the single link distance[a] clustering of the Manhattan distance (refer to Appendix D), the distances of sample pairs (1, 3) and (4, 5) are both 1, so they are first merged. We can cut the binary tree at any height to generate clustering results. In Figure 6.3, the binary tree is cut at a height of 2.5, resulting in clustering results $\{\{\{4, 5\}, \{1, 3\}\}, \{2\}\}$.

In hierarchical clustering, one important problem is how to measure the distance between two clusters. There are three common methods, as shown in Figure 6.4. Different methods may generate different clustering results.

1. **Single link:** It is also called the nearest neighbor distance, as shown in Figure 6.4(a). The distance between cluster G and cluster

[a]The distance between clusters is defined as the distance between the nearest samples of the two clusters.

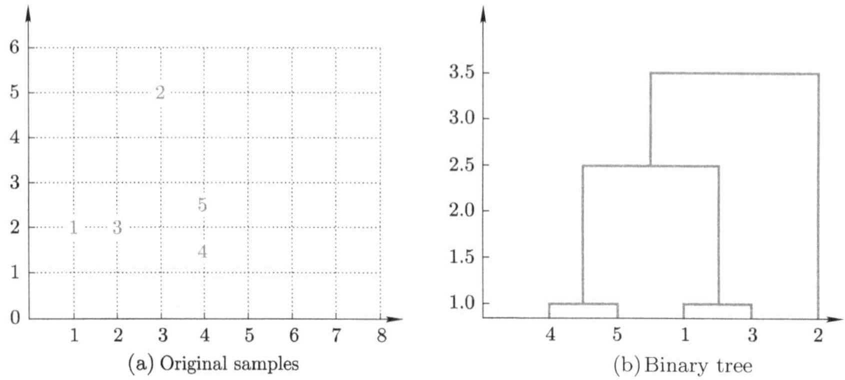

Figure 6.3 Visualization of hierarchical clustering by a binary tree.

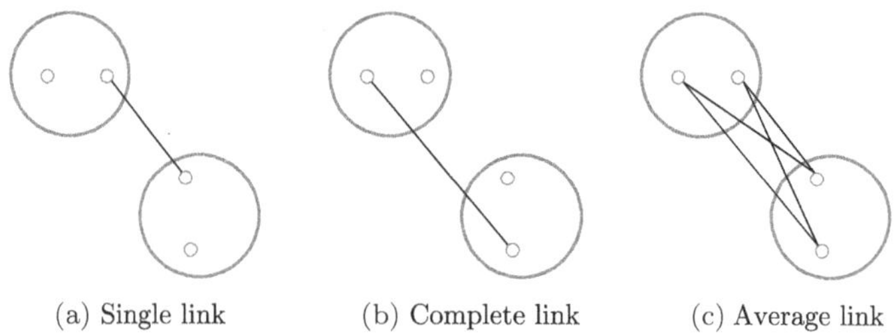

Figure 6.4 Three cluster distance measures in hierarchical clustering.

H is defined as the distance between the nearest members of the two clusters:

$$d_{sl}(G, H) = \min_{i \in G, i' \in H} d_{i,i'}; \tag{6.5}$$

2. **Complete link:** It is also called the farthest neighbor distance, as shown in Figure 6.4(b). The distance between cluster G and cluster H is defined as the distance between the farthest members of the two clusters:

$$d_{cl}(G, H) = \max_{i \in G, i' \in H} d_{i,i'}; \tag{6.6}$$

3. **Average link:** As shown in Figure 6.4(c), the distance between cluster G and cluster H is defined as the average distance of all

member pairs between two clusters:

$$d_{avg}(G,H) = \frac{1}{n_G n_H} \sum_{i \in G} \sum_{i' \in H} d_{i,i'}, \tag{6.7}$$

where n_G and n_H are the number of samples of cluster G and cluster H respectively.

In the single link method, two clusters can be merged when two members in the two clusters are close enough without considering the distances of other members, so the cluster formed by the single link method is highly likely to violate the compactness feature (i.e., the members in the cluster should be as similar as possible). If we define the cluster diameter as the maximum distance $d_G = \max_{i,i' \in G} d_{i,i'}$ between all members in the cluster, we can see that the single link method produces clusters with large diameters. Conversely, the complete link method is another extreme. Only when the distance between the joint members of the two clusters is relatively small, the two clusters can be merged, so the complete link method tends to generate compact clusters. The average link method is a compromise between single link and complete link, which can generate relatively compact clusters, and the distance between clusters is relatively far.

6.2.2. *Divisive Clustering*

Divisive clustering first treats all samples as a single cluster, and then recursively splits one existing cluster into two smaller clusters in a top-down manner. To split a cluster of n samples into two clusters, there are generally $2^{n-1} - 1$ choices. We often use the following heuristic methods:

1. **Bisecting K-means clustering:** Select the cluster with the largest radius, perform K-means clustering on the cluster and divide it into two sub-clusters. Repeat this process until the desired number of clusters is obtained.
2. **Minimum spanning tree method:** Consider each sample as a graph node, the distance between two samples is the weight of the edge. First we construct a minimum spanning tree based on this graph. Then, split the cluster into two clusters on the point of

the largest weight. Repeat this process until the desired number of clusters is obtained. In fact, the clustering results obtained by this method are consistent with the results obtained by single link agglomerative clustering.

3. **Distance analysis method:** Given that the starting cluster contains all samples, $G = \{1, \ldots, n\}$, we calculates the average distance between each sample $i \in G$ between all other samples $i' \in G$

$$d_i^G = \frac{1}{n_G} \sum_{i' \in G} d_{i,i'}. \tag{6.8}$$

Next, we remove the sample i^* with the largest average distance from G and move it into a new cluster H:

$$i^* = \operatorname*{argmax}_{i \in G} d_i^G, \quad G = G \backslash \{i^*\}, \quad H = \{i^*\}. \tag{6.9}$$

Following this rule, the samples are continuously removed from G until a certain condition is satisfied. In order to take into account the distance between the moving sample and the sample in H, the sample selection can be performed according to the following principles:

$$i^* = \operatorname*{argmax}_{i \in G} \quad (d_i^G - d_i^H), \tag{6.10}$$

where

$$d_i^H = \frac{1}{n_H} \sum_{i' \in H} d_{i,i'}. \tag{6.11}$$

The termination condition is: $(d_i^G - d_i^H) < 0$. Next, the same split is performed for G and H respectively until the termination condition is satisfied or each cluster contains only one sample.

There is no explicit global objective function to optimize in hierarchical clustering. The clustering process is gradually performed by local rules at different levels. Compared with K-means clustering, hierarchical clustering is more flexible. The disadvantage is that the spatial and time complexity are higher than K-means clustering.

6.3. Spectral Clustering

Spectral clustering solves the clustering problem from the perspective of graph cut. The basic idea is to firstly construct a weighted undirected graph $G = (V, E)$, where each node of V represents a sample, each edge of E represents the distance between two samples, forming a weight matrix $\boldsymbol{W}$. The elements of the matrix can be obtained by calculating the distance between each node and its nearest neighbors, which can ensure that G is sparse, and accelerate the calculation. For example, if $i \neq j$, $W_{ij} = \mathrm{e}^{-\|\boldsymbol{x}_i - \boldsymbol{x}_j\|/(2\sigma^2)}$; if $i = j$, $W_{ii} = 0$.

If we want to cut the graph into K clusters $\{C_1, C_2, \ldots, C_K\}$, a natural criterion is to minimize the objective function

$$\mathrm{cut}(C_1, \ldots, C_K) = \frac{1}{2} \sum_{k=1}^{K} g(C_k, \bar{C}_k), \tag{6.12}$$

where $\bar{C}_k = V \backslash C_k$ is a complement of C_k, $g(A, B) \triangleq \sum_{i \in A, j \in B} w_{ij}$. When $K = 2$, it is easy to solve the optimization problem and the optimal solution is to divide a single sample into one cluster, and the rest of the samples form another cluster. In order to ensure that each cluster is as large as possible, we choose to optimize the following objective function, also known as normalized cut:

$$\mathrm{Ncut}(C_1, \ldots, C_k) = \frac{1}{2} \frac{\mathrm{cut}(C_k, \bar{C}_k)}{\mathrm{vol}(C_k)}, \tag{6.13}$$

where $\mathrm{vol}(A) \triangleq \sum_{i \in A} d_i$, $d_i = \sum_{j=1}^{n} w_{ij}$ is the weighted degree of node i. This objective function makes the samples within the cluster as similar as possible, while the samples between the clusters are as different as possible.

The normalized cut problem can be seen as finding a binary vector $\boldsymbol{c}_i = (c_{i1}, c_{i2}, \ldots, c_{ik})$, $\boldsymbol{c}_{ik} \in \{0, 1\}$ (where $c_{ik} = 1$ means that sample i belongs to cluster k), to minimize objective function. Relax the binary constraint to a real value, i.e., $\boldsymbol{c}_i \in [0, 1]$, the problem can be solved as the eigenvector of the Laplacian matrix of the graph. Therefore, we generally call this kind of method as spectral clustering.[62]

An important element in spectral clustering is the graph Laplacian matrix. Let $\boldsymbol{D} = \mathrm{diag}(d_i)$ be a diagonal matrix, and each diagonal element is the weighted degree of each node. The graph Laplacian matrix is defined as

$$L \triangleq D - W. \tag{6.14}$$

This matrix has important properties. Because each row is summed to 1, 1 is one of the eigenvectors of $\boldsymbol{L}$, and the corresponding eigenvalue is 0. In addition, for any $\boldsymbol{f} \in \mathbb{R}^n$:

$$\begin{aligned} \boldsymbol{f}^{\mathrm{T}}\boldsymbol{L}\boldsymbol{f} &= \boldsymbol{f}^{\mathrm{T}}\boldsymbol{D}\boldsymbol{f} - \boldsymbol{f}^{\mathrm{T}}\boldsymbol{W}\boldsymbol{f} \\ &= \sum_i d_i f_i^2 - \sum_{i,j} f_i f_j w_{ij} \\ &= \frac{1}{2}\left(\sum_i d_i f_i^2 - 2\sum_{i,j} f_i f_j w_{ij} + \sum_j d_j f_j^2\right) \\ &= \frac{1}{2}\sum_{i,j} w_{ij}(f_i - f_j)^2. \end{aligned} \tag{6.15}$$

Thus $\boldsymbol{L}$ is a symmetric semi-definite matrix, and all its eigenvalues are non-negative real numbers.

The general steps of spectral clustering are as follows: Calculate the first K eigenvectors $\boldsymbol{u}_k$ of $\boldsymbol{L}$, let $\boldsymbol{U} = [\boldsymbol{u}_1, \ldots, \boldsymbol{u}_K] \in \mathbb{R}^{n\times K}$. Let $\boldsymbol{y}_i \in \mathbb{R}^K$ be the i-th row of matrix $\boldsymbol{U}$, and perform K-means clustering on the vector set $\{\boldsymbol{y}_i\}_{i=1}^n$. If $\boldsymbol{y}_i$ is grouped into the k-th cluster, the original sample i belongs to the k-th cluster, that is, the clustering result of the original dataset is the same as the clustering result of the vector set $\{\boldsymbol{y}_i\}_{i=1}^n$.

Spectral clustering is a clustering method that emerged around 2000 with good performance. As shown in Figure 6.5, spectral clustering can handle datasets of special shapes better than traditional K-means algorithm, such as the "half-moon" dataset (Figure 6.5(a), where the pentagram symbol indicates the centroid of the cluster) and the "circular" dataset (as shown in Figure 6.5(b)). There are many variants of spectral clustering. Interested readers are recommended to refer to the article[63] for further details.

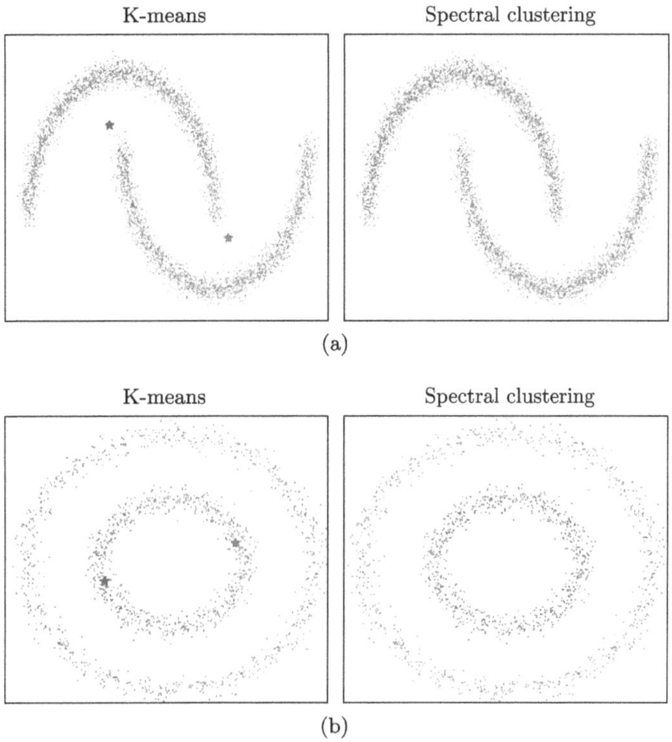

Figure 6.5 Comparison of K-means and spectral clustering. (a) Clustering results on "halfmoon" data. (b) Clustering results on "circular" data.

6.4. Density-Based Clustering

The basic assumption of density-based clustering is that the clustering results can be determined by the density of the sample distribution. In the sample space, each cluster consists of a group of dense samples, which are divided by low-density areas (noise). The goal of the algorithm is to filter out low-density areas to find dense samples. The density-based clustering method can find clusters of arbitrary shapes. Since this method can filter out noisy samples, it can handle data with noise and outliers well.

Below we will introduce the most commonly used density-based clustering method called Density-Based Spatial Clustering of Applications with Noise (DBSCAN).[64] DBSCAN can divide high-density areas into clusters and find clusters of arbitrary shape in the sample space with noise.

In the DBSCAN algorithm, we define the ε-neighborhood as the area centered on a given sample with a radius of ε. If the number of samples in the ε-neighbor of the given sample is not less than $N_{\min}$, then the sample is called a core point. The relationship between the samples can be described by the following concepts:

1. **Directly density-reachable:** For dataset $\boldsymbol{X}$, if sample $\boldsymbol{x}_j$ is within the ε-neighborhood of $\boldsymbol{x}_i$, and $\boldsymbol{x}_i$ is a core point, then sample $\boldsymbol{x}_j$ is directly density-reachable from sample $\boldsymbol{x}_i$;
2. **Density-reachable:** Given a sample sequence $\boldsymbol{p}_1, \boldsymbol{p}_2, \ldots, \boldsymbol{p}_m$, let $\boldsymbol{x}_i = \boldsymbol{p}_1, \boldsymbol{x}_j = \boldsymbol{p}_m$, if each sample $\boldsymbol{p}_i$ is directly density-reachable from $\boldsymbol{p}_{i-1}$, then sample $\boldsymbol{x}_j$ is density-reachable from sample $\boldsymbol{x}_i$. Note that density-reachable is unidirectional;
3. **Density-connected:** If there is a sample $\boldsymbol{p}$, which is density-reachable to both $\boldsymbol{x}_i$ and $\boldsymbol{x}_j$, then $\boldsymbol{x}_i$ and $\boldsymbol{x}_j$ are density-connected.

Based on these definitions, the general steps of DBSCAN are as follows:

1. Input two parameters: radius ε and the minimum number of samples required to form a high-density area $N_{\min}$. Mark all samples as unvisited;
2. Start with an arbitrarily unvisited sample, explore its ε-neighborhood. If there are enough samples in the ε-neighborhood (not less than $N_{\min}$), then create a new cluster. Otherwise, the sample is labeled as a noisy sample. If this sample is further found in the ε-neighborhood (with samples not less than $N_{\min}$) of another sample, then that sample will be added to this cluster too;
3. If a sample is located in a dense area of a cluster, the samples in its ε-neighborhood also belong to the same cluster. When these new samples are added to the cluster, if they are also in dense area, their ε-neighborhood samples will also be added to the cluster if they are also in dense area. This process will repeat until no more samples can be added. As such, a density-connected cluster is found. Then continue to explore the next unvisited sample to discover a new cluster or a noisy sample.

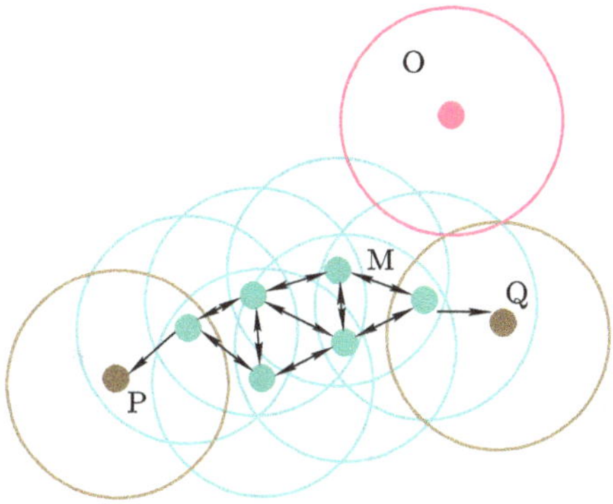

Figure 6.6 DBSCAN clustering algorithm.

In DBSCAN, clusters are formed by the connected density-reachable core points and samples in their ε-neighborhood. Figure 6.6 shows the basic idea of the DBSCAN algorithm, where $N_{\min} = 4$. Point M and other green points are core points. Because the ε-neighborhood (the green circle in the figure) area contains at least 4 points (including themselves) and they are mutually density-reachable, they form a cluster. Point P and Q are not core points, but they can be density-reachable from M through other core points, so they belong to the same cluster. Point O is a noisy point, which is neither a core point nor density-reachable from other points.

Compared to K-means, DBSCAN does not need to choose the number of clusters in advance. Other advantages of DBSCAN include: it can find clusters of any shape; it can effectively distinguish noisy samples and is insensitive to the order of the samples in the dataset. However, samples at the inter-cluster boundary will be grouped into the firstly detected cluster.

The main disadvantage of DBSCAN is that it is sensitive to the hyper-parameters $(\varepsilon, N_{\min})$. With $N_{\min}$ fixed, if ε is too large, most samples are grouped into the same cluster. If ε is too small, it will cause a cluster to split. With ε fixed, if $N_{\min}$ is too large, the samples in a cluster will be marked as noisy samples, and if $N_{\min}$ is too small, it will find a large number of core points. In practice, these two parameters need to be set according to empirical knowledge, such as the understanding of the data. If the results are not satisfactory, ε and $N_{\min}$ can be adjusted appropriately. After multiple runs and comparisons, the most suitable parameter values will be selected.

6.5. Summary

As a typical unsupervised learning method, clustering has many different algorithms. Because the evaluation criteria of clustering results varies, the clustering objective function can be designed from different angles or new clustering algorithms can be developed with heuristic criteria. In this chapter, we introduced four clustering algorithms. K-means is the most widely used algorithm in practice. To get better results, you can consider spectral clustering which requires more computing resources and time. However, the number of clusters needs to be determined in advance for both algorithms. If hierarchical structure is required of clusters, you can choose hierarchical clustering. There is no need to choose the number of clusters in advance and clustering results of arbitrary shapes can be generated in density-based clustering.

In addition to the clustering algorithms introduced in this chapter, another popular type of clustering algorithm emerged in recent years is the infinite mixture model. In this models, we do not need to set the number of clusters in advance but use the Dirichlet process to add a non-parametric prior to the number of clusters. The number of clusters increases with increasing sample size.[65–67]

6.6. Case Studies and Exercises

1. Vehicle clustering

There are many kinds of vehicles on the market, which makes it difficult for many consumers to choose. Manually comparison of vehicle of different companies, origins and categories is a very costly task. We can use clustering algorithm to cluster vehicles according to their parameters, so that cars with similar performance can be quickly identified. The provided dataset has 398 samples.[68] Each sample contains eight features, including the number of cylinders in the vehicle, engine displacement and vehicle origin. Please classify the cars into different groups using one of the clustering algorithms in this chapter.

2. Youth market segmentation

With the popularity of social networks such as Facebook and Twitter, more and more teenagers post messages on these platforms. These textual data can reflect their behaviors and tastes. Combining data of gender, age, number of friends and other user information on the social network platform, it is of great value to mine the youth market. Market segmentation can identify consumers with similar consumption preferences. Clustering is very suitable for accomplishing this task. Please use the given dataset extracted from a social network platform, which describes the basic information and interests of teenagers and use clustering algorithm to conduct youth market segmentation.

3. National happiness index clustering

The World Happiness Report is a landmark survey of the state of happiness in countries around the world. The first World Happiness Report was published in 2012. This case provides information on the happiness indicators surveyed from 157 countries and regions in the world in 2016, including national economic index, household index, health index and government trust index. Please cluster the countries according to the state of happiness and analyze the relationship between the clustering results and the region where the country is located.

Chapter 7

Association Rule Mining

7.1. Association Rule

When facing massive amount of data, is the computer able to automatically discover relationships hidden in the data? In the area of statistics, correlation tests are used to verify if there exist correlations between features. Pearson correlation coefficient is a statistical metric that reflects the linear correlation between features. Its value lies within the range of $[-1, 1]$, where the greater the absolute value is, the stronger the correlation between features will be.[a]

Although Pearson correlation coefficient can measure the strength of the correlation between features, it fails at explicitly indicating what the relationship is. In addition, the coefficient cannot explicate the relationship among multiple features. Association rule mining is a technique for explicitly mining the relationship among multiple features. Such a relationship can be expressed as an association rule in the following way:

$$\mathrm{A} \Rightarrow \mathrm{B}. \tag{7.1}$$

For example, when analyzing the market basket data, one can find that {male, diaper} $\Rightarrow$ {beer}. This rule means "if a customer is a male who buys diapers, he is also likely to buy beer". By using

[a]The calculation of Pearson correlation coefficient can be found in Appendix D.

this rule, market managers can optimize the placement of goods to increase sales.

In addition to market basket analysis, association rules are widely used in various areas such as communication, finance, transportation, health care, and user behavior analysis. For instance, in the field of health care, the relationship between different diagnostic methods can be discovered by analyzing the patients' historical diagnosis and treatment records. In the field of transportation, it is possible to optimize road planning and traffic management by analyzing the relationship of traffic jam in different road sections.

So, how do we discover such interesting rules from the data? And How do we define the rules quantitatively? There are two important metrics: importance and confidence. Importance is measured by the support degree $S(X)$ of all the features ($X = \mathrm{A} \cup \mathrm{B}$) contained in the association rules. The support can be calculated as the ratio of the samples covered by the association rules.

$$S(\mathrm{A} \Rightarrow \mathrm{B}) = S(X) = \frac{\text{Number of samples containing } X}{\text{Total number of samples}}. \tag{7.2}$$

Another metric is confidence, which is used to measure the credibility of the rules.

$$C(\mathrm{A} \Rightarrow \mathrm{B}) = \frac{S(\mathrm{A} \Rightarrow \mathrm{B})}{S(\mathrm{A})}. \tag{7.3}$$

To better explain support and confidence, we still use the above example of beer and diaper. The support of the rule, i.e., {male, diaper} $\Rightarrow$ {beer}, indicates the proportion of shopping records in which the *customer* is male and both beer and diapers were bought. The confidence of the rule indicates that among the shopping records in which a male customer purchased diapers, what percentage of the records in which beer is also purchased.

With support and confidence, we can quantitatively describe the rules that are required: given a dataset, discover all the rules with the support and confidence not less than the predefined minimum support degree *min_suppt* and the minimum confidence *min_conf*.

The rules that we are interested in should satisfy the conditions in terms of support and confidence. Such rules are called strong association rules. In general, this task can be divided into two steps:

1. Mine all rules that are not less than the *min_suppt*, named frequent rules.
2. Refine the frequent rules by filtering out the rules that are less than *min_conf*.

The key of association rule mining is to solve the problem of step 1. After the first step, through a simple test, the second step can be solved in a straightforward way. From Eq. (7.2), it can be observed that to mine the rules that meet with *min_suppt* is exactly equivalent to discovering a feature set (generally known as the itemset in data mining) that satisfies *min_suppt*. So, the key of association rule mining is to find out the frequent itemsets from the dataset. This is also the reason why association rule mining is also referred to as frequent itemset mining. Therefore, the general steps for the association rule mining are:

1. Find out all the frequent itemsets;
2. Generate frequent rules based on frequent itemsets;
3. Filter the rules based on some metrics, e.g., confidence.

Generally speaking, association rule mining can only deal with boolean feature (the value of the feature is either 1 or 0). If the feature is continuous or discrete with multiple values, preprocessing steps are needed, such as discretization and One-Hot encoding. Table 7.1 is a mobile phone review dataset. Each column of the dataset indicates whether the corresponding word appears in the review.

For ease of description, we use a feature set instead of a feature vector to represent each sample, and the elements in the feature set are all the features with the value of 1. In this case, each feature X_i of the sample is called an item. Each sample can be viewed as a collection of items, named itemset. If an itemset has k items, the set is called a k-itemset. Then each review in Table 7.1 can be represented as a set of words that appear in the review. For example, the first review is now represented by the itemset $\{X_1, X_2, X_3, X_4\}$, and it is a 4-itemset. The data are represented by the form of set, as shown in Table 7.2.

In Table 7.2, assume that *min_suppt* is set as 0.4 and *min_conf* is set as 0.6. The itemset $\{X_2, X_3\}$ is included in the first and third reviews, then the support $S(\{X_2, X_3\})$ is $0.4 \geqslant$ *min_suppt*, so the itemset is a frequent itemset. Since the itemset $\{X_2\}$ is included in

Table 7.1 Mobile phone review dataset.

Review ID	Function X_1	Speed X_2	Screen X_3	Touch sensitivity X_4	Customer service X_5
1	1	1	1	1	0
2	0	1	0	0	1
3	0	1	1	0	0
4	1	0	1	1	0
5	0	0	1	1	1

Table 7.2 The itemset representation of the mobile phone review dataset.

Review ID	Itemset
1	$\{X_1, X_2, X_3, X_4\}$
2	$\{X_2, X_5\}$
3	$\{X_2, X_3\}$
4	$\{X_1, X_3, X_4\}$
5	$\{X_3, X_4, X_5\}$

the reviews 1, 2 and 3, the support $S(\{X_2\})$ is $0.6 \geqslant$ *min_suppt*, so $\{X_2\}$ is also a frequent itemset.

Then, we test whether $\{X_2\} \Rightarrow \{X_3\}$ is a strong association rule. Its support is equal to that of itemset $\{X_2, X_3\}$, so it is a frequent rule. Then its confidence can be calculated according to Equation (7.3): $C(\{X_2\} \Rightarrow \{X_3\}) = \frac{S(\{X_2,X_3\})}{S(\{X_2\})} = \frac{0.4}{0.6} = 0.67 \geqslant \text{min\_conf}$. $\{X_2\} \Rightarrow \{X_3\}$ is a strong association rule.

From the above example, we can find that the key of association rule mining is to discover the frequent itemsets. Given a dataset, how can we find all the frequent itemsets hidden in the data? An intuitive idea is to enumerate all possible itemsets as the candidate itemsets, and then test whether their supports meet the pre-defined threshold. This idea is very simple but not efficiency. On one hand, with the increase of feature dimension d, the number of candidate itemsets will increase exponentially. For example, the number of k-itemset is

C_k^d, and the total number of itemsets is $\sum_{k=1}^{d} C_k^d$. On the other hand, for each candidate itemset, it needs to calculate its support based on the dataset. When the dataset is large, it will cost a large amount of computation.

In this chapter, we will introduce two typical association rule mining algorithms: Apriori and FP-Growth. Apriori uses a two-step strategy: (1) candidate itemsets generation; (2) filter out those itemsets that do not satisfy the test, i.e., starts from the frequent 1-itemset and gradually moves to the frequent 2-itemset, until all the frequent itemsets are discovered from the dataset. FP-Growth firstly compresses the original dataset, and then uses the "divide and conquer" strategy to find the frequent itemsets. Since it does not generate candidate itemsets, and the dataset is compressed, FP-Growth can be applied to large datasets, and it is more efficient than Apriori.

7.2. Apriori Algorithm

Apriori is proposed by Agrawal *et al.* in 1994 for frequent itemset mining.[69] Its basic idea is similar with what we introduced in the above section: generate candidate itemsets, and then check whether a candidate itemset satisfies the requirements of the support. The Apriori algorithm can improve computational efficiency, since it avoids enumerating all candidate itemsets by utilizing a property of support. The Apriori algorithm firstly finds frequent *1-itemsets*, and then finds frequent *2-itemsets* based on frequent *1-itemsets*. Similarly, the discovery of frequent *k-itemsets* is based on frequent $(k-1)$-*itemsets*. This is the meaning behind the name "A Priori".

7.2.1. *Apriori Property*

Why does the Apriori algorithm gradually mine frequent itemsets based on the size of the itemset? This strategy can successfully avoid generating massive useless candidate itemsets, which improves efficiency. Apriori mainly utilizes the property of the frequent itemset:

if an itemset A is not a frequent itemset, then none of its supersets[b] are frequent itemsets. This property is called the Apriori property.

According to the Apriori property, given a *k-itemset*, if any of its nonempty subsets is not frequent, then it can be directly determined that the *k-itemset* is not frequent without support checking. Therefore, all frequent $(k+1)$*-itemsets* must be the supersets of some frequent k-itemset. In this way, we can find all the frequent $(k+1)$*-itemsets* by only checking those $(k+1)$*-itemsets* generated by the frequent *k-itemsets* instead of enumerating them all.

7.2.2. *Apriori Algorithm*

The Apriori algorithm iterates from $k = 1$, and at step k, the task is to find all the frequent *k-itemsets*. Specifically, each step is divided into two sub-steps of generating candidate frequent itemsets and filtering out those infrequent itemsets. Firstly, we need to use the frequent $(k-1)$-itemsets L_{K-1} to generate the candidate itemsets C_K. Then we traverse the dataset once and filter out the infrequent k-itemsets from C_K to get frequent *k-itemsets*. The algorithm is shown in Figure 7.1.

We discuss how to generate candidate itemsets and filter infrequent itemsets, respectively.

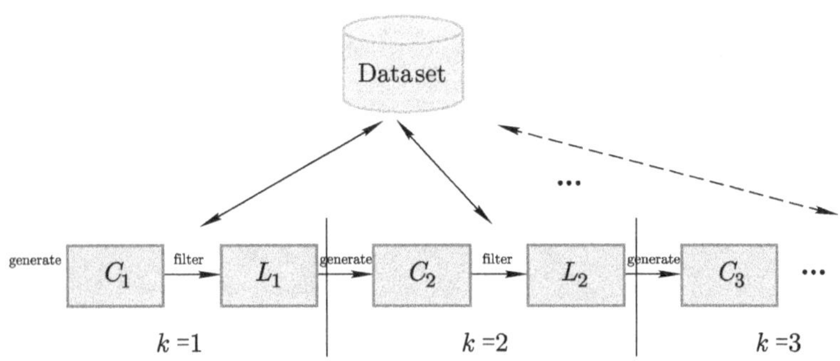

Figure 7.1 Apriori algorithm.

[b]If every element of set A is in set B, and B may contain elements that are not in A, then B is called the superset of A.

1. Candidate itemset generation

The merge of two *k-itemsets* can generate an itemset with the size not less than k. For instance, the *2-itemset* $\{X_1, X_2\}$ and *2-itemset* $\{X_3, X_4\}$ can be merged to obtain a *4-itemset* $\{X_1, X_2, X_3, X_4\}$. Similarly, the *2-itemset* $\{X_2, X_5\}$ and *2-itemset* $\{X_2, X_3\}$ can be merged to get *3-itemset* $\{X_2, X_3, X_5\}$.

In the generation process of the Apriori algorithm, we only need to obtain the $(k+1)$-itemset based on the k-itemset. If there is exactly one element different in two k-itemsets, a $(k+1)$-itemset can be obtained by merging them. Therefore, if we want to generate the candidate set C_{k+1} through frequent k-itemsets, we only need to merge the k-itemsets with only one different element in L_k. To increase efficiency, we need to sort features of the k-itemsets according to feature names. Thus, given two k-itemsets A and B, if the first $k-1$ elements are the same and only the k-th element is different, we merge A and B to generate a candidate $(k+1)$-itemset. For instance, the 2-itemset $\{X_1, X_2\}$ and the 2-itemset $\{X_3, X_4\}$ have two different elements, we do not need to merge them. By merging all the k-itemset pairs that satisfy the above condition, we can get the candidate itemsets C_{k+1}.

2. Filter candidate itemsets

After we get the candidate itemsets C_{k+1} through the generation step, we can traverse the dataset and calculate the support of each itemset in C_{k+1}. Then we can filter out the $(k+1)$-itemset whose support is less than *min_suppt* and get the frequent $(k+1)$-itemsets, i.e., L_{k+1}.

7.2.3. *Case Study of Apriori Algorithm*

We use the Apriori algorithm to mine frequent itemsets on the dataset shown in Table 7.2. Assume that the *min_suppt* is 0.6, and there are five amples. That means that a frequent itemset should appear in at least three samples. Firstly, each itemset that contains only one feature is a member of C_1:

$$C_1 = \{\{X_1\}, \{X_2\}, \{X_3\}, \{X_4\}, \{X_4\}\}.$$

Table 7.3 Support count of 1-itemset in mobile phone review dataset.

Itemset	Support count
X_1	2
X_2	3
X_3	4
X_4	3
X_5	2

Table 7.4 Support count of candidate 2-itemsets in mobile phone review dataset.

Itemset	Support count
$\{X_2, X_3\}$	2
$\{X_3, X_4\}$	3
$\{X_2, X_4\}$	1

Then we traverse the dataset and calculate the support of each itemset. The results are shown in Table 7.3. Since the supports of $\{X_1\}$ and $\{X_5\}$ are both smaller than *min_suppt*, we filter them out to get the frequent 1-itemset $L_1 : \{\{X_2\}, \{X_3\}, \{X_4\}\}$.

Then, we will generate candidate itemsets C_2 based on L_1. Note that $L_1 = \{\{X_2\}, \{X_3\}, \{X_4\}\}$, we can get $C_2 = \{\{X_2, X_3\}, \{X_3, X_4\}, \{X_2, X_4\}\}$ through simply merging 1-*itemsets*. Then we traverse the dataset and calculate the support of each candidate itemset in C_2, as shown in Table 7.4. It can be seen that only the support of $\{X_3, X_4\}$ is greater than *min_suppt*, so the frequent 2-itemset is $L_2 = \{\{X_3, X_4\}\}$. Since L_2 contains only one frequent itemset, the Apriori algorithm ends.

By merging L_1 and L_2, we get the frequent itemsets $\{\{X_2\}, \{X_3\}, \{X_4\}, \{X_3, X_4\}\}$ in mobile phone review dataset.

7.2.4. *Association Rules Generation*

Obtaining frequent itemsets from the dataset is just the first step in the discovery of association rules. We also need to generate frequent rules based on the frequent itemsets, and then find out strong association rules based on *confidence*.

In the mobile phone review dataset, *min_conf* is set as 0.5. We can generate two rules $X_3 \Rightarrow X_4$ and $X_4 \Rightarrow X_3$ according to the frequent itemset $\{X_3, X_4\}$. Then, we calculate the *confidence* based on the support in Table 7.3 and Table 7.4, respectively:

$$\begin{aligned} C(X_3 \Rightarrow X_4) &= \frac{S(\{X_3, X_4\})}{S(\{X_3\})} = \frac{3}{4} = 0.75, \\ C(X_4 \Rightarrow X_3) &= \frac{S(\{X_3, X_4\})}{S(\{X_4\})} = \frac{3}{3} = 1. \end{aligned} \tag{7.4}$$

It can be seen that the confidences of $X_3 \Rightarrow X_4$ and $X_4 \Rightarrow X_3$ are both greater than *min_conf*.

7.2.5. *Summary*

Apriori algorithm is one of the most representative achievements in the field of data mining, and it is also the most effective algorithm for association rule mining in the early days. However, it is not efficient. Especially when the dataset is very large, it cannot be loaded into memory at one time. The computational cost of traversing the dataset becomes the bottleneck of the performance. Apriori needs to traverse the dataset during each iteration, so its efficiency on large-scale datasets is very low. In the following section, we will introduce another algorithm for mining association rules: FP-Growth. FP-Growth only needs to traverse the dataset twice, which is quite efficient.

7.3. FP-Growth Algorithm

Instead of enumerating all itemsets, Apriori makes use of the monotonicity (the Apriori property) of frequent itemsets, which greatly reduces the number of candidate itemsets and improves efficiency. However, it may still generate a large number of useless candidate itemsets. For example, support that there are 1000 frequent 1-itemsets in a dataset, Apriori will generate at least 499,500 candidate 2-itemsets. It is likely that most of those candidate 2-itemsets are infrequent.

Another problem is that Apriori requires multiple traversals of the dataset. Association rules are usually mined from large-scale dataset. There are heavy I/O operations from the disk to traverse the dataset multiple times. Can we generate fewer candidate itemsets, or avoid generating useless candidate itemsets? Can we traverse the dataset less, instead of multiple times repeatedly? The FP-Growth algorithm is such a frequent itemset mining algorithm.

FP-Growth was proposed by Jiawei Han and Jian Pei in 2000.[70] FP-Growth designed a data structure to compress the original dataset, named FP-tree(Frequent Pattern tree). FP-tree stores itemsets and their supports in the original dataset. FP-tree is typically much smaller than the original dataset. When the original dataset is too large to fit in memory, we can use FP-tree to improve efficiency. Since it adopts the "divide and conquer" strategy, FP-Growth can be efficient even when the FP-tree can not fit in memory.

The FP-Growth algorithm mainly includes two steps: building FP-tree and frequent itemsets mining based on FP-tree.

7.3.1. *Build FP-tree*

FP-tree is used to mine frequent itemsets more efficiently. FP-tree stores a dataset in a tree structure, where the root of the tree is "null", and each node represents an item and its support. Firstly, the infrequent 1-itemsets in the original dataset will be filtered, because these itemsets cannot be contained in any frequent itemset. The remaining itemsets will be mapped to a path in the tree. Since samples may share common itemsets, the corresponding path in the tree can also be shared. To achieve a greater degree of compression itemsets are sorted in ascending order according to support.

Table 7.5 Product transaction dataset.

Transaction ID	Item	Sorted frequent itemset
100	$\{f, a, c, d, g, i, m, p\}$	$\{c, f, a, m, p\}$
200	$\{a, b, c, f, l, m, o\}$	$\{c, f, a, b, m\}$
300	$\{b, f, h, j, o\}$	$\{f, b\}$
400	$\{b, c, k, s, p\}$	$\{c, b, p\}$
500	$\{a, f, c, e, l, p, m, n\}$	$\{c, f, a, m, p\}$

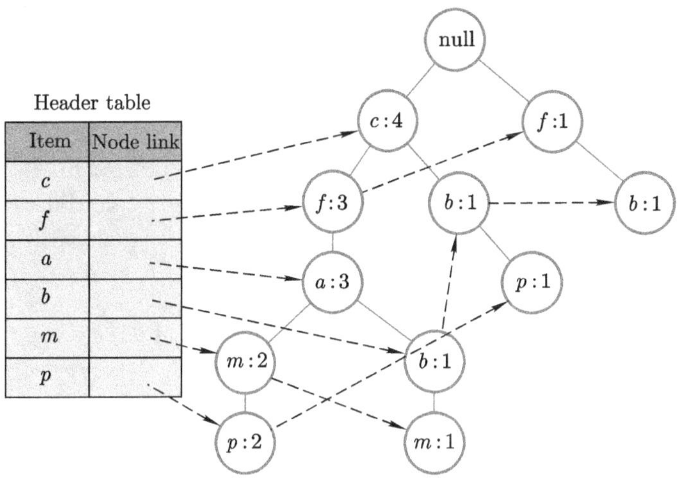

Figure 7.2 The FP-tree for the product transaction dataset.

We use a product transaction dataset as an example, as shown in Table 7.5. The dataset contains 5 transaction records. Assuming *min_suppt* is 0.6, the corresponding FP-tree is shown in Figure 7.2.

To mine frequent itemsets more efficiently, FP-tree maintains a header table. The table contains two columns: the first is item, which is sorted in a descending order according to support. The second column stores a linked list that connects the nodes with same items in the FP-tree.

Algorithm 5 shows the detailed steps of building FP-tree.

Algorithm 5 FP-tree construction algorithm

Input: A dataset D, support threshold *min_suppt*.
Output: FP-tree.
1: Traverse the dataset (the first time), calculate support for each item and sort items in descending order, filtering out the infrequent items according to *min_suppt* to get the frequent itemset L_1;
2: Create the root node of FP-tree, denoted by T, and mark the root node as "null". Create the header table and set the node link column to null. Traverse the dataset (the second time) as follows:
3: **For** each sample in D **do**
4: Extract the frequent items from the sample and sort them according to the order of items in L_1, denoted as P;
5: Insert P into the tree T and update the support count of the corresponding node;
6: Update the corresponding link in the header table.

Now, we show the process of FP-tree construction in detail. For the dataset in Table 7.5, we traverse the dataset, calculate the support of each item, and sort them by support in descending order. The results are as follows:

$$\{c:4, f:4, a:3, b:3, m:3, p:3, l:2, o:2, d:1, e:1, g:1, h:1, i:1, j:1, k:1, n:1\},$$

where $c:4$ indicates that the support count of item c is 4. Since *min_suppt* is 0.6 and the dataset size is 5, after filtering items whose support count is less than 3, the frequent 1-itemsets are obtained and denoted as $L_1 = \{c:4, f:4, a:3, b:3, m:3, p:3\}$. Then, create the FP-tree denoted as T, and build the root node as "null". Create the header table, and set the node link column as null, as shown in Figure 7.3.

For the first transaction $\{f, a, c, d, g, i, m, p\}$, the frequent items are extracted according to L_1 and sorted as $\{c, f, a, m, p\}$. Then insert each of these items into T. Since there are no items in the FP-tree, we create new nodes with support initialized to 1, and the corresponding node links will also be set, as shown in Figure 7.4.

For the second transaction, the sorted frequent itemset is $\{c, f, a, b, m\}$. It will be inserted into T with shared prefix. Note that the prefix "*cfa*" is the same as the first transaction, so the corresponding nodes can be shared, and the support of the corresponding node will be increased by 1. For "*bm*", the corresponding links of the

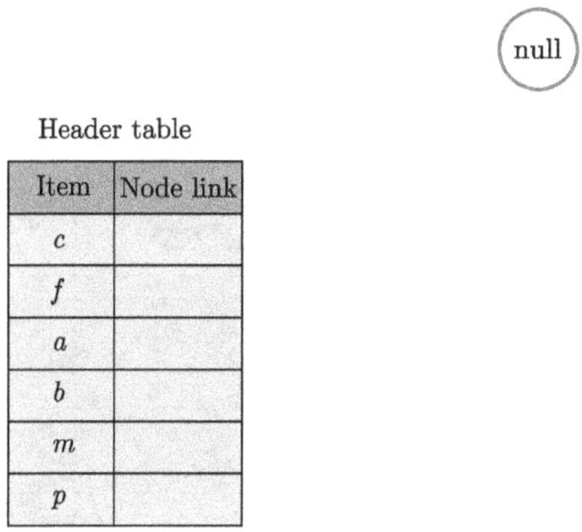

Figure 7.3 Build the root node.

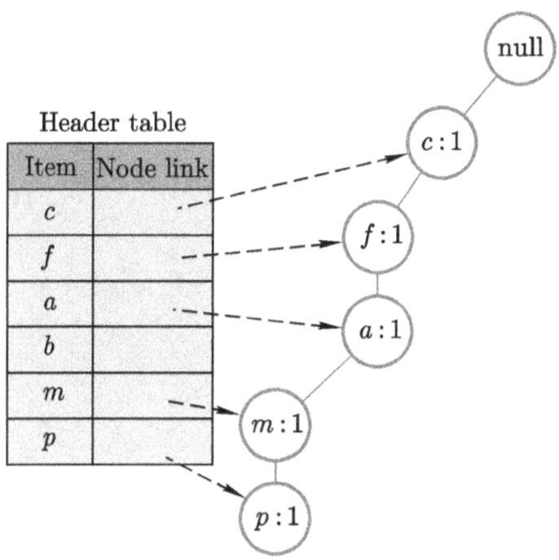

Figure 7.4 Insert $\{c, f, a, m, p\}$.

header table will be updated. For example, there are two "m" nodes in the FP-tree that are required to be linked. After traversing the second transaction, the FP-tree is shown in Figure 7.5.

Similarly, after traversing the third and fourth transaction, the FP-tree are shown in Figure 7.6 and Figure 7.7 respectively. After traversing the fifth transaction, we get the complete FP-tree in Figure 7.2.

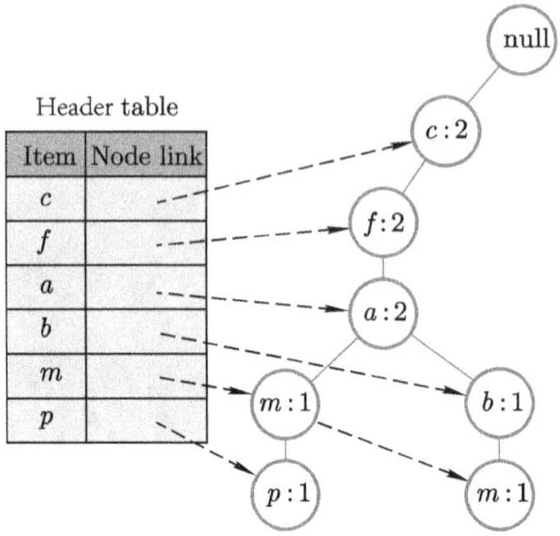

Figure 7.5 Insert $\{c, f, a, b, m\}$.

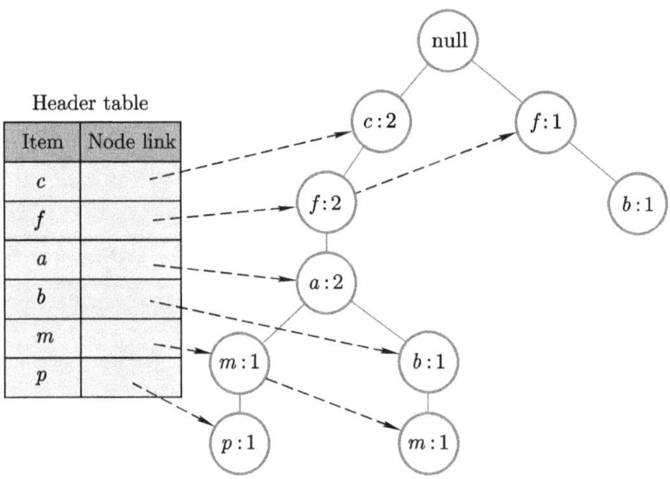

Figure 7.6 Insert $\{f, b\}$.

Since many items share nodes in the FP-tree, its size is usually much smaller than the original dataset. The support information is stored in each node of FP-tree. Therefore, we can use FP-tree instead of traversing the original dataset multiple times.

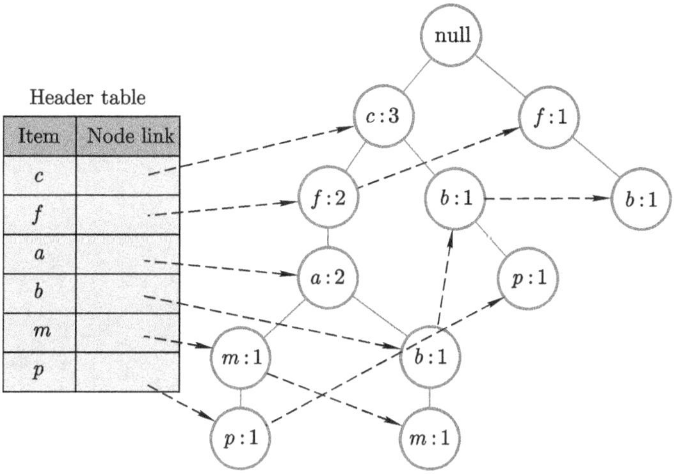

Figure 7.7 Insert $\{c, b, p\}$.

7.3.2. *FP-Growth Algorithm*

In this subsection, we will introduce how to mine frequent itemsets based on FP-tree. The algorithm is called the FP-Growth. The FP-Growth algorithm uses the "divide and conquer" strategy to divide the FP-tree into a series of conditional pattern base (CPB). Based on each conditional pattern base, we can construct a conditional FP-tree which is associated with a specific suffix itemset. In this way, we can mine frequent itemsets from conditional FP-trees independently.

1. Conditional pattern base (CPB)

The conditional pattern base can be considered as a sub-dataset projected from the FP-tree. For example, the process of constructing a conditional pattern base with suffix p from Figure 7.2 is: along the linked list of p in the header table, two paths *cfamp* and *cbp* from FP-tree will be found. Then a new itemset is constructed by items of the two paths (excluding node p). The support of the new itemset is set by the support of p in the corresponding path. Since the support of *cfam* is 2, and the support of *cb* is 1, the conditional pattern base with suffix p is $\{$*cfam*: 2, *cb*: 1$\}$. Similarly, we can obtain the condition pattern base with each item in the header table, as shown in Table 7.6.

Table 7.6 The conditional pattern bases for the product transaction dataset.

Item	CPB
c	$\{\}$
f	$\{c:3\}$
a	$\{cf:3\}$
b	$\{cfa:1, c:1\}$
m	$\{cfa:2, cfab:1\}$
p	$\{cfam:2, cb:1\}$

According to Table 7.6, we divide the original dataset into conditional pattern base for each item, which is equivalent to dividing a large dataset into small datasets. From these small datasets, we can mine frequent itemsets, which are suffixed with corresponding items. For example, all frequent itemsets with suffix p can be mined by mining their conditional pattern base $\{cfam : 2, cb : 1\}$, and construct FP-trees based on the CPB.

2. Conditional FP-tree

Based on the conditional pattern base, we can use Algorithm 5 to construct the corresponding conditional FP-tree. For example, the conditional pattern base for p is $\{cfam : 2, cb : 1\}$. Firstly, the infrequent items will be filtered out and we get $\{c : 3\}$, then the corresponding conditional FP-tree contains only the root node and node c. The conditional pattern base for m is $\{cfa : 2, cfab : 1\}$. Since the support of b is 1, the conditional FP-tree contains only one path $c \to f \to a$.

The procedure of FP-Growth is shown in Algorithm 6. If the conditional FP-tree contains only one path or a single path as the prefix, the frequent itemsets can be directly generated by combining the feature items in the path (line 2 to line 9). For example, the conditional FP-tree of m (support is 3) contains only one path $c \to f \to a$ so frequent itemsets {*cm*:3, *fm*: 3, *am*: 3, *cfm*:3, *cam*:3, f *am*:3, cf *am*:3} can be generated directly. When there are multiple paths, the common ancestor node of multiple paths are first found. Starting from this node, the conditional FP-tree is constructed, and frequent itemset mining is performed recursively (line 10 to line 15). Finally, join the above two sets of frequent itemsets (set(P) $\times$ set(Q)) and merge them with set(P) and set(Q) to get the complete frequent itemsets.

Algorithm 6 FP-Growth algorithm

Input: FP-tree corresponding to dataset D, support threshold min_suppt.
Output: Frequent itemsets.
1: **procedure** FP-GROWTH(Tree, α)
    $\triangleright$ Tree denotes a FP-tree, α is the current suffix
2:   **if** Tree contains a single path as prefix **then**
    $\triangleright$ Mining single path
3:     Assume that P is the prefix path of Tree,
4:     Mark the first branch node of Tree as "null" and the subtree with "null" as the root node as Q,
5:     **for** each non-empty combination β composed of feature items in path P **do**
6:       Generate a frequent itemset $\beta \cup \alpha$ with a support count equal to the support count of α.
7:     Set all frequent itemsets generated through P as set(P).
8:   **else**
9:     set Q = Tree.
10:   **for** each of the item a_i in Q **do**
    $\triangleright$ Mining multiple paths
11:     Generate frequent itemset $\beta = a_i \cup \alpha$ and its support count is set to the minimum support count of β,
12:     Construct the conditional FP-tree Tree$_\beta$ of β .
13:     **if** Tree$_\beta$ is not null **then**
14:       Run FP $-$ GROWTH(Tree$_\beta, \beta$).
15:   Set(Q) is the frequent itemsets generated through Q.
   return set(P) $\cup$ set(Q) $\cup$ (set(P) $\times$ set(Q))

7.3.3. *Association Rule Generation*

Similar to Apriori, FP-Growth generates frequent itemsets rather than association rules. Candidate association rules are a firstly generated by FP-Growth, which are further filtered through the predefined *min_conf* to obtain strong association rules.

7.3.4. *Summary*

FP-Growth converts the original dataset into a compressed tree structure by traversing it twice. Then the FP-tree is used to mine frequent itemsets. FP-tree is usually smaller than the original dataset, so FP-Growth tends to achieve higher performance compared to Apriori, which needs to traverse the original dataset multiple times. Secondly, the process of FP-tree mining also adopts the idea of "divide and conquer", which is implemented by recursively mining the conditional FP-tree. The size of the conditional FP-tree is further reduced compared to the FP-tree. In addition, FP-Growth does not generate useless candidate itemsets.

The FP-Growth algorithm also faces some disadvantages. Recall the fifth step in Algorithm 6. When there is a long path in the FP-tree, a great amount of combinations may be required. For example, suppose there is a path with the length 100, then the path will generate about 10^{30} frequent itemsets. Obviously, according to the Apriori property, we do not need to enumerate so many itemsets.

7.4. Case Studies Exercises

1. Movie recommendation

In the field of e-commerce, an effective recommendation algorithm will improve sales. For example, Taobao and JD use recommendation techniques to recommend products to potential customers. The online video websites can recommend movies based on the users' comments or ratings. This exercise provides a movie rating dataset from Grouplens, which records 671 user's ratings for 9,125 movies. The dataset consists of two parts, i.e., movie and rating. The first part records basic information of the movie, while the second part records 100004 comments. Each comment contains four features, as shown in Table 7.7.

Please find frequent itemsets and the association rules to recommend movies using this dataset.

2. Hobbies and habits survey

This exercise provides a dataset about youth hobbies and habits. The dataset is comprised of a questionnaire on young people's hobbies and habits, which was conducted by some university students majoring in statistics in 2013. Some statistics of the dataset are as follows:

- The dataset include *1010 questionnaires. There are 150 questions in each questionnaire.*
- The *columns.csv* file records the detailed description of the 150 questions.

Table 7.7 Feature description of the ratings dataset.

Feature	Description
userId	user ID
movieId	movie ID
rating	movie rating by the user
timestamp	rating time

- The dataset contains missing values, which are not provided by the surveyor.
- The dataset contains both numeric and discrete features.
- Numerical features represent the degree of approval, which is gradual increased from 1 to 5.

The 150 questions can be divided into different categories: music preferences, movie preferences, hobbies and interests, hygiene habits, personality characteristics, outlook on life, consumption habits, and personal information.

Please use a association rule algorithm to mine the relationships between the surveyed items.

3. Shopping records

When referring to association rules, the most typical case is the "beer and diapers" shopping case. Wal-Mart firstly applied association rule algorithm to find the combinations of goods that a customer might purchase together, and then put them on shelves nearby to increase sales.

This case provides a dataset of shopping records,[73] with a total of 9,835 samples. Each sample records the products purchased by one customer at the mall. Please use an association rule mining algorithm to find frequent itemsets and association rules, so as to provide an optimal product placement strategy for the mall.

Chapter 8

Dimensionality Reduction

Dimensionality reduction is the process of reducing the number of features of data. Why do we want to reduce feature dimension? First of all, the data we encounter is usually high-dimensional. Although the original data is of high dimension, the "intrinsic" dimension of the data may be low. Let's see an example. Suppose we need to measure customer value using mobile communication data. Communication data includes user's access time, package price, monthly call charge, monthly data usage, monthly call duration, amount of arrears and number of months in arrears. The monthly call charge, monthly data usage, and monthly call duration indicate the user's spending power. The amount of arrears and the number of months in arrears are also correlated, and both indicate the user arrears index. When evaluating a user's value, we tend not to be interested in specific features, but rather to be more abstract about spending power, arrears index and user's loyalty. Compared to the original seven dimensions, we can use the three dimensions of spending power arrears index and user's loyalty more intuitively.

An important challenge brought by high-dimension to data analysis is the curse of dimensionality, that is, the model complexity and computation cost increase exponentially with the number of features. Dimensionality reduction is one way to overcome the issue of the curse of dimensionality. It can reduce both model complexity and training time. In addition, dimensionality reduction can also be used as a way of feature extraction.

The methods of dimensionality reduction can be divided into two classes: *linear methods* and *nonlinear methods*. The former includes principal component analysis, linear discriminant analysis, etc. The latter includes multi-dimensional scaling, locally linear embedding, etc. In this chapter, we focus on four dimensionality reduction algorithms. Other algorithms are briefly introduced in Section 8.5.

8.1. Principal Component Analysis

Principal component analysis (PCA) is a linear dimensionality reduction method proposed by Karl Pearson in 1901.[74] It has been widely applied in the fields of face recognition and image compression. In principal component analysis, the importance of information is expressed by variance. Therefore, its basic idea is to construct a series of linear combinations of original features to form low-dimensional features. With the low-dimensional features, we can not only remove correlations of original features, but also retain variance of the high-dimensional data.

We can understand PCA with a simple example. Suppose that the original dataset contains two features x_1 and x_2, as shown in Figure 8.1. We need to reduce the dataset to one dimension, that is, find a direction in which the variance of the original dataset is maximized after projection. Intuitively, the direction of the line in Figure 8.1 is what we need. This is the first expression of PCA. We can also understand PCA from the perspective of minimizing the data reconstruction error. In the subsequent analysis of this section, it will be found that these two are actually equivalent.

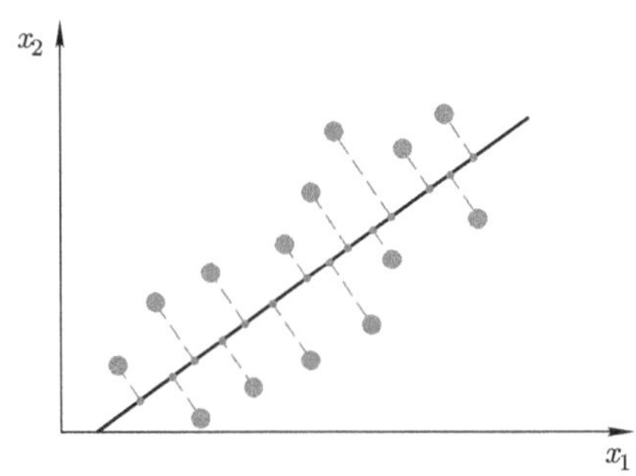

Figure 8.1 PCA example.

8.1.1. *PCA Algorithm*

Suppose that we have a dataset $D = \{\boldsymbol{x}_1, \boldsymbol{x}_2, \ldots, \boldsymbol{x}_n\}$. Each sample is represented as a d-dimension vector, and each dimension is a continuous feature. The dataset D can also be represented as a $n \times d$ matrix $\boldsymbol{X}$. For ease of description, we assume that the mean of each feature is zero. Now we need to linearly reduce the d-dimensional data to l dimensions ($l < d$). We can use a $d \times l$ matrix $\boldsymbol{W}$ to represent the dimensionality reduction process. The data after dimensionality reduction can be expressed as

$$\boldsymbol{Y} = \boldsymbol{X}\boldsymbol{W}. \tag{8.1}$$

The variance of the data after dimensionality reduction is

$$\begin{aligned} \mathrm{Var}(\boldsymbol{Y}) &= \frac{1}{n-1}\mathrm{tr}(\boldsymbol{Y}^{\mathrm{T}}\boldsymbol{Y}) \\ &= \frac{1}{n-1}\mathrm{tr}(\boldsymbol{W}^{\mathrm{T}}\boldsymbol{X}^{\mathrm{T}}\boldsymbol{X}\boldsymbol{W}) \\ &= \mathrm{tr}\left(\boldsymbol{W}^{\mathrm{T}}\frac{1}{n-1}\boldsymbol{X}^{\mathrm{T}}\boldsymbol{X}\boldsymbol{W}\right). \end{aligned} \tag{8.2}$$

Denote the covariance matrix of $\boldsymbol{X}$ by $\boldsymbol{\Sigma} = \frac{1}{n-1}\boldsymbol{X}^{\mathrm{T}}\boldsymbol{X}$. Then the variance after dimensionality reduction is $\mathrm{tr}(\boldsymbol{W}^{\mathrm{T}}\boldsymbol{\Sigma}\boldsymbol{W})$. Considering that the objective of PCA is to maximize the variance of the data after dimensionality reduction, and the features after dimensionality reduction are uncorrelated, the optimization problem can be formulated as

$$\max_{\boldsymbol{W}} \ \mathrm{tr}(\boldsymbol{W}^{\mathrm{T}}\boldsymbol{\Sigma}\boldsymbol{W}), \quad \text{s.t.} \quad \boldsymbol{w}_i^{\mathrm{T}}\boldsymbol{w}_i = 1, \quad i \in \{1, 2, \ldots, l\}. \tag{8.3}$$

Using the Lagrangian method, we can transform the above constrained optimization problem into an unconstrained optimization problem. The corresponding Lagrangian function is

$$L(\boldsymbol{W}, \boldsymbol{\lambda}) = \mathrm{tr}(\boldsymbol{W}^{\mathrm{T}}\boldsymbol{\Sigma}\boldsymbol{W}) - \sum_{i=1}^{l}\lambda_i(\boldsymbol{w}_i^{\mathrm{T}}\boldsymbol{w}_i - 1), \tag{8.4}$$

where λ_i is the Lagrange multiplier. Taking the partial derivative with respect to $\boldsymbol{w}_i$ and letting the derivative equal to zero, we can get

$$\boldsymbol{\Sigma w}_i = \lambda_i \boldsymbol{w}_i. \tag{8.5}$$

It can be seen that $\boldsymbol{w}_i$ is the eigenvector of the covariance matrix $\boldsymbol{\Sigma}$, and λ_i is the corresponding eigenvalue. Multiply both sides of Eq. (8.5) by $\boldsymbol{w}_i^{\mathrm{T}}$, and notice that $\boldsymbol{w}_i^{\mathrm{T}}\boldsymbol{w}_i = 1$, we will get

$$\boldsymbol{w}_i^{\mathrm{T}}\boldsymbol{\Sigma w}_i = \lambda_i \boldsymbol{w}_i^{\mathrm{T}}\boldsymbol{w}_i = \lambda_i. \tag{8.6}$$

Substituting $\mathrm{tr}(\boldsymbol{W}^{\mathrm{T}}\boldsymbol{\Sigma W}) = \sum_{i=1}^{l} \boldsymbol{w}_i^{\mathrm{T}}\boldsymbol{\Sigma w}_i$ to Eq. (8.6), we further have:

$$\mathrm{tr}(\boldsymbol{W}^{\mathrm{T}}\boldsymbol{\Sigma W}) = \sum_{i}^{l} \lambda_i. \tag{8.7}$$

In PCA, the variance after projection is equal to the sum of the eigenvalues of the covariance matrix of the dataset $\boldsymbol{X}$. To maximize variance, we only need to first obtain the eigenvectors and eigenvalues of $\boldsymbol{\Sigma}$, and then take the eigenvectors corresponding to the l largest eigenvalues. The detailed procedure of the PCA algorithm is shown in Algorithm 7.

Algorithm 7 PCA algorithm

Input: Data matrix $\boldsymbol{X}$, dimension after dimensionality reduction l.
Output: Transformation matrix $\boldsymbol{W} = (\boldsymbol{w}_1, \boldsymbol{w}_2, \cdots, \boldsymbol{w}_l)$.

1: Centralize each sample $\boldsymbol{x}_i$ in $\boldsymbol{X}$:

$$\boldsymbol{x}_i \leftarrow \boldsymbol{x}_i - \boldsymbol{m},$$

where $\boldsymbol{m} = \frac{1}{n}\sum_{j=1}^{n} \boldsymbol{x}_j$ is the sample mean;

2: Compute the covariance matrix $\boldsymbol{\Sigma} = \dfrac{1}{n-1}\boldsymbol{X}^{\mathrm{T}}\boldsymbol{X}$;

3: Perform eigenvalue decomposition on the covariance matrix $\boldsymbol{\Sigma}$ and sort the eigenvalues in descending order: $\lambda_1 \geqslant \lambda_2 \geqslant \cdots \geqslant \lambda_d$;

4: Take the eigenvectors $\boldsymbol{w}_1, \boldsymbol{w}_2, \cdots, \boldsymbol{w}_l$ corresponding to the l largest eigenvalues to form the transformation matrix $\boldsymbol{W}$.

8.1.2. *Summary*

Principal component analysis aims to maximumly retrain the variance of data. In practice, we can choose the dimension l according to the variance we need to retain. It can be seen from Eq. (8.6) that the variance after the dimensionality reduction is determined by the corresponding eigenvalues of the covariance matrix. Suppose we have sorted the eigenvalues in descending order. If we choose the dimension l, the variance retained ratio after dimensionality reduction is

$$\mathrm{Var}_{\mathrm{ratio}}(l) = \frac{\sum_{i=1}^{l} \lambda_i}{\sum_{j=1}^{d} \lambda_j}. \tag{8.8}$$

In practice, we can first set a variance ratio threshold (for example, 80%), and then determine the most suitable dimension l according to Eq. (8.8).

From the perspective of data reconstruction, the data after dimensionality reduction $\boldsymbol{Y} = \boldsymbol{X}\boldsymbol{W}$ is obtained through the matrix $\boldsymbol{W}$. We can get the reconstructed data $\boldsymbol{X}\boldsymbol{W}\boldsymbol{W}^{\mathrm{T}}$ through $\boldsymbol{W}^{\mathrm{T}}$. The reconstruction error is

$$\|\boldsymbol{X} - \boldsymbol{X}\boldsymbol{W}\boldsymbol{W}^{\mathrm{T}}\|_{\mathrm{F}}^2. \tag{8.9}$$

We can prove that minimizing (8.9) is equivalent to maximizing the variance after dimensionality reduction.

As a linear dimensionality reduction method, each feature of low-dimensional space in PCA can be written as a weighted sum of high-dimensional features, which is convenient for understanding low-dimensional data. In addition, PCA is convex and has a global optimal solution. PCA is computationally efficient, with time complexity of $O(d^3)$ and space complexity of $O(d^2)$.

When there is a linear structure in the data and the variance is large, PCA is usually the preferred dimensionality reduction method. However, maximizing variance is not always an appropriate objective. For example, when the sample contains a category label, we are more interested in predicting the label. At this point, finding the low-dimensional representation of the data to better distinguish different categories in low-dimensional space becomes our main optimization objective.

8.2. Linear Discriminant Analysis

In the previous section we discussed principal component analysis, a linear dimensionality reduction method that aims at maximizing the

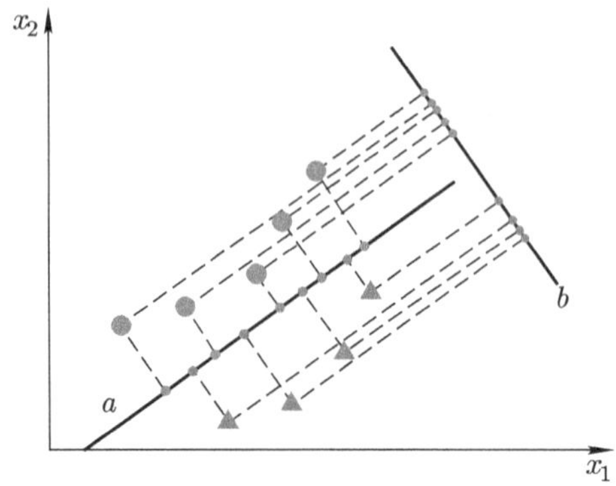

Figure 8.2 An example to show the principle of LDA.

variance. However, variance may not fully represent the information in the data. As shown in Figure 8.2, the samples belong to two categories. We hope to linearly reduce the data to one-dimension, so that the data can be well classified in one-dimensional space. Using PCA, we get the direction a in Figure 8.2. Obviously, after dimensionality reduction the two categories cannot be distinguished very well.

The linear discriminant analysis we introduced in this section can solve the above problem well. Linear discriminant analysis (LDA) is a typical supervised linear dimensionality reduction method. It was proposed by Ronald Fisher in 1936 and therefore was also known as Fisher discriminant analysis.[75] LDA has been widely applied in dimensionality reduction, face recognition and other fields.

Instead of maximizing variance, LDA uses the category information of the samples to find a linear low-dimensional representation of the data, making the low-dimensional representation easier to classify. For the data shown in Figure 8.2, LDA can find direction b, which is easy to perform classification.

8.2.1. *The Optimization Objective of LDA*

Suppose the dataset is $D = \{(\boldsymbol{x}_1, y_1), (\boldsymbol{x}_2, y_2), \ldots, (\boldsymbol{x}_n, y_n)\}$. Each sample $\boldsymbol{x}_i$ is a d-dimension vector, and the sample label y_i has a value range of $\{1, 2, \ldots, C\}$. The dataset contains n samples, and the number of samples in class C is denoted as n_c, thus $n = \sum_{i=1}^{C} n_i$. To facilitate the subsequent description, we represent the features of the dataset as a $n \times d$ matrix $\boldsymbol{X}$. Now, assume that the data is reduced to l dimension through the $d \times l$ transformation matrix $\boldsymbol{W}$, i.e.,

$$\boldsymbol{Z} = \boldsymbol{X}\boldsymbol{W}. \tag{8.10}$$

LDA exploits the category labels of the samples to find a linear low-dimensional representation. This objective can be quantified from two perspectives. The first perspective is to make samples of the same category as close as possible after dimensionality reduction. This can be quantified using the variance of the samples within the category, also known as within-class scatter. The second perspective is to make samples of different categories as far as possible after dimensionality reduction. This can be quantified using the variance of the sample means of different categories, also known as between-class scatter.

The calculation of within-class scatter and between-class scatter requires calculation of sample mean before and after dimensionality reduction and the sample mean of each category. The sample mean of the dataset D and sample mean in class c are

$$\boldsymbol{m} = \frac{1}{n}\sum_{i=1}^{n} \boldsymbol{x}_i, \tag{8.11}$$

$$\boldsymbol{m}_c = \frac{1}{n_c}\sum_{i=1}^{n_c} \boldsymbol{x}_i. \tag{8.12}$$

The within-class scatter of the class c before dimensionality reduction is

$$\boldsymbol{S}_c = \sum_{i=1}^{n_c} (\boldsymbol{x}_i - \boldsymbol{m}_c)(\boldsymbol{x}_i - \boldsymbol{m}_c)^{\mathrm{T}}. \tag{8.13}$$

Then within-class scatter of the whole dataset before dimensionality reduction is

$$\boldsymbol{S}_w = \sum_{c=1}^{C} \frac{n_c}{n} \boldsymbol{S}_c. \tag{8.14}$$

Suppose we use $\boldsymbol{w}$ to project the data into a new direction ($\boldsymbol{w}$ is any column of $\boldsymbol{W}$). Then in this direction, the within-class scatter of the whole dataset is $\boldsymbol{w}^{\mathrm{T}}\boldsymbol{S}_w\boldsymbol{w}$.

The between-class scatter is used to measure the degree of dispersion between different categories of data. It can be calculated using the mean of each class, $\boldsymbol{m}_c$, and the mean of the dataset, $\boldsymbol{m}$. The calculation of the between-class scatter matrix $\boldsymbol{S}_b$ is

$$\boldsymbol{S}_b = \sum_{c=1}^{C} \frac{n_c}{n}(\boldsymbol{m}_c - \boldsymbol{m})(\boldsymbol{m}_c - \boldsymbol{m})^{\mathrm{T}}. \tag{8.15}$$

Suppose we use $\boldsymbol{w}$ to project the data into a new direction. In this direction, the between-class scatter is $\boldsymbol{w}^{\mathrm{T}}\boldsymbol{S}_b\boldsymbol{w}$.

For a specific projection direction $\boldsymbol{w}$, the within-class scatter after projection is $\boldsymbol{w}^{\mathrm{T}}\boldsymbol{S}_w\boldsymbol{w}$, and the between-class scatter is $\boldsymbol{w}^{\mathrm{T}}\boldsymbol{S}_b\boldsymbol{w}$. The objective of LDA is to make the projected data easier to classify, requiring $\boldsymbol{w}^{\mathrm{T}}\boldsymbol{S}_w\boldsymbol{w}$ to be as small as possible and $\boldsymbol{w}^{\mathrm{T}}\boldsymbol{S}_b\boldsymbol{w}$ as large as possible. Our optimization objective is

$$\max_{\boldsymbol{w}} \quad \frac{\boldsymbol{w}^{\mathrm{T}}\boldsymbol{S}_b\boldsymbol{w}}{\boldsymbol{w}^{\mathrm{T}}\boldsymbol{S}_w\boldsymbol{w}}. \tag{8.16}$$

Observing (8.16), we find that scaling the value of each dimension of $\boldsymbol{w}$ does not change the value of the objective function. This also means that the main concern when finding $\boldsymbol{w}$ is not its specific value, but its direction. So we can add constraints to $\boldsymbol{w}$, such as $\boldsymbol{w}^{\mathrm{T}}\boldsymbol{S}_w\boldsymbol{w} = 1$. Then the optimization problem is equivalent to

$$\max_{\boldsymbol{w}} \quad \boldsymbol{w}^{\mathrm{T}}\boldsymbol{S}_b\boldsymbol{w}, \quad \text{s.t.} \quad \boldsymbol{w}^{\mathrm{T}}\boldsymbol{S}_w\boldsymbol{w} = 1. \tag{8.17}$$

8.2.2. *The Solution of LDA*

We can use the Lagrangian method to solve the problem (8.17), and the corresponding Lagrangian function is

$$L(\boldsymbol{w}, \lambda) = \boldsymbol{w}^{\mathrm{T}}\boldsymbol{S}_b\boldsymbol{w} - \lambda(\boldsymbol{w}^{\mathrm{T}}\boldsymbol{S}_w\boldsymbol{w} - 1), \tag{8.18}$$

where λ is the Lagrange multiplier. Take the partial derivative with respect to $\boldsymbol{w}$:

$$\frac{\partial L(\boldsymbol{w}, \lambda)}{\partial \boldsymbol{w}} = 2\boldsymbol{S}_b\boldsymbol{w} - 2\lambda\boldsymbol{S}_w\boldsymbol{w}, \tag{8.19}$$

let Eq. (8.19) equal to zero, then

$$\boldsymbol{S}_b\boldsymbol{w} = \lambda\boldsymbol{S}_w\boldsymbol{w}. \tag{8.20}$$

Multiply both sides of Eq. (8.20) by $\boldsymbol{S}_w^{-1}$, we can get

$$\boldsymbol{S}_w^{-1}\boldsymbol{S}_b\boldsymbol{w} = \lambda\boldsymbol{w}. \tag{8.21}$$

Observing (8.21), we can see that the projection direction $\boldsymbol{w}$ is the eigenvector of the matrix $\boldsymbol{S}_w^{-1}\boldsymbol{S}_b$, and λ is the corresponding eigenvalue. There are multiple eigenvalues of the matrix. So which one is our objective? Let's analyze the optimization objective again. According to (8.17), we need to maximize $\boldsymbol{w}^{\mathrm{T}}\boldsymbol{S}_b\boldsymbol{w}$. Substituting Eq. (8.20), we have

$$\boldsymbol{w}^{\mathrm{T}}\boldsymbol{S}_b\boldsymbol{w} = \boldsymbol{w}^{\mathrm{T}}\lambda\boldsymbol{S}_w\boldsymbol{w} = \lambda\boldsymbol{w}^{\mathrm{T}}\boldsymbol{S}_w\boldsymbol{w}. \tag{8.22}$$

Since we have the assumption $\boldsymbol{w}^{\mathrm{T}}\boldsymbol{S}_w\boldsymbol{w} = 1$, we can get

$$\boldsymbol{w}^{\mathrm{T}}\boldsymbol{S}_b\boldsymbol{w} = \lambda, \tag{8.23}$$

that is, the optimization problem (8.17) is equivalent to

$$\max_{\boldsymbol{w}} \ \lambda, \quad \text{s.t.} \quad \boldsymbol{S}_w^{-1}\boldsymbol{S}_b\boldsymbol{w} = \lambda\boldsymbol{w}. \tag{8.24}$$

Therefore, the solution $\boldsymbol{w}$ is the eigenvector corresponding to the largest eigenvalue of the matrix $\boldsymbol{S}_w^{-1}\boldsymbol{S}_b$. If we want to use LDA to reduce the data to l dimensions, we only need to choose the first l largest eigenvalues of the matrix $\boldsymbol{S}_w^{-1}\boldsymbol{S}_b$ and the corresponding eigenvectors. Combining the l projection vectors $(\boldsymbol{w}_1, \boldsymbol{w}_2, \ldots, \boldsymbol{w}_l)$ by column, we get the transformation matrix $\boldsymbol{W}$. The detailed procedure of the LDA algorithm is shown in Algorithm 8.

Algorithm 8 LDA algorithm

Input: Dataset $D = \{(\boldsymbol{x}_1, y_1), (\boldsymbol{x}_2, y_2), \cdots, (\boldsymbol{x}_n, y_n)\}$, the dimension l.
Output: Transformation matrix $\boldsymbol{W} = (\boldsymbol{w}_1, \boldsymbol{w}_2, \cdots, \boldsymbol{w}_l)$.

1: Calculate the mean of the dataset $\boldsymbol{m}$ and the mean of each class $\boldsymbol{m}_c$:

$$\boldsymbol{m} = \frac{1}{n}\sum_{i=1}^{n}\boldsymbol{x}_i, \quad \boldsymbol{m}_c = \frac{1}{n_c}\sum_{i=1}^{n_c}\boldsymbol{x}_i;$$

2: Calculate the within-class scatter matrix $\boldsymbol{S}_w = \sum_{c=1}^{C}\frac{n_c}{n}\boldsymbol{S}_c$, where $\boldsymbol{S}_c$ is the within-class scatter matrix of the c-class samples, calculated according to (8.13);

3: Compute the between-class scatter matrix $\boldsymbol{S}_b$:

$$\boldsymbol{S}_b = \sum_{c=1}^{C}\frac{n_c}{n}(\boldsymbol{m}_c - \boldsymbol{m})(\boldsymbol{m}_c - \boldsymbol{m})^{\mathrm{T}};$$

4: Compute the matrix $\boldsymbol{S}_w^{-1}\boldsymbol{S}_b$, perform eigenvalue decomposition and then sort the eigenvalues of the matrix $\boldsymbol{S}_w^{-1}\boldsymbol{S}_b$ in descending order;

5: Select the eigenvectors $(\boldsymbol{w}_1, \boldsymbol{w}_2, \cdots, \boldsymbol{w}_l)$ corresponding to the l largest eigenvalues to form the transformation matrix $\boldsymbol{W}$.

8.2.3. *Summary*

Linear discriminant analysis is a supervised linear dimensionality reduction method. It can be used both as a dimensionality reduction method and as a classification algorithm. LDA can find the features that are better for the classification, so it can also be used as a feature selection method for classification problems.

LDA algorithm finds the transformation matrix $\boldsymbol{W}$ by eigenvalue decomposition of the matrix $\boldsymbol{S}_w^{-1}\boldsymbol{S}_b$. In the previous section, we assumed that $\boldsymbol{S}_w$ is invertible. What happens if the matrix $\boldsymbol{S}_w$ is non-invertible? For example, when the sample size n_c for each class in the dataset D is much smaller than the sample dimension d, the matrix $\boldsymbol{S}_w$ is non-invertible. In this case, we can adjust the matrix $\boldsymbol{S}_w$ as follows:

$$\boldsymbol{S}_w = (1 - \lambda)\boldsymbol{S}_w + \lambda\boldsymbol{I}, \tag{8.25}$$

where $\lambda > 0$ is a shrinkage hyper-parameter and $\boldsymbol{I}$ is the identity matrix. This algorithm is also called shrinkage discriminant analysis.[76]

As typical representatives of linear dimensionality reduction methods, what is the difference between LDA and PCA? First of all, their basic ideas are different. PCA selects the direction in which the sample projection has the largest variance, maximumly retaining internal information of data. LDA exploits the labels so that the sample distance between different categories after projection is maximized and the sample distance of the same category is minimized. Secondly, the learning schema is different. PCA belongs to unsupervised learning and has a wider range of applications. However, it cannot be guaranteed that data is easy to analyze after dimensionality reduction. LDA is a supervised learning method with the ability of data classification and dimensionality reduction.

To intuitively understand the differences between PCA and LDA, we use a optical character recognition dataset to compare the dimensionality reduction results.[46] The dataset contains 20,000 samples of 26 English capital letters. Each sample represents a rectangular area in the optical image that contains only a single character. Each sample contains 16 features and a label feature *letter*. For comparison purpose, we select a subset of the three letters B, P, and H. Figure 8.3 and Figure 8.4 show the dimensionality reduction results

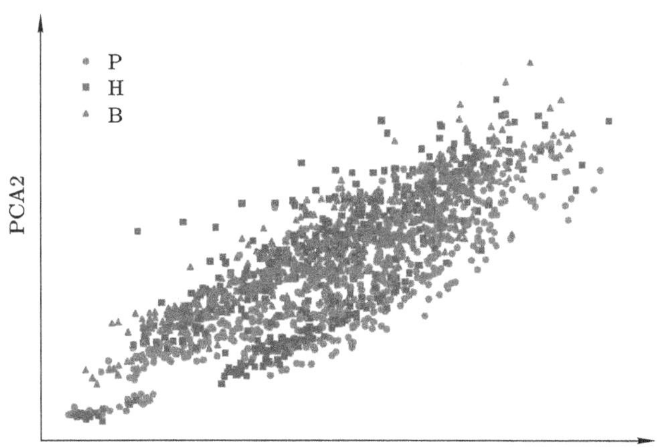

Figure 8.3 Dimensionality reduction result of PCA.

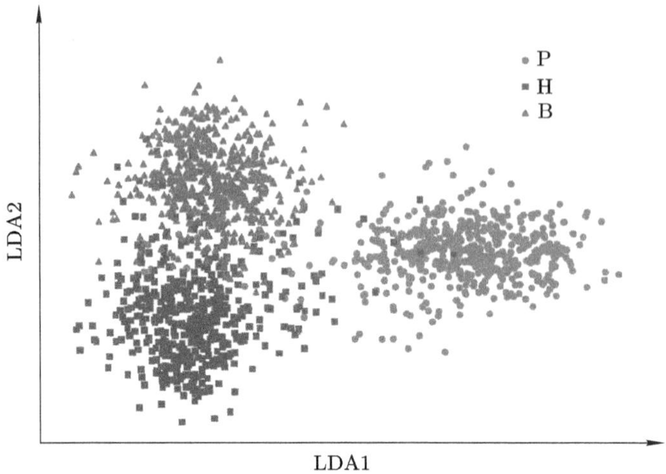

Figure 8.4 Dimensionality reduction result of LDA.

(reduction to two dimensions). Intuitively, the variance after dimensionality reduction with PCA is large, but the samples of different letters are difficult to distinguish. And after dimensionality reduction with LDA, the samples of different letters can be distinguished well.

8.3. Multi-dimensional Scaling

Multi-dimensional Scaling (MDS) is another dimension reduction method that aims to find low-dimensional representation of data so that similarity between samples before and after dimensionality reduction is preserved.[77,78] In practice, we may not be able to directly observe the features of each sample, but only the distance or similarity between samples. For example, in the field of advertising and cognitive psychology, we often get some survey data, including questions about the preferences or similarities of different subjects. These questions may be in the form of "which product of B and C do you think is closer to A "or" rate the similarity of product A and B using a number between 0 and 10".

MDS only uses the distance information between samples to find the feature representation of each sample. With that feature representation, the distance of two samples is as close as possible to their distance in the original space. Since MDS can retain distance information, the results can be visualized on a scatter plot when dimension is reduced to 1, 2 and 3.

8.3.1. *The Optimization Objective of MDS*

Suppose the dataset contains n samples. $\boldsymbol{D}$ is the $n \times n$ distance matrix, whose element d_{ij} denotes the distance between sample i and j. If sample i is represented using a d dimension feature vector $\boldsymbol{x}_i$, our dataset can also be represented by $n \times d$ matrix $\boldsymbol{X}$. $\boldsymbol{D}$ is calculated from the d-dimension feature vectors of the samples. The goal of the MDS is to find the low-dimensional representation (l dimensions) of each sample so that the sample distance after dimensionality reduction is as consistent as possible with the distance matrix $\boldsymbol{D}$. Suppose $\boldsymbol{z}_i$ is the low-dimensional representation of $\boldsymbol{x}_i$ the Euclidean distance between sample $\boldsymbol{z}_j$ and $\boldsymbol{z}_i$ is $\|\boldsymbol{z}_i - \boldsymbol{z}_j\|_2$. The optimization objective of MDS is then

$$\min_{\boldsymbol{z}_1,\boldsymbol{z}_2,\cdots,\boldsymbol{z}_n} \sum_{i \neq j} \left(\|\boldsymbol{z}_i - \boldsymbol{z}_j\|_2 - d_{ij}\right)^2. \tag{8.26}$$

8.3.2. *The Solution of MDS*

Since MDS aims to retain distance information of samples, translation or rotation of all the samples does not affect the result. In the discussion below, we can assume that sample mean is zero, that is, $\sum_{i=1}^n \boldsymbol{x}_i = \boldsymbol{0}$. We can construct a new matrix $\boldsymbol{B} = \boldsymbol{X}\boldsymbol{X}^{\mathrm{T}}$ from the feature matrix $\boldsymbol{X}$. Then each element in $\boldsymbol{B}$ can be represented as $b_{ij} = \boldsymbol{x}_i^{\mathrm{T}}\boldsymbol{x}_j$.

Assuming the data matrix $\boldsymbol{X}$ can perfectly retain the distance information, d_{ij} can be calculated directly by $\boldsymbol{x}_i$ and $\boldsymbol{x}_j$:

$$\begin{aligned} d_{ij}^2 &= (\boldsymbol{x}_i - \boldsymbol{x}_j)^{\mathrm{T}}(\boldsymbol{x}_i - \boldsymbol{x}_j) = \boldsymbol{x}_i^{\mathrm{T}}\boldsymbol{x}_i + \boldsymbol{x}_j^{\mathrm{T}}\boldsymbol{x}_j - 2\boldsymbol{x}_i^{\mathrm{T}}\boldsymbol{x}_j \\ &= b_{ii} + b_{jj} - 2b_{ij}. \end{aligned} \tag{8.27}$$

This gives the relationship between the elements of matrix $\boldsymbol{D}$ and the elements of matrix $\boldsymbol{B}$. Further, we want to directly calculate each element of b_{ij} in $\boldsymbol{B}$ using the elements in $\boldsymbol{D}$. This way we can calculate $\boldsymbol{B}$ directly through $\boldsymbol{D}$ and after we get $\boldsymbol{B}$, calculate $\boldsymbol{X}$.

Now let's see how to calculate the element b_{ij} in $\boldsymbol{B}$ from the distance matrix. In Eq. (8.27), take sum of both sides of the equation

$$\sum_{i=1}^{n} d_{ij}^2 = \sum_{i=1}^{n} b_{ii} + \sum_{i=1}^{n} b_{jj} - 2\sum_{i=1}^{n} b_{ij}. \tag{8.28}$$

Denote the first item on the right side of (8.28) as $T = \sum_{i=1}^{n} b_{ii}$. T is actually the sum of diagonal elements of matrix $\boldsymbol{B}$, also known as the *trace* of matrix $\boldsymbol{B}$. Since the sample mean is zero, the third term on the right side of Eq. (8.28) is equal to zero ($\sum_{i=1}^{n} b_{ij} = \sum_{i=1}^{n} \boldsymbol{x}_i^{\mathrm{T}} \boldsymbol{x}_j = (\sum_{i=1}^{n} \boldsymbol{x}_i)^{\mathrm{T}} \boldsymbol{x}_j = \boldsymbol{0}^{\mathrm{T}} \boldsymbol{x}_j = 0$). Equation (8.28) can be rewritten as

$$\sum_{i=1}^{n} d_{ij}^2 = T + n b_{jj}. \tag{8.29}$$

Further summing the j on both sides of Eq. (8.29), we obtain

$$\sum_{i=1}^{n}\sum_{j=1}^{n} d_{ij}^2 = \sum_{j=1}^{n} T + n\sum_{j=1}^{n} b_{jj} = 2nT. \tag{8.30}$$

Then substitute Eq. (8.30) into Eq. (8.29)

$$b_{jj} = \frac{1}{n}\sum_{i=1}^{n} d_{ij}^2 - \frac{1}{2n^2}\sum_{i=1}^{n}\sum_{j=1}^{n} d_{ij}^2. \tag{8.31}$$

Similarly, one can get the calculation of b_{ii}:

$$b_{ii} = \frac{1}{n}\sum_{j=1}^{n} d_{ij}^2 - \frac{1}{2n^2}\sum_{i=1}^{n}\sum_{j=1}^{n} d_{ij}^2. \tag{8.32}$$

Substituting Eq. (8.31) and Eq. (8.32) into Eq. (8.27), the formula for calculating matrix $\boldsymbol{B}$ from distance matrix $\boldsymbol{D}$ is obtained

$$b_{ij} = \frac{1}{2}\left(\frac{1}{n}\sum_{i=1}^{n} d_{ij}^2 + \frac{1}{n}\sum_{j=1}^{n} d_{ij}^2 - \frac{1}{n^2}\sum_{i=1}^{n}\sum_{j=1}^{n} d_{ij}^2 - d_{ij}^2\right). \tag{8.33}$$

So after getting matrix $\boldsymbol{B}$, how to obtain feature representation of the samples? Notice that matrix $\boldsymbol{B}$ is a $n \times n$ square matrix, we can perform eigenvalue decomposition

$$\boldsymbol{B} = \boldsymbol{U}\boldsymbol{\Lambda}\boldsymbol{U}^{\mathrm{T}}, \tag{8.34}$$

where $\boldsymbol{\Lambda}$ is a diagonal matrix of eigenvalues, assuming $\lambda_1 \geqslant \lambda_2 \geqslant \cdots \geqslant \lambda_n$. The i-th column of $\boldsymbol{U}$ is the eigenvector corresponding to the eigenvalue λ_i, and satisfies $\boldsymbol{u}_i^{\mathrm{T}}\boldsymbol{u}_i = 1$.

Assuming that the distance matrix $\boldsymbol{D}$ is calculated from the $n \times d$ matrix $\boldsymbol{X}$, then $\boldsymbol{B}$ has only d non-zero eigenvalues. At this point, denote $\boldsymbol{\Lambda}_d$ as the diagonal matrix consisting of the first d eigenvalues and by $\boldsymbol{U}_d$ as the matrix consisting of the first d eigenvectors, then $\boldsymbol{B}$ can be represented as:

$$\boldsymbol{B} = \boldsymbol{U}_d\boldsymbol{\Lambda}_d\boldsymbol{U}_d^{\mathrm{T}}. \tag{8.35}$$

Since $\boldsymbol{B} = \boldsymbol{X}\boldsymbol{X}^{\mathrm{T}}$, the data matrix is

$$\boldsymbol{X} = \boldsymbol{U}_d\boldsymbol{\Lambda}_d^{1/2}. \tag{8.36}$$

Suppose we need to reduce the data to the l dimension ($l < d$), then we can take the first l eigenvalues of the matrix $\boldsymbol{B}$ and their corresponding eigenvectors. The $n \times l$ matrix $\boldsymbol{Z}$ of the data after dimensionality reduction can be calculated by:

$$\boldsymbol{Z} = \boldsymbol{U}_l\boldsymbol{\Lambda}_l^{1/2}. \tag{8.37}$$

The solution obtained by Eq. (8.37) is also the solution to the optimization problem (8.28). The detailed procedure of the MDS algorithm is shown in Algorithm 9.

Algorithm 9 MDS algorithm

Input: Distance matrix $\boldsymbol{D}$, dimension l.
Output: The $n \times l$ data matrix $\boldsymbol{Z}$ after dimensionality reduction.

1: Calculate matrix $\boldsymbol{B}$ from the distance matrix $\boldsymbol{D}$, where b_{ij} is calculated according to equation (8.33);
2: Perform eigenvalue decomposition on the matrix $\boldsymbol{B}$: $\boldsymbol{B} = \boldsymbol{U}\boldsymbol{\Lambda}\boldsymbol{U}^{\mathrm{T}}$;
3: Take the first l largest eigenvalues to form the diagonal matrix $\boldsymbol{\Lambda}_l$, and combine the columns with the l corresponding eigenvectors into $\boldsymbol{U}_l$;
4: The data matrix after dimension reduction is calculated as: $\boldsymbol{Z} = \boldsymbol{U}_l\boldsymbol{\Lambda}_l^{1/2}$.

8.3.3. *Practical Example*

In this section we present an example to demonstrate the performance of MDS. Table 8.1 shows a distance dataset between several cities in China. The dataset does not directly give information about the city itself. We can use MDS to get the feature representation of each city. Using the distance between cities as the input to Algorithm 9, and set $l = 2$, we can get the coordinates of each city in the 2D plane. The coordinates are shown in Figure 8.5.[a] It can be seen that the distance information between cities can used to reconstruct the location of each city.

8.3.4. *Summary*

In this section we have introduced metric MDS, that is for any sample i, sample j and sample k, the distance matrix satisfies the following conditions:

$$d_{ik} + d_{jk} \geqslant d_{ij}. \tag{8.38}$$

For metric MDS, if we choose Euclidean distance to calculate the distance between samples, the low-dimensional representation will be consistent with PCA. When inequality (8.38) is not satisfied, MDS no longer aims to retain the sample distance, but tries to retain the order of the distances between the samples. This method is called non-metric MDS.[79] If the distance matrix $\boldsymbol{D}$ satisfies $d_{ik} < d_{jk}$, non-metric MDS tries to keep this order relationship after the dimensionality reduction. Suppose we use Euclidean distance, $\|\boldsymbol{z}_i - \boldsymbol{z}_k\|_2 < \|\boldsymbol{z}_j - \boldsymbol{z}_k\|_2$ will hold after dimensionality reduction.

In this section, we assume that the distance between samples is calculated using the Euclidean distance. Other distances can also be used. For example, the Isomap algorithm (we will discuss in Section 8.5) uses Euclidean distance only for nearest samples. Samples with distances less than a threshold are then connected by edges to form a graph between samples. The distance of the distant samples is calculated by the length of shortest path between the two samples in the graphs.

[a]We have rotated and translated the coordinates after dimensionality reduction.

Table 8.1 The distance between several cities in China (unit: km).

	Beijing	Shanghai	Harbin	Urumqi	Guiyang	Lanzhou	Fuzhou	Lhasa	Guangzhou	Wuhan
Beijing	0	1064	1055	2417	1734	1187	1558	2563	1888	1049
Shanghai	1064	0	1675	3268	1527	1717	610	2902	1213	683
Harbin	1055	1675	0	3061	2769	2192	2282	3558	2791	1992
Urumqi	2417	3268	3061	0	2571	1624	3463	1602	3281	2766
Guiyang	1734	1527	2769	2571	0	1087	1256	1560	763	870
Lanzhou	1187	1717	2192	1624	1087	0	1840	1376	1699	1148
Fuzhou	1558	610	2282	3463	1256	1840	0	2786	693	698
Lhasa	2563	2902	3558	1602	1560	1376	2786	0	2311	2227
Guangzhou	1888	1213	2971	3281	763	1699	693	2311	0	839
Wuhan	1049	683	1992	2766	870	1148	698	2227	839	0

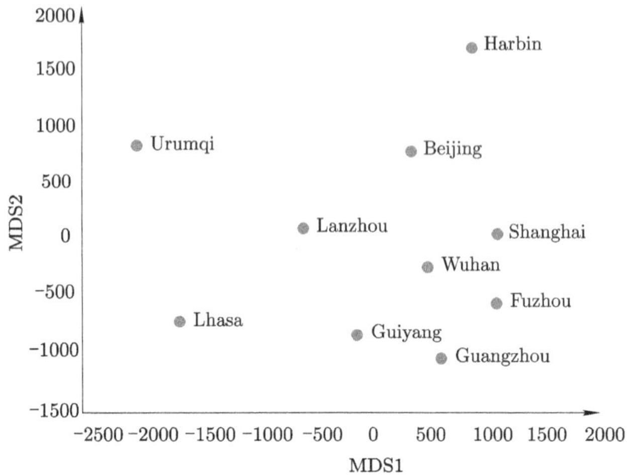

Figure 8.5 Results of multi-dimensional scaling to map cities to two-dimensional space.

In addition to dimensionality reduction, MDS is used in other fields. For example, in social network analysis, when the shortest path between nodes is regarded as the distance between nodes and l is set as 2, MDS can be applied as a graph layout algorithm.[80] In chemistry, MDS is used in molecular conformation to analyze the molecular structure.[81]

8.4. Locally Linear Embedding

MDS obtains a low-dimensional representation by retaining the sample distance of the high-dimensional data. Sample distance in high-dimensional space may not describe data characteristics very well. For example, when data is distributed over a manifold in high-dimensional space, the distance can not reflect the characteristics. Manifold refers to a space with Euclidean space properties locally, which is a generalization of curves and surfaces in Euclidean space.

In this section we introduce the locally linear embedding (LLE) algorithm, assuming that data is distributed in a low-dimensional manifold of high-dimensional space.[82,83] The basic idea of LLE is that there is a local linear relationship in the data and the local linear relationship needs to be retained after dimensionality reduction.

The locally linear embedding algorithm consists of three steps. The first step is to find neighbor samples. Calculate the Euclidean distance between samples and find the nearest neighbors for each sample. The second step is locally linear reconstruction. Assuming that each sample can be expressed in a linear combination of its neighbors, the weights can be obtained by the least squares. The third step is to find a low-dimensional representation. In the low-dimensional representation, the relationship between a sample and its neighbors can still be expressed using the weights obtained in the second step.

8.4.1. *LLE Algorithm*

Suppose that the dataset $D = \{\boldsymbol{x}_1, \boldsymbol{x}_2, \ldots, \boldsymbol{x}_n\}$ contains n samples, and the dimension of each sample is d. The dataset can also be represented as a $n \times d$ matrix $\boldsymbol{X}$. For each sample $\boldsymbol{x}_i$, we choose the k nearest neighbors to linearly reconstruct $\boldsymbol{x}_i$ (usually choosing $k \leqslant d$). Our objective is to reduce the data to l dimension ($l < d$). The d-dimension sample $\boldsymbol{x}_i$ is transformed to the l-dimension sample $\boldsymbol{y}_i$. The dataset after dimensionality reduction can be represented as a $n \times l$ matrix $\boldsymbol{Y}$. With the above notation, the three steps of the LLE algorithm can be formalized as

1. For each sample $\boldsymbol{x}_i$ in the dataset, use KNN algorithm to find k neighbors.
2. Assume that the linear relationship between the samples is represented by the $n \times n$ matrix $\boldsymbol{W}$, whose element w_{ij} represents the coefficient of sample j when reconstructing sample i. To calculate $\boldsymbol{W}$, we need to minimize the reconstruction error of the samples

$$\min_{\boldsymbol{W}} \sum_{i=1}^{n} \|\boldsymbol{x}_i - \sum_{j} w_{ij}\boldsymbol{x}_j\|_2^2. \tag{8.39}$$

3. Use the weight matrix $\boldsymbol{W}$ to find the low-dimensional representation $\boldsymbol{Y}$. Similarly, we need to minimize reconstruction error of low-dimensional samples

$$\min_{\boldsymbol{Y}} \sum_{i=1}^{n} \left\|\boldsymbol{y}_i - \sum_{j} w_{ij}\boldsymbol{y}_j\right\|_2^2. \tag{8.40}$$

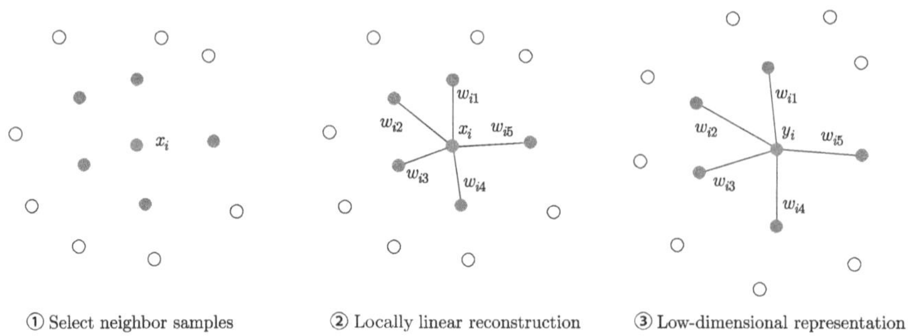

Figure 8.6 The three steps of LLE.

We can use Figure 8.6 to illustrate the procedure of LLE algorithm. For a particular sample $\boldsymbol{x}_i$, find its five nearest samples by calculating its Euclidean distances to the other samples. Suppose these five samples are $\{\boldsymbol{x}_1, \boldsymbol{x}_2, \ldots, \boldsymbol{x}_5\}$. Now, use the neighbors to get the linear reconstruction of $\boldsymbol{x}_i$, that is, $\hat{\boldsymbol{x}}_i = \sum_{j=1}^{5} w_{ij}\boldsymbol{x}_j$. The reconstruction weights $\boldsymbol{w}_i = \{w_{i1}, w_{i2}, \ldots, w_{i5}\}$ can be obtained by minimizing the reconstruction error $\|\boldsymbol{x}_i - \hat{\boldsymbol{x}}_i\|^2$. The reconstruction weights can be used to obtain a low-dimensional representation of each sample. As shown in the right sub-figure in Figure 8.6, we get the low-dimensional representation $\boldsymbol{y}_i$ of the sample $\boldsymbol{x}_i$.

In LLE the selection of neighbor samples can be done by KNN introduced in Section 4.2. However, the choice of k will affect the solution to the local linear reconstruction step, which we will discuss in the next section. The following section focuses on how to perform locally linear reconstruction and find low-dimensional representation.

8.4.2. *Locally Linear Reconstruction*

The optimization objective of the locally linear reconstruction step is shown in (8.39). There are some constraints of the parameter $\boldsymbol{W}$. First, the sample is only linearly reconstructed from the k nearest neighbors. We denote the nearest neighbors of $\boldsymbol{x}_i$ as $N(i)$. So if the sample $\boldsymbol{x}_j$ belongs to the nearest neighbors of $\boldsymbol{x}_i$, then $w_{ij} > 0$, otherwise $w_{ij} = 0$. We can further add a constraint $\sum_{j=1}^{n} w_{ij} = 1$, which does not affect the optimization objective. We found that all constraints are only for a single sample, that is, constraints are added to the row vector of the parameter matrix $\boldsymbol{W}$. Therefore, the problem of (8.39) can be decomposed into linear reconstruction problems for each sample:

$$\min_{\boldsymbol{w}_i} \|\boldsymbol{x}_i - \sum_j w_{ij}\boldsymbol{x}_j\|_2^2, \quad \text{s.t.} \quad \sum_j w_{ij} = 1, \quad w_{ij} = 0, \quad j \notin N(i). \tag{8.41}$$

Denote the above objective function as $\text{RSS}(\boldsymbol{w}_i)$. Taking into account that $\sum_{j=1}^{n} w_{ij} = 1$, the objective function can be rewritten as

$$\begin{aligned}\text{RSS}(\boldsymbol{w}_i) &= \left\|\boldsymbol{x}_i - \sum_{j\in N(i)} w_{ij}\boldsymbol{x}_j\right\|_2^2 = \left\|\sum_j w_{ij}\boldsymbol{x}_i - \sum_j w_{ij}\boldsymbol{x}_j\right\|_2^2 \\ &= \left\|\sum_j w_{ij}(\boldsymbol{x}_i - \boldsymbol{x}_j)\right\|_2^2. \end{aligned}\tag{8.42}$$

Since each sample has only k nearest neighbors, we can treat $\boldsymbol{w}_i$ as a *column vector of* k dimension. We can also introduce a $k \times d$ matrix $\boldsymbol{Z}$ with its j column as $\boldsymbol{z}_j = \boldsymbol{x}_i - \boldsymbol{x}_j$. At this point, the objective function can be written as $\text{RSS}(\boldsymbol{w}_i) = \boldsymbol{w}_i^{\mathrm{T}}\boldsymbol{Z}\boldsymbol{Z}^{\mathrm{T}}\boldsymbol{w}_i$. For ease of derivation, the notation $\boldsymbol{G}_i = \boldsymbol{Z}\boldsymbol{Z}^{\mathrm{T}}$ can be further introduced, which is commonly referred to as the *Gram matrix*. Now, the locally linear reconstruction optimization problem for sample $\boldsymbol{x}_i$ becomes

$$\min_{\boldsymbol{w}_i} \quad \text{RSS}(\boldsymbol{w}_i) = \boldsymbol{w}_i^{\mathrm{T}}\boldsymbol{G}_i\boldsymbol{w}_i, \quad \text{s.t.} \qquad \boldsymbol{w}_i^{\mathrm{T}}\mathbf{1} = 1, \tag{8.43}$$

where $\mathbf{1}$ is the vector with all the elements equal to 1.

This problem can be solved using the Lagrangian method, and the corresponding Lagrangian function is

$$L(\boldsymbol{w}_i, \alpha) = \boldsymbol{w}_i^{\mathrm{T}}\boldsymbol{G}_i\boldsymbol{w}_i - \alpha(\boldsymbol{w}_i^{\mathrm{T}}\mathbf{1} - 1), \tag{8.44}$$

where α is the Lagrangian multiplier. Taking the partial derivative with respect to $\boldsymbol{w}_i$ and make it equal to zero, that is

$$\frac{\partial L(\boldsymbol{w}_i, \alpha)}{\partial \boldsymbol{w}_i} = 2\boldsymbol{G}_i\boldsymbol{w}_i - \alpha\mathbf{1} = \mathbf{0}. \tag{8.45}$$

So we get the solution

$$\boldsymbol{w}_i = \frac{\alpha}{2}\boldsymbol{G}_i^{-1}\mathbf{1}. \tag{8.46}$$

Considering the constraint $\boldsymbol{w}_i^{\mathrm{T}}\mathbf{1} = 1$, we can first calculate $\boldsymbol{G}_i^{-1}\mathbf{1}$ and then normalize it to get $\boldsymbol{w}_i$. When the number of nearest neighbors is greater than the original data dimension ($k > d$), the matrix $\boldsymbol{G}_i$ is not a full rank matrix and is non-invertible. We need to add a regularization term to the objective function, i.e.,

$$\mathrm{RSS}(\boldsymbol{w}_i) = \boldsymbol{w}_i^{\mathrm{T}}\boldsymbol{G}_i\boldsymbol{w}_i + \lambda\boldsymbol{w}_i^{\mathrm{T}}\boldsymbol{w}_i, \tag{8.47}$$

where λ is the regularization coefficient. Then we get the solution

$$\boldsymbol{w}_i = \frac{\alpha}{2}(\boldsymbol{G}_i + \lambda\boldsymbol{I})^{-1}\mathbf{1}. \tag{8.48}$$

For each sample, we can get its reconstruction weight vector $\boldsymbol{w}_i$ according to Eq. (8.46) or Eq. (8.48). The local reconstruction matrix $\boldsymbol{W}$ can be obtained by combining the n weight vectors of the samples in the dataset.

8.4.3. *Low-Dimensional Representation*

After obtaining $\boldsymbol{W}$, we map sample $\boldsymbol{x}_i$ to $\boldsymbol{y}_i$ in the low-dimensional space by LLE. Since LLE guarantees that the locally linear relationship $\boldsymbol{W}$ remains the same, we can still obtain a low-dimensional representation by minimizing the reconstruction error of the data, as shown in Eq. (8.40). Denote the objective function as $\phi(\boldsymbol{Y}) = \sum_{i=1}^{n}\|\boldsymbol{y}_i - \sum_j w_{ij}\boldsymbol{y}_j\|^2$. By introducing a new matrix $\boldsymbol{M} = (\boldsymbol{I} - \boldsymbol{W})^{\mathrm{T}}(\boldsymbol{I} - \boldsymbol{W})$, the objective function can be written in the following form:

$$\phi(\boldsymbol{Y}) = \mathrm{tr}(\boldsymbol{Y}^{\mathrm{T}}\boldsymbol{M}\boldsymbol{Y}). \tag{8.49}$$

Observing Eq. (8.40) and Eq. (8.49), we find that if we translate the low-dimensional data, it does not affect the optimization objective. So we can add a constraint to $\boldsymbol{Y}$, assuming the sample has a mean of zero, i.e., $\sum_{i=1}^{n}\boldsymbol{y}_i = \mathbf{0}$. We further assume that after dimensionality reduction the features are independent of each other, and the variance of each feature is 1. Then we need to add the constraint $\frac{1}{n}\boldsymbol{Y}^{\mathrm{T}}\boldsymbol{Y} = \boldsymbol{I}$. Now, finding a low-dimensional representation is equivalent to

$$\min_{\boldsymbol{Y}} \quad \phi(\boldsymbol{Y}) = \mathrm{tr}(\boldsymbol{Y}^{\mathrm{T}}\boldsymbol{M}\boldsymbol{Y}), \quad \text{s.t.} \quad \sum_{i=1}^{n} \boldsymbol{y}_i = \boldsymbol{0}, \quad \frac{1}{n}\boldsymbol{Y}^{\mathrm{T}}\boldsymbol{Y} = \boldsymbol{I}. \tag{8.50}$$

We can also use the Lagrangian method to solve the above optimization problem. The eigenvalue decomposition is performed on $\boldsymbol{M}$, and the eigenvectors corresponding to the smallest l eigenvalues are chosen to form matrix $\boldsymbol{Y}$. The minimum eigenvalue of $\boldsymbol{M}$ is very close to zero. Therefore, in practice, the smallest $l+1$ eigenvalues are usually taken, then the minimum eigenvalue is deleted, and the eigenvectors corresponding to the remaining l eigenvalues are selected to form the matrix $\boldsymbol{Y}$ by column.

8.4.4. *Summary*

The idea of LLE is simple, finding a low-dimensional representation of the data by retaining the locally linear relationship of high-dimensional data. Although it utilizes the local linearity characteristic, it is a nonlinear dimensionality reduction method. LLE contains only two hyper-parameters to be tuned: the number of nearest neighbors k and the regularization coefficient λ. LLE algorithm has a global optimal solution. These advantages make LLE widely used in image recognition, image classification and data visualization.

LLE also has some weaknesses. When the sample size is small, the sample and its nearest neighbors can be far away, resulting in poor performance. LLE is sensitive to noise and outliers. In PCA and LDA, there is a transformation formula $\boldsymbol{Y} = \boldsymbol{X}\boldsymbol{W}$ between low-dimensional and high-dimensional data, but there is no such explicit mapping in LLE. We cannot directly obtain the low-dimensional representation for new samples. In addition, LLE assumes that the samples are distributed in a single smooth manifold, which is not satisfied in classification problems.

8.5. Other Dimensionality Reduction Methods

Dimensionality reduction is an important research topic in data science. It is an important way to overcome the issue of ss curse of dimension. In addition to the four methods introduced above, researchers have proposed many other dimensionality reduction methods. Typical examples include kernel principal component analysis, isometric mapping, diffusion map, t-SNE, etc. Here, we briefly introduce the basic idea of them.

Kernel principal component analysis (Kernel PCA, KPCA) uses the kernel method to compensate for the weakness that traditional PCA can only perform linear dimensionality reduction.[84] The idea of KPCA is to map the original dataset $\boldsymbol{X}$ to a higher dimensional space by nonlinear mapping ϕ, and use PCA for dimensionality reduction of the samples $\phi(\boldsymbol{X})$. KPCA is effectively applied in signal denoising[85] and face recognition.[86]

Isometric mapping (Isomap) is an improved algorithm based on MDS.[87] For nearest neighbors, Isomap uses Euclidean distance to calculate sample similarity, and then connects the adjacent samples with edges to form a sample graph. For long-distance samples, the similarity is calculated by the shortest path between the nodes. Finally, MDS is used to obtain low-dimensional representation. Isomap can preserve global structure of data, but it is sensitive to noise and computationally complex.

Diffusion map is a nonlinear dimensionality reduction method based on dynamic systems.[88,89] It is to measure sample similarity by constructing a diffusion map using the diffusion distance in the graph. Unlike Isomap, which uses the shortest path to measure sample similarity, diffusion map essentially considers all paths between samples. Diffusion map avoids operations such as matrix decomposition and is less sensitive to noise.

t-distributed Stochastic Neighbor Embedding (t-SNE) is a nonlinear dimensionality reduction method for visualizing high-dimensional data.[90,91] t-SNE uses the conditional probability between two samples to calculate similarity, and retain this conditional probability information as much as possible during dimensionality reduction.

More dimensionality reduction methods include auto-encoders,[92] Sammon mapping,[93] and Laplacian eigenmaps.[94] A comparative analysis of different nonlinear dimensionality reduction methods, as well as their performance for practical tasks, can be found in [95].

8.6. Case Studies and Exercises

1. Principal component analysis for the diagnosis of malignant mesothelioma
Malignant mesotheliomas are very aggressive pleural tumors. The main factors of the disease are: asbestos exposure time, sputum virus 40 (SV40) infection, genetic predisposition and smoking habits. In this case, the malignant mesothelioma dataset was selected.[54] Starting from the 34 related features of malignant mesothelioma, we can make a diagnosis of the disease. Please use the principal component analysis method to reduce feature size and construct a classifier for the diagnosis of malignant mesothelioma.

2. Principal component analysis for face recognition
Face recognition is a biometric recognition technology based on human facial features for identification. A classic algorithm for solving the face recognition problem is the eigenface, which is essentially the principal component analysis dimensionality reduction. This exercise provides a dataset of some famous politicians photos, a total of 1288 photos.[96] Please use principal component analysis to generate a set of eigenfaces, and then use the SVM algorithm to classify the images after dimensionality reduction (that is, determine whether the images belong to a certain politician).

3. Handwritten digital image dimensionality reduction
This exercise use a handwritten digital image dataset.[47] The dataset contains 1797 8×8 pixel handwritten digital images. The target feature stores the digit each image represents.

Please use the dimensionality reduction methods in this chapter to map the image data into a two-dimensional space and visualize it.

Chapter 9

Feature Selection

In the era of big data, we face the problem of dealing with high-dimensional data in many industries. However, high-dimensional data brings many challenges to data analysis. Firstly, more features means more time for model training. Secondly, with the increase of the number of features, model complexity will increase while model generalization ability will decrease.

In this chapter, we will discuss how to select a subset of features from all the features of a dataset. This process is called feature selection. Here we emphasize the difference between feature selection and dimension reduction introduced in Chapter 8. They both prevent overfitting by reducing the number of features. However, dimension reduction constructs new features that can better represent the original data while feature selection selects a subset of the features.

9.1. General Process of Feature Selection

Dash and Liu proposed a basic flowchart for feature selection as shown in Figure 9.1.[97] In this flowchart, a typical feature selection algorithm consists of the following four basic steps:

1. Subset generation. Generate candidate feature subset based on a certain strategy;
2. Subset evaluation. Use a certain metric to evaluate the feature subset;

3. Stopping criterion. Decide when the feature selection algorithm stops;
4. Subset validation. Verify the effectiveness of the feature subset.

Subset generation is essentially a searching process, which could start with an empty set, a randomly generated feature subset, or the entire feature set. The first strategy is called forward search, which iteratively adds features. The second is called backward search, which iteratively removes features. The third strategy is called bi-directional search, which starts with a random subset of features and combines forward and backward search strategies. In each iteration, new relevant features will be added while redundant features will be removed.

Subset evaluation is an iterative process. An evaluation metric will be calculated by assessing the new feature subset at each iteration. The metric value will be compared with the historical optimal value. If it is better, it will become the new optimal subset.

There are two ways of determining stopping criterion: one is based on subset generation, while the other is based on subset evaluation. The stopping criteria based on subset generation, are as follows:

1. Reach the pre-defined number of features;
2. Reach the pre-defined number of iterations.

The stopping criteria based on subset evaluation are as follows:

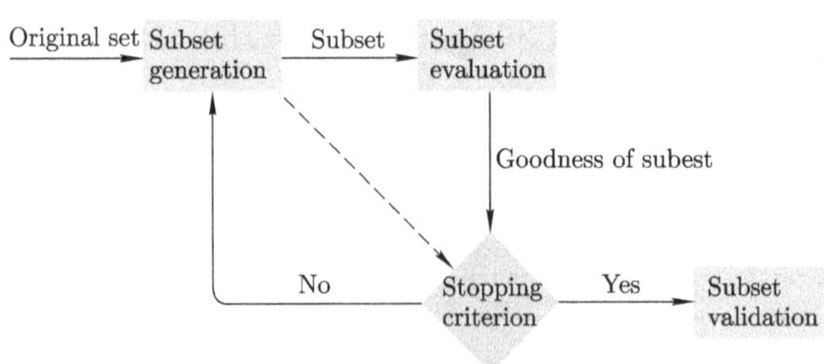

Figure 9.1 Feature selection flowchart.

1. The evaluation metric can no longer increase despite adding or removing features;
2. The evaluation metric reaches the pre-defined threshold, which denotes that the training model is good enough.

The last step of feature selection is subset validation, that is, comparing the performance between the selected feature subset and the original feature subset, to determine whether the feature subset will be retained.

Essentially, feature selection is an optimization problem. Theoretically speaking, we can list all possible feature subsets, then evaluate each feature subset, and select the best one. Suppose we have d features, then the total number of feature subsets is 2^d. The time complexity of enumerating method is $O(2^d)$. For example, if we use a breadth-first strategy to traverse the feature subspace to enumerate all the combinations, the computation amount is huge and the practiability is poor. Therefore, researchers have proposed other strategies such as heuristic search and random search, to achieve both higher efficiency and better quality.

Greedy algorithm is the most commonly used heuristic search strategy. To achieve an approximate optimal solution, greedy algorithm adopts the optimal solution under the current conditions at each iteration, without considering the historical or future selections. Greedy algorithm can be divided into three types according to different searching strategies: forward search algorithm, backward search algorithm and bidirectional search algorithm.

In the forward search algorithm, each feature is considered as a candidate subset, and the optimal subset involving one feature will be determined firstly. Then, new feature will be added into the optimal one-feature subset to obtain optimal two-feature subset. The iteration will continue until the evaluation metric cannot be improved. Backward search starts from the complete feature set, and features are iteratively removed according to the subset evaluation result to find the optimal subset. Bidirectional search combines the above two algorithms, which not only adds but also removes features.

Note that the greedy algorithm only considers local optimal subset at each iteration. For example, in the third iteration if $\{x_5\}$ is better than $\{x_6\}$, then the optimal subset is $\{x_1, x_2, x_5\}$, even if

$\{x_1, x_2, x_6, x_8\}$ is better than $\{x_1, x_2, x_5, x_8\}$ at next iteration. However, such problems are hard to avoid without enumerating search.

9.2. Feature Selection Method

Feature selection method can be divided into three categories: filtering method, wrapper method and embedding method.

9.2.1. *Filtering Method*

In filtering methods, feature selection and model training are independent processes. The key idea is to examine the correlation between candidate features and the target, and assess each feature according to the correlation. Finally, the features with higher correlations are selected.

Usually, a statistical metric is used to measure the correlation with the target feature. In general, the metric is a vector that contains the importance of each feature in the subset, and the importance of the subset is the sum of all elements of the vector. We can set a threshold and any feature whose importance is greater than the threshold will be selected. Some common used metrics include:

1. **Pearson correlation coefficient:** It is a classic metric to determine the linear correlation between a feature $\boldsymbol{x}$ and the target $\boldsymbol{y}$, which can be computed by the following formula:

$$\mathrm{Corr}(\boldsymbol{x}, \boldsymbol{y}) = \frac{\mathrm{Cov}(\boldsymbol{x}, \boldsymbol{y})}{\sigma_{\boldsymbol{x}}\sigma_{\boldsymbol{y}}}, \tag{9.1}$$

   where $\mathrm{Cov}(\boldsymbol{x}, \boldsymbol{y})$ is the covariance of $\boldsymbol{x}$ and $\boldsymbol{y}$, and $\sigma_{\boldsymbol{x}}$ is the standard deviation of $\boldsymbol{x}$.
2. **Information entropy and information gain:** For discrete features, we can use information gain as the metric. Assume that the target feature Y has a value range of $\{1, 2, \ldots, C\}$, where the

probability of the occurrence of c is denoted as p_c then the information entropy of Y can be defined as

$$H(Y) = -\sum_{c=1}^{C} p_c \log_2 p_c, \tag{9.2}$$

since entropy comes from the lower bound with the length of the binary code, it usually takes 2 as the base. If another variable X is given with value range of $\{1, 2, \ldots, K\}$, then the conditional entropy of Y is represented as

$$H(Y|X) = \sum_{k=1}^{K} p(X = k) H(Y|X = k), \tag{9.3}$$

the information gain of Y with respect to X should be defined as

$$\mathrm{IG}(Y|X) = H(Y) - H(Y|X). \tag{9.4}$$

Information entropy and information gain have also been used in decision tree (see Section 4.3.1). For feature subset X_1 and X_2, if $\mathrm{IG}(Y|X_1) > \mathrm{IG}(Y|X_2)$, then subset X_1 will be selected.

3. **Distance:** A good feature subset will bring samples of the same class closer and samples of different classes farther apart. So, the distance between samples is utilized to evaluate the feature subset. Commonly used distances include Euclidean distance, Mahalanobis distance and KL divergence (see Appendix D).

9.2.2. *Wrapper Method*

The wrapper method combines the feature selection process with the model training process. At each round, a model is trained by the feature subset, and the feature subset is evaluated based on the model's performance. In this way, we can get a feature subset with good prediction performance. However, the feature subset is not universally applicable to different models, so we need to repeatedly execute the feature selection process for different models. Moreover, due to the repeated model training and test for each candidate subset, the computation complexity will become very high, especially when the dataset is large.

The wrapper method treats the learner as a "black box", so various learning algorithms can be used to evaluate features, such as decision tree, Bayesian classifier, and support vector machine. The evaluation metrics could also be flexible, such as the accuracy, recall, AUC area, Akaike information criterion, and Bayesian information criteria. Here, we briefly introduce two metrics:

1. Akaike information criterion (AIC)

$$\mathrm{AIC} = 2d - 2\ln(\hat{L}); \tag{9.5}$$

2. Bayesian information criterion (BIC)

$$\mathrm{BIC} = d\ln(n) - 2\ln(\hat{L}), \tag{9.6}$$

where d represents the number of parameters, $\hat{L}$ represents the maximum value of the likelihood function, and n represents the number of samples. AIC and BIC are used to balance model complexity (the number of parameters) and model performance (using the maximum likelihood). We tend to choose model or feature subset with smaller AIC and BIC. Unlike AIC, in BIC, the penalty term will increase as the sample size increases. When the sample size increases the feature with low correlation to the target feature can be removed to reduce model complexity.

Recursive feature elimination (RFE) is a popular wrapper-based feature selection method. RFE recursively removes features by training models. At each iteration, each feature is assigned a weight during the model training process, and the feature with the smallest weight will be eliminated. The above processing will be repeatedly executed until the number of features meets our requirements. Note that the stability of RFE has a close relationship with the model performance.

9.2.3. *Embedding Method*

In embedding method, feature selection and model training are performed simultaneously. When model training ends, feature subset are also obtained simultaneously. Decision tree is one of most typical embedding-based feature selection methods. During the tree generation process, each node should select a feature for splitting. The order of splitting features indicates the importance of features in a decision tree. Therefore, we can use the decision tree algorithm to select features.

Below we introduce several common embedding-based feature selection methods.

1. Regularization

Regularization is a way to overcome overfitting by adding penalty to the model. It has been introduced in Section 3.2. LASSO has the effect of making the coefficients of certain features to be 0. Thus, it can also be used for feature selection.

According to the characteristics of the problem, we can also use regularizations with other norms for feature selection, such as elastic net.[29] It has a grouping effect in the feature selection process, that is, the coefficients of the highly correlated features tend to be equal.

2. Tree-based model

Tree-based models, such as random forest, GBDT, etc., can measure the importance of features so that features can be selected based on their weights. In a tree structure, the more important feature is, the closer it is to the root. We can use the following metrics to measure the importance of features.

1. **Mean decrease impurity (MDI):** In a decision tree, each node is considered as a condition that is decided by a feature, and each node aims to divide the dataset into two parts according to node feature. We can use impurity to decide the feature of each node. For classification, Gini index or information gain are usually used, while for regression, variance or least-squares are usually used. For decision tree, feature importance is computed during training by the impurity decrease. For random forest, the average impurity decrease of a feature in different decision trees can be used to measure its importance.
2. **Mean decrease accuracy (MDA):** Another metric for feature selection is to directly measure the impact of each feature on the

model performance. This method will measure the influence on the model accuracy by changing the order of the feature values. Obviously, for those non-important features, there will be less influence on the accuracy when changing the order of feature values. On the contrary, for those important features, the changes in the order of feature values will greatly decrease the accuracy.

9.3. Unsupervised Feature Selection

The aforementioned methods exploit the target feature, that is, to use the correlation between features and target for feature selection. This kind of methods can be referred to as supervised method. However, in practice most datasets are unlabeled. Unsupervised feature selection method has received extensive attention.

There are mainly two types of unsupervised feature selection methods. The first one retains the similarity or manifold structure to remove redundant features; the second one predicts the category label for each sample by clustering (also referred to as "pseudo-label"), and then perform supervised feature selection, as shown in Figure 9.2.

Unsupervised feature selection methods can also be divided into filtering method and embedding methods according to whether feature selection and model training are performed simultaneously. Laplacian score and SPEC mentioned below are representatives of filtering-based unsupervised feature selection methods, while MCFS is the representative of embedding-based unsupervised feature selection methods. In the following, we list several unsupervised feature selection methods:

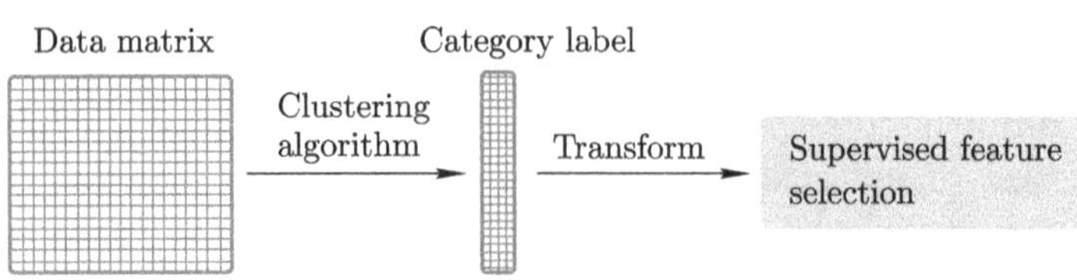

Figure 9.2 Unsupervised feature selection based on clustering.

1. **Laplacian score[98]:** This method selects feature subsets that are able to retain the manifold structure. Each feature has a corresponding Laplacian score, which can be used to sort the features for selection;
2. **SPEC[99]:** This method is an extension of Laplacian score. It uses spectral analysis methods for feature selection;
3. **Similarity preserving feature selection (SPFS)[100]:** This method selects feature subset maximumly retaining the similarity between samples;
4. **Multi-cluster feature selection (MCFS)[101]:** This method uses a spectral clustering method with L_1 regularization for feature selection;
5. **Nonnegative discriminative unsupervised feature selection (NDFS)[102]:** This method combines non-negative spectral analysis, L_1 regularization, and L_2 regularization.

Please refer to the literature for more introduction to unsupervised feature selections and comparison between different methods in practical applications.[101,103]

9.4. Summary

Feature selection is an important branch in data science, and it is also an important problem often faced by data scientists in practice. In the era of big data, high-dimensional data becomes increasingly common, and it brings higher demands on the effectiveness and efficiency for feature selection. As to high-dimensional data, the filtering method and wrapper method gradually fade out of sight due to their low efficiency. For most real scenarios, we use embedding methods for feature selection, which can not only improve efficiency, but also avoid generating feature subsets.

9.5. Case Studies and Exercises

1. Course evaluation

Course evaluation is an effective feedback on the teaching quality and the popularity of course content. The school management department can make adjustments to the courses and teachers according to the feedback to improve the quality of education. In Gazi university in Turkey, a questionnaire was designed for course investigation. In the questionnaire, students are required to answer 28 questions that are related to the courses and teachers. For each question, students are asked to select a number from {1, 2, 3, 4, 5} as the answer. The value of the number represents the degree of approval towards the specific question. Some questions are shown as follows:

Q1: The semester course content, teaching method and evaluation system were provided at the start.
Q9: I greatly enjoyed the class and was eager to actively participate during the lectures.
Q17: The instructor arrived on time for classes.

There are 5,820 samples in the dataset. Each sample consists of 33 features, among which 28 features are the feedbacks of the questionnaire, while the other features are the information of the course and teacher, including teacher ID, course ID, the number of class, the attendance rate and the difficulty level of the course.

Please use the feature selection methods introduced in this chapter to identify key features that affect the attendance rate and the popularity of the course.

2. Bank telemarketing

Nowadays, business organizations will adopt a variety of marketing tools to promote their products and services, such as telemarketing, television advertising, and plane publicity. The aim is to introduce products to the potential customers more accurately and effectively to increase revenue and profits.

In this exercise, we use a dataset which records the information of promoting the time deposit product by means of telemarketing in a Portuguese bank.[101] The dataset contains 45,211 samples, and each sample includes 17 features. Among these features, the last feature records whether the customer purchase the product, while the other 16 features can be roughly divided into three categories: customer information, marketing campaign information and socio-economic environment information. Customer information includes age, job, marital status, education level, housing, loan, etc. Marketing campaign information includes contact communication type, last contact time, etc. Socio-economic environment information includes the employment variation rate, consumer price index, consumer confidence index, etc.

Please use the feature selection methods to identify the key features and build a model to predict whether the customer will purchase the product.

Chapter 10

EM Algorithm

10.1. EM Algorithm

Expectation-maximization (EM) algorithm aims to find the maximum likelihood estimation of the parameters in a probabilistic model. In 1950, Ceppellini *et al.* first applied the EM algorithm to the estimation of gene frequency, which was further promoted in hidden Markov model by Hartley and Baum *et al.*[a] Probabilistic model depends on unobserved latent variables, the EM algorithm is designed to estimate these latent variables. The EM algorithm uses incomplete information to estimate parameters of probabilistic model. The derivation of the EM algorithm is described as follows.

Our goal is to find the optimal parameters that maximize the log-likelihood function. However, the likelihood function now contains a latent variable $\boldsymbol{z}$. Given the observable variables $\boldsymbol{X} = \{\boldsymbol{x}_1, \boldsymbol{x}_2, \ldots, \boldsymbol{x}_n\}$ and latent variables $\boldsymbol{Z}$, the logarithm of the likelihood function $p(\boldsymbol{X}|\boldsymbol{\theta})$ are

$$\sum_i \ln p(\boldsymbol{x}_i|\boldsymbol{\theta}) = \sum_i \ln \sum_{\boldsymbol{z}_i} p(\boldsymbol{x}_i, \boldsymbol{z}_i|\boldsymbol{\theta}). \tag{10.1}$$

It is assumed that each observed sample $\boldsymbol{x}_i$ has its corresponding latent variable $\boldsymbol{z}_i$. If we solve the maximization problem by using

[a]See Section 11.2 of this book for hidden Markov models.

the traditional method, we first partially derive the log-likelihood function, and then use the gradient method to optimize or to solve the equation with the partial derivative equaling to zero. This traditional method can hardly obtain the optimal parameter $\boldsymbol{\theta}$. Because in the log-likelihood function, the latent variables $\boldsymbol{z}_i$ should be summed up first and then taken logarithm, which makes it difficult to obtain an analytical solution.

Therefore, we consider making some mathematical adjustments:

$$\begin{aligned}\sum_i \ln p(\boldsymbol{x}_i|\boldsymbol{\theta}) &= \sum_i \ln \sum_{\boldsymbol{z}_i} p(\boldsymbol{x}_i, \boldsymbol{z}_i|\boldsymbol{\theta}) \\ &= \sum_i \ln \sum_{\boldsymbol{z}_i} Q_i(\boldsymbol{z}_i)\frac{p(\boldsymbol{x}_i, \boldsymbol{z}_i|\boldsymbol{\theta})}{Q_i(\boldsymbol{z}_i)}, \end{aligned} \tag{10.2}$$

where $Q_i(\boldsymbol{z}_i|\boldsymbol{\theta})$ is introduced to represent the probability distribution of latent variable $\boldsymbol{z_i}$ which is usually denoted as $Q_i(\boldsymbol{z}_i)$. $Q_i(\boldsymbol{z}_i)$ satisfies the conditions $\sum_{\boldsymbol{z}} Q_i(\boldsymbol{z}) = 1$ and $Q_i(\boldsymbol{z}) > 0$.

Next, by using the Jensen's inequality (see Appendix C), we can get:

$$\sum_i \ln \sum_{\boldsymbol{z}_i} Q_i(\boldsymbol{z}_i)\frac{p(\boldsymbol{x}_i, \boldsymbol{z}_i|\boldsymbol{\theta})}{Q_i(\boldsymbol{z}_i)} \geqslant \sum_i \sum_{\boldsymbol{z}_i} Q_i(\boldsymbol{z}_i) \ln \frac{p(\boldsymbol{x}_i, \boldsymbol{z}_i|\boldsymbol{\theta})}{Q_i(\boldsymbol{z}_i)}. \tag{10.3}$$

That is

$$\sum_i \ln \left(E_{Q_i(\boldsymbol{z}_i)} \left(\frac{p(\boldsymbol{x}_i, \boldsymbol{z}_i|\boldsymbol{\theta})}{Q_i(\boldsymbol{z}_i)} \right) \right) \geqslant \sum_i E_{Q_i(\boldsymbol{z}_i)} \left(\ln \left(\frac{p(\boldsymbol{x}_i, \boldsymbol{z}_i|\boldsymbol{\theta})}{Q_i(\boldsymbol{z}_i)} \right) \right). \tag{10.4}$$

$\frac{p(\boldsymbol{x}_i, \boldsymbol{z}_i|\boldsymbol{\theta})}{Q_i(\boldsymbol{z}_i)}$ is an independent variable. Since the $\ln(\cdot)$ function is a concave function, Jensen's inequality direction should be reversed, that is, $f(E(X)) \geqslant E(f(X))$, so as to hold the above inequality. The right term of Eq. (10.4) is called the lower bound of the log-likelihood function.

Note that our goal is to maximize the log-likelihood function, that is, to maximize the left-hand of Eq. (10.4). To this end, we need to find the condition that makes left-hand equals to the right hand, and then maximize the right-hand under that condition. It is because when the lower bound of the likelihood function is maximized that, the value of the likelihood function will become larger. Therefore, we can establish the lower bound of the log-likelihood function with respect to the current parameter $\boldsymbol{\theta}^{(t)}$, that is, determining $Q_i(\boldsymbol{z}_i)$ (the Expectation step, referred to as E step), and the lower bound is the expectation of $\ln \frac{p(\boldsymbol{x}_i, \boldsymbol{z}_i|\boldsymbol{\theta})}{Q_i(\boldsymbol{z}_i)}$ under the distribution $Q_i(\boldsymbol{z}_i|\boldsymbol{\theta}^{(t)})$ denoted as $\mathrm{LB}(\boldsymbol{\theta}, \boldsymbol{\theta}^{(t)})$. Then new parameter $\boldsymbol{\theta}^{(t+1)}$ will be determined to maximize the lower bound (this step is Maximization step, referred to as M step). This two-step iterative process will be executed repeatedly until the extremum of the lower bound approaches the extremum of the log-likelihood function. So, this algorithm is called EM algorithm, which is the maximum expectation algorithm. Figure 10.1 illustrates the diagram of the EM algorithm. The green line is a log-likelihood function, where E step constructs its lower bound (i.e., the pink line in Figure 10.1), while M step maximizes the lower bound, estimates new parameters, and the iteration ends until the maximum value of the likelihood function is approached.

When does the Jensen' inequality take the equal sign? To meet the condition of equality, it holds that:

$$\frac{p(\boldsymbol{x}_i, \boldsymbol{z}_i|\boldsymbol{\theta})}{Q_i(\boldsymbol{z}_i)} = c, \tag{10.5}$$

where c is a constant. According to Eq. (10.5) $Q_i(\boldsymbol{z}_i)$ can be represented by $p(\boldsymbol{x}_i, \boldsymbol{z}_i|\boldsymbol{\theta})$. Since $\sum_{\boldsymbol{z}_i} Q_i(\boldsymbol{z}_i) = 1$, we can get $\sum_{\boldsymbol{z}_i} p(\boldsymbol{x}_i, \boldsymbol{z}_i|\boldsymbol{\theta}) = c$. Then we get:

$$Q_i(\boldsymbol{z}_i) = \frac{p(\boldsymbol{x}_i, \boldsymbol{z}_i|\boldsymbol{\theta})}{\sum_{\boldsymbol{z}_i} p(\boldsymbol{x}_i, \boldsymbol{z}_i|\boldsymbol{\theta})} = \frac{p(\boldsymbol{x}_i, \boldsymbol{z}_i|\boldsymbol{\theta})}{p(\boldsymbol{x}_i|\boldsymbol{\theta})} = p(\boldsymbol{z}_i|\boldsymbol{x}_i, \boldsymbol{\theta}). \tag{10.6}$$

We can see that to obtain a good lower bound of the likelihood function, we just need to set $Q_i(\boldsymbol{z}_i)$ equals to the conditional distribution of $\boldsymbol{z}_i$ given $\boldsymbol{x}_i$ and the parameter θ.

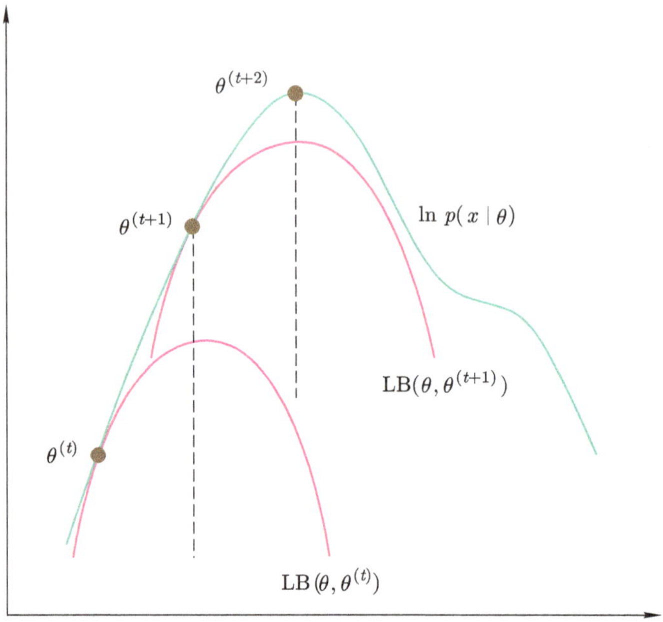

Figure 10.1 Schematic diagram of the EM algorithm.

Therefore, the E-step can also be described as: set the probability distribution $Q_i(\boldsymbol{z}_i) = p(\boldsymbol{z}_i|\boldsymbol{x}_i, \theta)$ to the lower bound of the log-likelihood function. Thus, we can obtain the complete steps of the EM algorithm as:

1. Initialize the parameter $\boldsymbol{\theta}^{(0)}$
2. E-step: For each sample i calculate

$$Q_i(\boldsymbol{z}_i|\boldsymbol{\theta}^{(t)}) = p(\boldsymbol{z}_i|\boldsymbol{x}_i, \boldsymbol{\theta}^{(t)}), \tag{10.7}$$

to find the lower bound of the log-likelihood function;
3. M-step: Maximize the logarithm likelihood function

$$\boldsymbol{\theta}^{(t+1)} := \mathrm{argmax}_{\boldsymbol{\theta}} \quad \sum_i \sum_{\boldsymbol{z}_i} Q_i(\boldsymbol{z}_i|\boldsymbol{\theta}^{(t)}) \ln \frac{p(\boldsymbol{x}_i, \boldsymbol{z}_i|\boldsymbol{\theta})}{Q_i(\boldsymbol{z}_i|\boldsymbol{\theta}^{(t)})}; \tag{10.8}$$

4. Repeat step 2 and 3 until convergence.

The termination condition can be pre-defined. For example, select a sufficiently small positive number ε, and set the termination condition as $|\boldsymbol{\theta}^{(i+1)} - \boldsymbol{\theta}^{(i)}| < \varepsilon$. Although EM maximizes the lower bound of the likelihood function at each iteration, it can only guarantee a local optimal solution. The initialization of parameters will affect the final result, so in practice, multiple runs are required to select the best result.

10.2. Application of EM: Gaussian Mixture Model

One application of the EM algorithm is to estimate the latent parameters of the Gaussian mixture model. We introduce the Gaussian mixture model first.

Suppose a dataset is generated from multiple normal distributions with different mean values and expectations. The proportions of distribution and their parameters are not known. The dataset sample is produced by the combination of multiple small datasets — the probability of which small dataset each sample belongs to is subject to a multinomial distribution. Moreover, the samples belonging to each small dataset all come from normal distributions with unknown parameters.

For example, there are 100 students in the class, and their heights are subject to two normal distributions with respect to different genders. Overall speaking, the problem about the mixed heights of boys and girls is a typical Gaussian mixture model. More specifically, there are only two categories: boys and girls, which are subject to binomial distributions. For each category, it is a Gaussian distribution with individual parameters. Then, the latent variables $\boldsymbol{z} = (z_1, z_2)$, where $z_1 = 1$ denotes boys and $z_2 = 1$ denotes girls. We will introduce how to solve this problem.

To do this, a K-dimensional binary random vector $\boldsymbol{z}$ is introduced as a latent variable. One-hot encoding strategy is adopted for assigning the value of each random variable, where only one-dimensional z_k is 1 and the rest is 0. In this way, the vector z_k will indicate that it belongs to the k class. Note that $z_k \in \{0, 1\}$, $\sum_k z_k = 1$. We will continue to convert the joint distribution of the observable variable $\boldsymbol{x}$ and the latent variable $\boldsymbol{z}$ into the product of the marginal distribution $p(\boldsymbol{z})$ and the conditional probability distribution $p(\boldsymbol{x}|\boldsymbol{z})$

$$p(\boldsymbol{x}, \boldsymbol{z}) = p(\boldsymbol{x}|\boldsymbol{z})p(\boldsymbol{z}), \tag{10.9}$$

where the marginal distribution of $\boldsymbol{z}$ is determined by the hybrid parameter π_k, i.e.,

$$p(z_k = 1) = \pi_k, \tag{10.10}$$

where $\pi_k \in [0, 1]$ and $\sum_{k=1}^{K} \pi_k = 1$. So, the marginal distribution can be written by multiple distribution

$$p(\boldsymbol{z}) = \prod_{k=1}^{K} \pi_k^{z_k}, \tag{10.11}$$

The conditional distribution can be written as a normal distribution, and describe the distribution that each small dataset satisfies.

$$p(\boldsymbol{x}|z_k = 1) = N(\boldsymbol{x}|\boldsymbol{\mu}_k, \boldsymbol{\Sigma}_k), \tag{10.12}$$

and its equivalent form:

$$p(\boldsymbol{x}|\boldsymbol{z}) = \prod_{k=1}^{K} N(\boldsymbol{x}|\boldsymbol{\mu}_k, \boldsymbol{\Sigma}_k)^{z_k}. \tag{10.13}$$

Then, sum up all the possible states of $\boldsymbol{z}$ to obtain the marginal distribution of the observable variable $\boldsymbol{x}$

$$p(\boldsymbol{x}) = \sum_{\boldsymbol{z}} p(\boldsymbol{z})p(\boldsymbol{x}|\boldsymbol{z}) = \sum_{k=1}^{K} \pi_k N(\boldsymbol{x}|\boldsymbol{\mu}_k, \boldsymbol{\Sigma}_k). \tag{10.14}$$

Obviously, the distributions of samples are a linear combination of K Gaussian models with some proportion, named Gaussian mixture model. $\boldsymbol{x}$ denotes the observable variable and $\boldsymbol{z}$ denotes the latent variable. For this kind of problems, the EM algorithm can be adopted to perform maximum likelihood estimation for parameter estimation.

Next, a probability distribution $Q_i(\boldsymbol{z}) = p(\boldsymbol{z}|\boldsymbol{x}|\boldsymbol{\theta}) = \gamma_n(z_k)$ is introduced to represent the conditional probability of the latent variable $\boldsymbol{z}$ with respect to $\boldsymbol{x}$, i.e., the probability of $z_k = 1$. Bayesian formula is used to solve this problem.

$$\begin{aligned}\gamma_n(z_k) = p(z_k = 1|\boldsymbol{x}_n) &= \frac{p(z_k = 1)p(\boldsymbol{x}|z_k = 1)}{\sum_{j=1}^{K} p(z_j = 1)p(\boldsymbol{x}|z_j = 1)} \\ &= \frac{\pi_k N(\boldsymbol{x}_n|\boldsymbol{\mu}_k, \boldsymbol{\Sigma}_k)}{\sum_{j=1}^{K} \pi_j N(\boldsymbol{x}_n|\boldsymbol{\mu}_j, \boldsymbol{\Sigma}_j)},\end{aligned} \tag{10.15}$$

where π_k is considered as the prior probability of $z_k = 1$, and $\gamma(z_k)$ is considered as "post-probability" of $\boldsymbol{x}$. In other words, $\gamma(z_k)$ can be interpreted as "how much this class contributes to the specific sample" or "for this sample, how much is formed by this class".

Then, write the function of the maximum likelihood to identify the maximum likelihood.

$$\ln p\left(\{\boldsymbol{x}_n\}_{n=1}^{N}|\boldsymbol{\theta}\right) = \sum_{n=1}^{N}\sum_{k=1}^{K} \pi_k N(\boldsymbol{x}_n|\boldsymbol{\mu}_k, \boldsymbol{\Sigma}_k), \tag{10.16}$$

where the parameter set $\boldsymbol{\theta} = \{\boldsymbol{\pi}, \{\boldsymbol{\mu}_k\}_{k=1}^{K}, \{\boldsymbol{\Sigma}_k\}_{k=1}^{K}\}$. In this way, the two steps of EM are

1. **E step:** Calculate $\gamma_n(z_k)$ according to formula (10.15):

$$\gamma_n(z_k) = p(z_k = 1|\boldsymbol{x}_n, \boldsymbol{\theta}^{(t)}) = \frac{\pi_k N(\boldsymbol{x}_n|\boldsymbol{\mu}_k, \boldsymbol{\Sigma}_k)}{\sum_{j=1}^{K} \pi_j N(\boldsymbol{x}_n|\boldsymbol{\mu}_j, \boldsymbol{\Sigma}_j)}; \tag{10.17}$$

2. **M step:** Take $\boldsymbol{\theta}$ as a variable to maximize the lower bound $\mathrm{LB}(\boldsymbol{\theta}, \boldsymbol{\theta}^{(t)})$ of the likelihood function.

$$\begin{aligned}\mathrm{LB}(\boldsymbol{\theta}, \boldsymbol{\theta}^{(t)}) &= E\left(\sum_n \ln p(\boldsymbol{x}_n, \boldsymbol{z}_n)\right) \\ &= \sum_n E\left(\ln\left(\prod_{k=1}^{K} \pi_k p(\boldsymbol{x}_n|\boldsymbol{\theta}_k)^{z_k}\right)\right) \\ &= \sum_n\sum_k E(z_k)\ln\left(\pi_k p(\boldsymbol{x}_n|\boldsymbol{\theta}_k)\right)\end{aligned}$$

$$= \sum_n \sum_k p(z_k = 1|\boldsymbol{x}_n, \boldsymbol{\theta}^{(t)}) \ln (\pi_k p(\boldsymbol{x}_n|\boldsymbol{\theta}_k))$$
$$= \sum_n \sum_k \gamma_n(z_k) \ln \pi_k + \sum_n \sum_k \gamma_n(z_k) \ln p(\boldsymbol{x}_n|\boldsymbol{\theta}_k). \tag{10.18}$$

The solution of this equation can be analytically obtained by means of logarithmic partial derivatives. The analytical solution is

$$\pi_k = \frac{1}{N} \sum_{n=1}^{N} \gamma_n(z_k), \tag{10.19}$$

$$\boldsymbol{\mu}_k = \frac{\sum_{n=1}^{N} \gamma_n(z_k) \boldsymbol{x}_n}{\sum_{n=1}^{N} \gamma_n(z_k)}, \tag{10.20}$$

$$\boldsymbol{\Sigma}_k = \frac{\sum_{n=1}^{N} \gamma_n(z_k)(\boldsymbol{x}_n - \boldsymbol{\mu}_k)(\boldsymbol{x}_n - \boldsymbol{\mu}_k)^{\mathrm{T}}}{\sum_{n=1}^{N} \gamma_n(z_k)}. \tag{10.21}$$

Until now, the parameters $\boldsymbol{\theta}_k = (\boldsymbol{\mu}_k, \boldsymbol{\Sigma}_k)$ of K Gaussian distributions in the Gaussian mixture model are obtained. Back to the height issue, we actually get an estimation of whether a sample is a boy or a girl, and the average height and variance of boys and girls, respectively.

10.3. Summary

EM algorithm is an important method to solve the maximum likelihood parameter estimation problem when there are invisible latent variables. This chapter demonstrates how to apply EM algorithm into the parameter estimation of Gaussian mixture model. The EM algorithm is widely used in various applications, which can be applied not only to the Gaussian mixture model, but also to the hidden Markov model and the Latent Dirichlet Allocation model. It will become a very powerful numerical calculation tool when combined with other technologies.

Table 10.1 Features of the building energy dataset.

Feature name	Feature description
x_1	Relative compactness
x_2	Surface area
x_3	Wall area
x_4	Roof area
x_5	Overall height
x_6	Orientation
x_7	Glazing area
x_8	Distribution of glazing area
y_1	Heating load
y_2	Cooling load

In the next chapter, we will continue to introduce the applications of the EM algorithm. However, it should be mentioned that as not all the lower bound of the log likelihood function can be calculated, EM algorithm is not suitable for all the issues about parameter estimations with latent variables. For example, the parameters of the well-known restricted Boltzmann machines (RBM) have no such properties, therefore, EM algorithm is not available.

10.4. Case Studies and Exercises

1. Energy efficiency of building

Nowadays, building is one of the most important sources of energy consumption. Buildings heavily rely on lighting, heating, air conditioning, and other facilities to provide people with comfortable environments for working and living. In order to study energy efficiency, researchers simulated 12 different architectural exteriors. Buildings are differentiated by the areas of glass covering, the locations of glass, and the directions of the building, etc.

In this case, a dataset of building energy is provided, with 10 features and 768 samples. The heating load and cooling load of buildings are the key features. Ten features of the dataset are listed in Table 10.1.

Table 10.2 Features of the wheat grain dataset.

Feature name	Feature description	Example
x_1	area	15.26
x_2	perimeter	14.84
x_3	compactness	0.871
x_4	length of kernel	5.763
x_5	width of kernel	3.312
x_6	asymmetry coefficient	2.221
x_7	length of kernel groove	5.22
Label	variety of wheat, 1: Kama, 2: Rosa, 3: Canadian	1

Please use the Gaussian mixture model to cluster the buildings with different energy consumption based on the 10 features of the dataset.

2. Wheat variety classification

Wheat is a kind of grass plant widely planted all over the world. It is rich in starch, protein and many microelements. As one of the staple foods of human beings, wheat can be used to make bread, beer and many other foods.

In this exercise, a dataset consisted of wheat's internal structure is provided with 8 features and 210 samples as shown in Table 10.2. Among them, there are 70 samples for each type of wheat. The dataset was collected by Polish researchers, and it was comprised with high-quality image of the internal structure of three kinds of wheat grain (Kama, Rosa, and Canadian) that are obtained by soft X-ray technology.

Based on the given dataset, please use the Gaussian mixture model to classify and predict the wheat variety.

3. Survey on dress style

In summer, ladies love to wear skirts. Different collections of skirt styles and colors will reveal their unique charms. The dataset of this case contains the information on the dress style, the material, the color, the size, and the price, etc. Based on the above information, the salesman will decide whether to recommend a specific type of dress to a customer. There are 14 features and 501 samples in the dataset.

Please use Gaussian mixture model to determine whether the skirt should be recommended based on the dataset. Since the feature are all nominal types, preprocessing is required, such as feature encoding, etc.

Chapter 11

Probabilistic Graphical Model

11.1. Overview of Probabilistic Graphical Model

The vast majority of tasks in data science involve the core problem of how to infer the variables we are interested in based on observed data or variables. For example, we could observe various physical indicators of patients from intelligent medical diagnosis, including blood pressure, blood sugar, white blood cell count and red blood cell count, so we wanted to speculate the likelihood of the patient having leukemia. In natural language generation, the word with the highest possibility is inferred from the first few words of a complete sentence. The probabilistic model considers the task of analyzing the relationship between these variables as a problem of modeling the probability distribution of random variables. Suppose we analyze three variables $\boldsymbol{x}, \boldsymbol{y}, \boldsymbol{\theta}$, where $\boldsymbol{x}$ is the observed data, $\boldsymbol{y}$ is the variable we are interested in, and $\boldsymbol{\theta}$ is the parameter representing the model. The question we are interested in is how to estimate the probability distribution of $\boldsymbol{y}$ given the value of the $\boldsymbol{x}, \boldsymbol{\theta}$ variable or only the value of $\boldsymbol{x}$, which is $p(\boldsymbol{y}|\boldsymbol{x}, \boldsymbol{\theta})$ or $p(\boldsymbol{y}|\boldsymbol{x})$. There are two modeling strategies for this problem.

1. Model the joint probability distribution and use the joint distribution to infer $p(\boldsymbol{y}|\boldsymbol{x}, \boldsymbol{\theta})$ and $p(\boldsymbol{y}|\boldsymbol{x})$ based on the dependencies between variables or probability rules. The method of modeling the joint probability distribution of all variables is called a generative model. According to the joint distribution, we sample the probability distribution, so this model explains the mechanism of

the observed data. A typical generative model is the naive Bayes classifier. It models the joint distribution of the observed sample $\boldsymbol{x}$ and the corresponding sample label y as:

$$p(y, \boldsymbol{x}) = p(\boldsymbol{x}|y)p(y) = p(y)\prod_{j=1}^{d} p(x_j|y).$$

The hidden Markov model is also a generative model, which is described in Section 11.2.

2. Model $p(\boldsymbol{y}|\boldsymbol{x}, \boldsymbol{\theta})$ directly and analyze the influence of $\boldsymbol{x}$ and $\boldsymbol{\theta}$ on the variable $\boldsymbol{y}$, this method of conditional probability distribution modeling is called the discriminative model. This kind of model is directly concerned with the prediction task $p(\boldsymbol{y}|\boldsymbol{x}, \boldsymbol{\theta})$, and can not be used to generate data, or explain the mechanism of data generation. A typical discriminative model is logistic regression which directly models the conditional probability distribution:

$$p(y|\boldsymbol{x}, \boldsymbol{\theta}) = \frac{1}{1 + \mathrm{e}^{-\boldsymbol{\theta}^{\mathrm{T}}\boldsymbol{x}}}.$$

The CRF model introduced in Section 11.3 is also a discriminative model.

The advantage of generative model is that it provides a mechanism for how data is generated and it can generate new data based on the learned model. Its disadvantage is that it is generally difficult to observe the distribution of the variable $\boldsymbol{x}$ because it involves how to accurately capture the dependencies between features. The naive Bayes classifier is based on a strong assumption that features are conditional independent, making this generative model simple. Although the discriminative model cannot produce data, it has the advantage of avoiding modeling the dependencies between observed variables and is directly concerned with the prediction task. These two types of models have their own strengths in practical applications.

When the number of variables increases, joint distribution modeling makes the process of conditional probability inference complicated. The probabilistic graphical model provides an intuitive and concise framework to represent the dependencies between variables and the joint probability distribution. The graphical model consists of the nodes representing the variables and the edges describing the relationships between the nodes. The graphical model can be divided into two categories depending on whether the edge has direction: directed graphical model and undirected graphical model.

11.1.1. *Directed Graphical Model*

Directed graphical model is also known as Bayesian network or belief network. Edges in the network use directed arrows to indicate that child nodes are dependent on the parent node. Figure 11.1 shows a directed graphical model of five variables. This is equivalent to some sort of "topological sorting" on the relationship between variables with arrows. Under this sorting, we assume that each node (variable) only "depends" on its directly connected parent rather than all predecessor nodes. The following conditional independence holds:

$$x_s \perp\!\!\!\perp \boldsymbol{x}_{\mathrm{pred}(s)\backslash \mathrm{pa}(s)} | \boldsymbol{x}_{\mathrm{pa}(s)}, \tag{11.1}$$

where x_s is a child node, $\boldsymbol{x}_{\mathrm{pa}(s)}$ is the parent node of s, and $\boldsymbol{x}_{\mathrm{pred}(s)}$ is all the predecessor nodes of s. Thus the joint distribution of the five variables in Figure 11.1 is

$$p(\boldsymbol{x}) = p(x_1)p(x_2|x_1)p(x_3|x_1)p(x_4|x_1)p(x_5|x_2,x_4). \tag{11.2}$$

We can use the directed graphical model to represent the naive Bayesian classifier. The classifier assumes that the features are conditionally independent under a given classification label. As shown

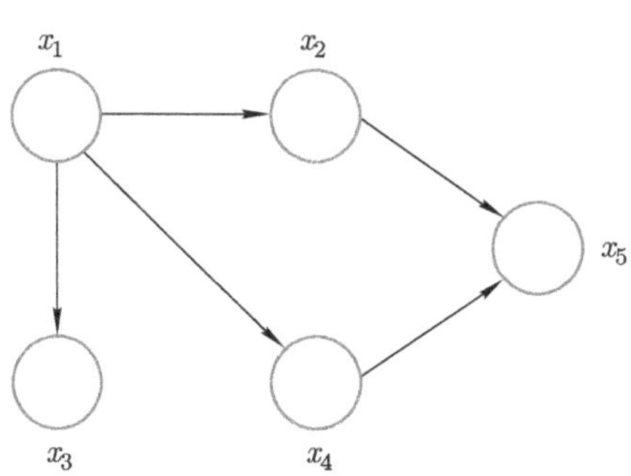

Figure 11.1 Directed graphical model of five variables.

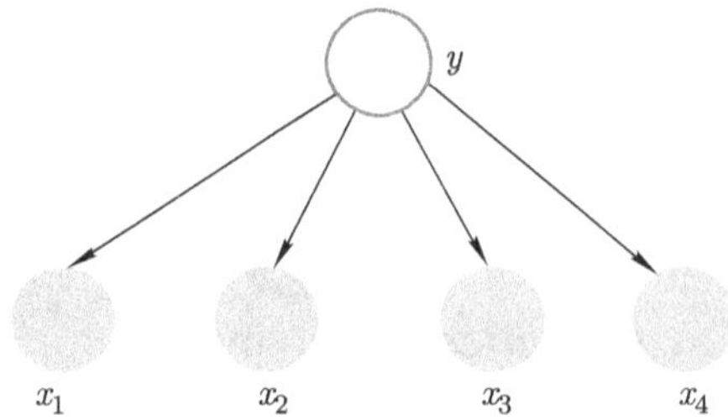

Figure 11.2 The directed graphical representation of naive Bayes.

in Figure 11.2, the joint probability distribution is

$$p(y, \boldsymbol{x}) = p(y) \prod_{j=1}^{d} p(x_j|y). \tag{11.3}$$

11.1.2. *Undirected Graphical Model*

Undirected graphical model, also known as Markov Random Field (MRF), overcomes the shortcomings of the directed graphical model that a directed dependency must be added to the parent and child nodes. It uses undirected edge to connect two nodes. There is no distinction between parent node and child node, which make more sense in some applications. For example, when using a probabilistic graphical model to model an image, we treat each pixel as a node (variable). Undirected edges between adjacent pixels makes more sense and are more reasonable than directed edges.

Since MRF does not have directional edge, we can't represent the joint distribution with a simple conditional probability like directed graphical model. It replace the role of conditional probability with the potential function of the clique. A clique is a collection of nodes in an undirected graph such that every two different nodes in the clique are directly connected. If a clique is not a true subset of any other clique, then the clique is called a maximal clique. As shown in Figure 11.3, the cliques consisting of two nodes are $\{x_1, x_2\}, \{x_1, x_3\}, \{x_1, x_4\}, \{x_2, x_4\}, \{x_2, x_5\}, \{x_4, x_5\}$. The cliques consisting of three nodes are

$$\{x_1, x_2, x_4\}, \{x_2, x_4, x_5\}.$$

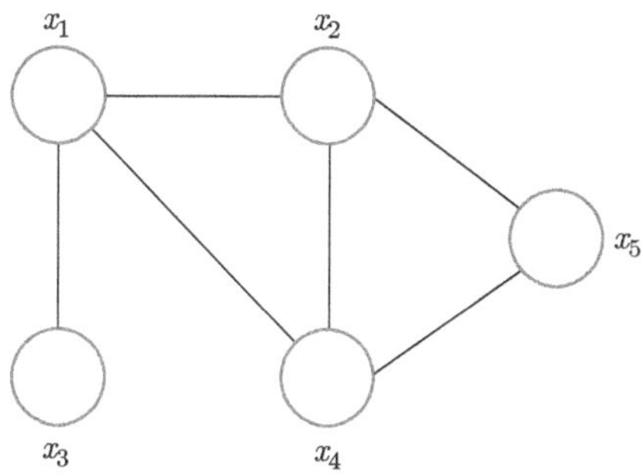

Figure 11.3 Undirected graphical model of five variables.

Thus, the maximal cliques include $\{x_1, x_3\}, \{x_1, x_2, x_4\}, \{x_2, x_4, x_5\}$.

We use c to represent a clique, and $\boldsymbol{x}_c$ to represent variables in the clique c. The Hammersley–Clifford theorem tells us that the joint distribution of all variables is proportional to the product of all potential function of the maximal clique $\phi_c(\boldsymbol{x}_c)$.[105,106]

$$p(\boldsymbol{x}) = \frac{1}{Z} \prod_{c \in \mathcal{C}} \phi_c(\boldsymbol{x}_c), \tag{11.4}$$

where $Z = \sum_{\boldsymbol{x}} \prod_{c \in \mathcal{C}} \phi_c(\boldsymbol{x}_c)$ is a normalized constant (here we assume $\boldsymbol{x}$ is a discrete variable, thus Z is the sum form; If $\boldsymbol{x}$ is a continuous variable, it is the integral form). Z is also called the partition function, which is used to normalize the probability distribution. However, the calculation of the partition function is difficult. If each variable has M values, then M^n addition operations are required to calculate Z. In Figure 11.3, the joint distribution of the five variables is

$$p(\boldsymbol{x}) = \frac{1}{Z} \phi_{13}(x_1, x_3) \phi_{124}(x_1, x_2, x_4) \phi_{245}(x_2, x_4, x_5). \tag{11.5}$$

The potential function is generally expressed in the following exponential form

$$\phi_c(\boldsymbol{x}_c) = \mathrm{e}^{\boldsymbol{\theta}_c^{\mathrm{T}} \boldsymbol{f}_c(\boldsymbol{x}_c)}, \tag{11.6}$$

where $\boldsymbol{f}_c(\boldsymbol{x}_c)$ is an application-specific feature vector derived from clique c. $\boldsymbol{\theta}$ represents the parameter about the clique. So the logarithm of the joint probability is

$$\ln p(\boldsymbol{x}) = \sum_c \boldsymbol{\theta}_c^{\mathrm{T}} \boldsymbol{f}_c(\boldsymbol{x}_c) - \ln Z(\boldsymbol{\theta}), \tag{11.7}$$

which is generally called maximum entropy or log-linear model.

Below we will introduce two probabilistic graphical models widely used in data science: hidden Markov model and conditional random field. They are typical examples of directed graphical model and undirected graphical model, respectively.

11.2. Hidden Markov Model

Hidden Markov Model (HMM) is a directed graphical model for time series data. HMM is widely used for many tasks in the field of speech recognition and natural language processing, such as word segmentation, part-of-speech tagging, named entity recognition, etc. The graphical representation of HMM is shown in Figure 11.4.

The variables of HMM are divided into two types: hidden variables and observed variables. The hidden variable y_i represents the state of the system at time i, and the observed variable x_i represents the observed value at time i. x is conditionally independent given y. The

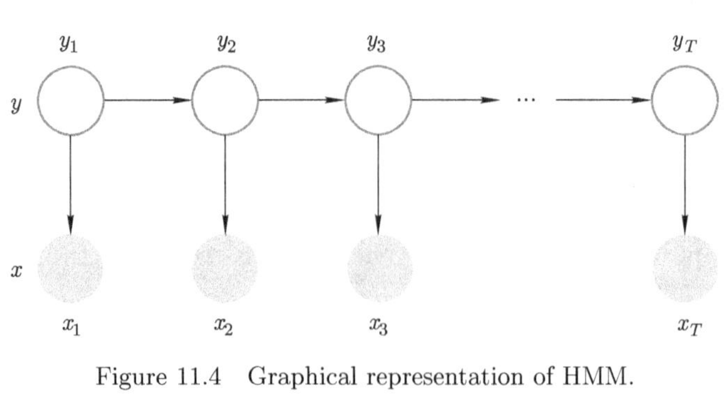

Figure 11.4 Graphical representation of HMM.

joint probability distribution of hidden and observed variables is

$$p(\boldsymbol{x}_{1:T}, \boldsymbol{y}_{1:T}) = p(y_1)p(x_1|y_1)\prod_{t=2}^{T} p(x_t|y_t)p(y_t|y_{t-1}), \tag{11.8}$$

where $\boldsymbol{x}_{1:T} = (x_1, x_2, \ldots, x_T)$, $\boldsymbol{y}_{1:T} = (y_1, y_2, \ldots, y_T)$.

The value space of hidden variable is $\mathcal{Y} = \{s_1, s_2, \ldots, s_s\}$ and the value space of observed variable is $\mathcal{X} = \{o_1, o_2, \ldots, o_O\}$. HMM mainly contains the following three parameters:

1. **Initial state probability $\boldsymbol{\pi}$:** The probability of each state in the initial state $t = 1$ is denoted as $\boldsymbol{\pi} = (\pi_1, \pi_2, \ldots, \pi_S)$, where $\pi_i = p(y_i = s_i)$.
2. **Output observation probability $\boldsymbol{B}$:** The probability that the model will obtain individual observations under a certain state. $p(x_t = o_j|y_t = s_i)$, $1 \leqslant i \leqslant S, 1 \leqslant j \leqslant O$, the matrix is denoted as $\boldsymbol{B} = [b_i(j)]_{S\times O}$.
3. **State transition probability $\boldsymbol{A}$:** Transition probability between hidden states $p(y_{t+1} = s_j|y_t = s_i)$, $1 \leqslant i, j \leqslant S$, which is represented by the matrix $\boldsymbol{A} = [a_{ij}]_{S\times S}$.

For convenience, we denote the three parameters of HMM as a unified mark $\boldsymbol{\lambda} = \{\boldsymbol{A}, \boldsymbol{B}, \boldsymbol{\pi}\}$. Given $\boldsymbol{\lambda}$, the generation process of the sequence $\{x_1, x_2, \ldots, x_T\}$ is as f.

1. Set $t = 1$, and select the initial state y_1 according to the initial probability distribution π;
2. Generate x_t according to state y_t and output observation probability $\boldsymbol{B}$;
3. Generate state y_{t+1} according to state y_t and state transition probability $\boldsymbol{A}$;
4. Set $t = t + 1$ if $t < T$, and go to step 2; otherwise stop.

HMM is a powerful tool for modeling sequence data. We can also assign actual meaning to the implicit state variable $\boldsymbol{y}_{1:T}$ in real applications. Some examples are:

1. **Speech recognition:** x_t represents the feature extracted from the speech signals. The hidden variable y_t represents the spoken word or word. The transition model $p(y_{t+1} = s_j|y_t = s_i)$ is the

language model, while the observation model $p(x_t = o_j | y_t = s_i)$ is the acoustic model. Refer to the literature for more details;[107]

2. **Behavior recognition:** x_t represents the feature extracted from the video frame image. The hidden variable y_t represents activity type of the character (such as running, walking, sitting, etc.). Refer to the literature;[108]
3. **Gene search:** x_t represents the type of the DNA molecule base: adenine (A), cytosine (C), guanine (G), and thymine (T). The hidden variable y_t represents whether this base is in the effective gene coding region.[109]

When applying the HMM, we will encounter the following three basic
problems:

4. **The estimation problem:** Given the parameters $\boldsymbol{\lambda} = \{\boldsymbol{A}, \boldsymbol{B}, \boldsymbol{\pi}\}$ and the observed sequence $\boldsymbol{x}_{1:T} = \{x_1, x_2, \ldots, x_T\}$, estimate the probability of the observed sequence $p(\boldsymbol{x}_{1:T} | \boldsymbol{\lambda})$;
5. **The decoding problem:** Given the parameters $\boldsymbol{\lambda} = \{\boldsymbol{A}, \boldsymbol{B}, \boldsymbol{\pi}\}$ and the observed sequence $\boldsymbol{x}_{1:T} = \{x_1, x_2, \ldots, x_T\}$, infer the most likely state sequence $\boldsymbol{y}_{1:T}$, that is, estimate the probability $p(\boldsymbol{y}_{1:T} | \boldsymbol{x}_{1:T})$;
6. **The learning problem:** Given the observed sequence $\boldsymbol{x}_{1:T} = \{x_1, x_2, \ldots, x_T\}$, estimate the model parameter $\boldsymbol{\lambda} = \{\boldsymbol{A}, \boldsymbol{B}, \boldsymbol{\pi}\}$, that is, solve the maximum likelihood problem arg $\max_{\boldsymbol{\lambda}} p(\boldsymbol{x}_{1:T} | \boldsymbol{\lambda})$.

The first problem is estimation and the model that best matches the given observed sequence can be selected through its solution. The second problem is finding the state sequence most likely to generate an observed sequence. Speech recognition, Chinese word segmentation, and named entity recognition can all be regarded as the decoding problem. The third problem is learning. The learning problem is relatively simple given a labeled training set. However, the state sequence is difficult to obtain in practice so the EM algorithm is usually used to estimate the model parameters.

The corresponding algorithms for solving the three problems of HMM are: forward and backward algorithm, Viterbi algorithm and Baum–Welch algorithm.

11.2.1. *The Estimation Problem: Forward and Backward Algorithm*

In HMM, the generation process of the observed sequence $\boldsymbol{x}_{1:T}$ is: Suppose there is a state sequence $\boldsymbol{y}_{1:T}$ of the same length as the observed sequence, generate observations at each point time, according to the state and state transition matrix $\boldsymbol{A}$ of the current time. Therefore, given an observed sequence, the probability is equal to the sum of the probabilities of generating the observed sequence under all possible state sequences.

$$p(\boldsymbol{x}_{1:T}|\boldsymbol{\lambda}) = \sum_{\boldsymbol{y}_{1:T}} p(\boldsymbol{x}_{1:T}, \boldsymbol{y}_{1:T}|\boldsymbol{\lambda}) = \sum_{\boldsymbol{y}_{1:T}} p(\boldsymbol{x}_{1:T}|\boldsymbol{y}_{1:T}, \boldsymbol{\lambda})p(\boldsymbol{y}_{1:T}|\boldsymbol{\lambda}). \tag{11.9}$$

With the conditional independence assumption, we can get

$$p(\boldsymbol{x}_{1:T}|\boldsymbol{y}_{1:T}, \boldsymbol{\lambda}) = \prod_{t=1}^{T} p(x_t|y_t, \boldsymbol{\lambda}), \tag{11.10}$$

and

$$p(\boldsymbol{y}_{1:T}|\boldsymbol{\lambda}) = p(y_1)\prod_{t=2}^{T} p(y_t|y_{t-1}, \boldsymbol{\lambda}). \tag{11.11}$$

A straightforward method to calculate $p(\boldsymbol{x}_{1:T}|\boldsymbol{\lambda})$ is to enumerate each possible hidden state sequence $\boldsymbol{y}_{1:T}$. However, this is infeasible. A hidden state sequence of length T has S^t values. Even if the hidden state has only three values, the hidden state sequence of length 15 has 14348907 possible values. A more efficient method for calculating the probability of an observed sequence is to use the idea of dynamic programming, which is solved by a recursive method called the forward algorithm. A forward variable $\alpha_t(i)$ is needed in the forward algorithm, which means the joint probability of the observed sequence $\{x_1, x_2, \ldots, x_t\}$ and the hidden state at time t is s_i, expressed as

$$\alpha_t(i) = p(x_1, x_2, \ldots, x_t, y_t = s_i|\boldsymbol{\lambda}). \tag{11.12}$$

With the forward variable $\alpha_t(i)$, the probability of observing the sequence $\boldsymbol{x}_{1:T}$ is

$$p(\boldsymbol{x}_{1:T}|\boldsymbol{\lambda}) = \sum_{i=1}^{S} \alpha_T(i). \tag{11.13}$$

It can be seen that as long as we can calculate $\alpha_T(i)$, $p(\boldsymbol{x}_{1:T}|\boldsymbol{\lambda})$ can be calculated by simple summation. Fortunately, the calculation of $\alpha_t(i)$ can be performed efficiently using recursive methods. The algorithm is as follows:

Algorithm 10 Forward algorithm

1: Calculate the initial value of the forward variable $\alpha_1(i) = \pi_i b_i(x_1), i = 1, \ldots, S$;
2: Recursively calculate the forward variables:

$$\alpha_{t+1}(j) = \left[\sum_{i=1}^{S} \alpha_t(i) a_{ij}\right] b_j(x_{t+1}), \quad 1 \leqslant t \leqslant T-1, 1 \leqslant j \leqslant S;$$

3: Calculate the probability of observed sequence $p(\boldsymbol{x}_{1:T}|\boldsymbol{\lambda}) = \sum_{i=1}^{S} \alpha_T(i)$.

The backward algorithm can also be used to efficiently solve the estimation problem of HMM. The difference is that a backward variable $\beta_t(i) = p(x_T, x_{T-1}, \ldots, x_{t+1}, y_t = s_i|\boldsymbol{\lambda})$ needs to be defined, which means the joint probability of observing the sequence $x_T, x_{T-1}, \ldots, x_{t+1}$ and the hidden state at time t is s_i.

The time complexity of the forward and backward algorithm is $O(S^2T)$, while the time complexity of the exhaustive method is $O(S^T)$.

11.2.2. *The Decoding Problem: Viterbi Algorithm*

The second problem of the HMM is the decoding problem, that is, given the observed sequence $\boldsymbol{x}_{1:T}$ and the model parameter $\boldsymbol{\lambda}$, find the most likely state sequence $\boldsymbol{y}_{1:T}$. The decoding problem is widely applied in speech recognition and natural language processing. For example, the part-of-speech tagging task can be seen as a typical decoding problem. Given a sequence of sentence words, we need to infer the corresponding part-of-speech sequence. In gene search, given the base sequence, the most likely sequence corresponding to whether the nucleotide is involved in the gene coding is obtained. The decoding problem can be expressed formally as

$$\arg\max_{\boldsymbol{y}_{1:T}} \quad p(\boldsymbol{y}_{1:T}|\boldsymbol{x}_{1:T}, \boldsymbol{\lambda}). \tag{11.14}$$

By applying the Bayes' theorem, we can know

$$p(\boldsymbol{y}_{1:T}|\boldsymbol{x}_{1:T}, \boldsymbol{\lambda}) = \frac{p(\boldsymbol{x}_{1:T}, \boldsymbol{y}_{1:T}|\boldsymbol{\lambda})}{p(\boldsymbol{x}_{1:T}|\boldsymbol{\lambda})}, \tag{11.15}$$

So the decoding problem is equivalent to

$$\arg\max_{\boldsymbol{y}_{1:T}} \quad p(\boldsymbol{x}_{1:T}, \boldsymbol{y}_{1:T}|\boldsymbol{\lambda}). \tag{11.16}$$

The essence of the decoding problem is to obtain the state sequence $\boldsymbol{y}_{1:T}$ given the observed sequence $\boldsymbol{x}_{1:T}$ and the parameter $\boldsymbol{\lambda}$, so that $p(\boldsymbol{x}_{1:T}, \boldsymbol{y}_{1:T}|\boldsymbol{\lambda})$ is maximized. The Viterbi algorithm is a classic algorithm for solving the decoding problem of HMM. Its basic principle is similar to that of the forward and backward algorithm. It uses the idea of dynamic programming. Unlike the forward and backward algorithm, only the maximum probability value of the observed sequence at the current time and the state value of the corresponding current time are recorded at each time t. Denote $\delta_t(i)$ as the maximum probability of the observed sequence $x_1, x_2, \ldots, x_t$ is generated when the state sequence $y_1, y_2, \ldots, y_t$ and $y_t = s_i$:

$$\delta_t(i) = \max_{y_1, y_2, \cdots, y_{t-1}} \quad p(y_1, y_2, \ldots, y_{t-1}, y_t = s_i, x_1, x_2, \ldots, x_t|\boldsymbol{\lambda}).$$

$\delta_t(i)$ can be efficiently calculated according to Eq. (11.8):

$$\delta_{t+1}(j) = [\max_i \delta_t(i) a_{ij}] \cdot b_j(x_{t+1}).$$

We also need a variable to record the most likely sequence of states. $\Psi_t(i)$ is used to represent the state at the previous time when the state at time t is s_i, and s_t^* to represent the optimal state sequence at time t. The Viterbi algorithm for solving the optimal state sequence is as follows:

Algorithm 11 Viterbi algorithm

1: Calculate initial value:

$$\delta_t(i) = \pi_i b_i(x_1), \quad i = 1, \ldots, S,$$
$$\Psi_t(i) = 0, \quad i = 1, \ldots, S;$$

2: Iteratively calculate intermediate variables:
- Calculate $\delta_t(j)$:

$$\delta_t(j) = \max_i \, [\delta_{t-1}(i) a_{ij}] \, b_j(x_t), \quad 2 \leqslant t \leqslant T, 1 \leqslant j \leqslant S;$$

- Calculate $\Psi_t(j)$:

$$\Psi_t(j) = \arg\max_i \, [\delta_{t-1}(i) a_{ij}], \quad 2 \leqslant t \leqslant T, 1 \leqslant j \leqslant S;$$

3: Calculate termination state: $s_T^* = \arg\max_i \, [\delta_T(i)]$;
4: $s_t^* = \Psi_{t+1}(s_{t+1}^*), \quad t = T-1, T-2, \ldots, 1.$

The time complexity of the Viterbi algorithm is $O(S^2T)$. The space complexity is $O(S^T)$. The Viterbi algorithm returns only one of the most optimal state sequence. We can extend it to return multiple state sequences.[109] We can then reorder these state sequences by certain methods to select one. This reordering approach is widely used in speech recognition. For example, considering the sentence "你很漂亮 (Ni hen piao liang)", the state sequence output by the system may be "米很漂亮 (Mi hen piao liang)", or "你狠，漂亮 (Ni hen piao liang)", here we can use the global context to reorder these possible sequences to get the best sequence.

11.2.3. *The Learning Problem: Baum–Welch Algorithm*

The third problem of HMM is the problem of learning model parameters. The goal is to find the $\boldsymbol{\lambda}$ that maximizes the value of $p(\boldsymbol{x}_{1:T}|\boldsymbol{\lambda})$. For models with hidden variables, the learning of parameters is difficult. The EM algorithm is often used to perform parameter learning. Specific to the HMM model, the classical algorithm for parameter learning is often called the Baum–Welch algorithm. The Baum–Welch algorithm uses the idea of recursion to get *the local maximum of* $p(\boldsymbol{x}_{1:T}|\boldsymbol{\lambda})$, thus obtaining the model parameter $\boldsymbol{\lambda} = \{\boldsymbol{A}, \boldsymbol{B}, \boldsymbol{\pi}\}$.

First we define the Baum–Welch variable $\varepsilon_t(i,j) = p(y_t = s_i|y_{t+1} = s_j|\boldsymbol{x}_{1:T}, \boldsymbol{\lambda})$, which means given model parameter $\boldsymbol{\lambda}$ and the observed sequence $\boldsymbol{x}_{1:T}$, the probability that the state at time t is s_i and the state at time $t+1$ is s_j. According to the definitions of the forward variable α and the backward variable β, $\varepsilon_t(i,j)$ can be expressed as

$$\begin{aligned}\varepsilon_t(i,j) &= \frac{p(y_t = s_i, y_{t+1} = s_j, X|\boldsymbol{\lambda})}{p(X|\boldsymbol{\lambda})}\\ &= \frac{\alpha_t(i)a_{ij}b_j(x_{t+1})\beta_{t+1}(j)}{p(\boldsymbol{x}_{1:T}|\boldsymbol{\lambda})}\\ &= \frac{\alpha_t(i)a_{ij}b_j(x_{t+1})\beta_{t+1}(j)}{\sum_i\sum_j \alpha_t(i)a_{ij}b_j(x_{t+1})\beta_{t+1}(j)}.\end{aligned} \tag{11.17}$$

The variable $\gamma_t(i)$ is further defined, which means that given model parameter $\boldsymbol{\lambda}$ and the observed sequence $\boldsymbol{x}$, the probability of state at time t is s_i.

$$\gamma_t(i) = \sum_{j=1}^{S} \varepsilon_t(i,j). \tag{11.18}$$

The relevant markings and their corresponding meanings are as follows:

1. $\varepsilon_t(i,j)$: Given the model parameter $\boldsymbol{\lambda}$ and the observed sequence $\boldsymbol{x}_{1:T}$, the probability that the state at time t is s_i and the state at time $t+1$ is s_j;
2. $\gamma_t(i)$: Given the model parameter $\boldsymbol{\lambda}$ and the observed sequence $\boldsymbol{x}_{1:T}$, the probability of the state at time t is s_i;
3. $\sum_{t=1}^{T-1}\gamma_t(i)$ The expected value of the number of state transitions issued from state s_i;
4. $\sum_{t=1}^{T-1}\varepsilon_t(i,j)$: The expected value of the number of transitions from the state s_i to the state s_j.

The parameter $\boldsymbol{\lambda} = \{\boldsymbol{A}, \boldsymbol{B}, \boldsymbol{\pi}\}$ of the HMM model can be estimated using the following methods.

$$\pi_i = \text{Probability of initial state } s_i = \gamma_1(i); \tag{11.19}$$

$$\begin{aligned} a_{ij} &= \frac{\text{Expected value of transitions from state } s_i \text{ to } s_j}{\text{Expected value of state transitions from state } s_i} \\ &= \frac{\sum_{t=1}^{T-1} \varepsilon_t(i,j)}{\sum_{t=1}^{T-1} \gamma_t(i)}; \end{aligned} \tag{11.20}$$

$$\begin{aligned} b_j(o_k) &= \frac{\text{Expected value of } o_k \text{ observed at state } s_j}{\text{Expected value at state } s_j \text{ observed at state } s_j} \\ &= \frac{\sum_{t=1}^{T} (\gamma_t(j) \cdot I(x_t = o_k))}{\sum_{t=1}^{T} \gamma_t(j)}. \end{aligned} \tag{11.21}$$

The Baum–Welch algorithm is the process of calculating the $\varepsilon_t(i,j)$ and model parameters $\boldsymbol{\lambda}$ using an iterative method. This algorithm is a special case of the EM algorithm, in which the process of calculating $\varepsilon_t(i,j)$ is equivalent to the E-step. The process of calculating the model parameter $\boldsymbol{\lambda}$ is equivalent to the M-step. The Baum-Welch algorithm is shown in Algorithm 12.

We can guarantee that the updated parameter $\boldsymbol{\lambda}^{\text{new}}$ can continuously improve the likelihood of the HMM model until the algorithm converges, i.e., $p(\boldsymbol{x}_{1:T}|\boldsymbol{\lambda}^{\text{new}}) \geqslant p(\boldsymbol{x}_{1:T}|\boldsymbol{\lambda}^{\text{old}})$. Therefore, the parameter estimates returned by the Baum–Welch algorithm are also the maximum likelihood estimates of the HMM model. Eq. (11.19), (11.20) and (11.21) can be obtained directly by maximizing the following auxiliary functions.

$$\text{LB}(\boldsymbol{\lambda}, \boldsymbol{\lambda}^{\text{old}}) = \sum_{\boldsymbol{y}_{1:T}} p(\boldsymbol{y}_{1:T}|\boldsymbol{x}_{1:T}, \boldsymbol{\lambda}^{\text{old}}) \ln p(\boldsymbol{x}_{1:T}, \boldsymbol{y}_{1:T}|\boldsymbol{\lambda}). \tag{11.22}$$

Due to the characteristics of the EM algorithm itself, only the local optimal solution can be obtained. Different parameter initialization may result in different local optimal solutions. We often adopt the following strategies to initialize parameters:

Algorithm 12 Baum–Welch algorithm

1: Calculate the initial value: $\boldsymbol{A}$ and $\boldsymbol{B}$ are probability matrices satisfying $a_{i0} = 0, b_0(j) = 0$ $(i = 1, \ldots, S,\ j = 1, \ldots, O)$;

2: Repeat the following steps until convergence:

- (**E-step**) Run the forward and backward algorithms to calculate α and β, then compute $\varepsilon_t(i, j)$ and $\gamma_t(i)$:

$$\varepsilon_t(i,j) = \frac{\alpha_t(i)a_{ij}b_j(x_{t+1})\beta_{t+1}(j)}{\sum_i \sum_j \alpha_t(i)a_{ij}b_j(x_{t+1})\beta_{t+1}(j)},$$

$$\gamma_t(i) = \sum_{j=1}^{S} \varepsilon_t(i,j);$$

- (**M**-step) Re-estimate the model parameters:

$$\pi_i = \gamma_1(i),$$

$$a_{ij} = \frac{\sum_{t=1}^{T-1} \varepsilon_t(i,j)}{\sum_{t=1}^{T-1} \gamma_t(i)},$$

$$b_j(o_k) = \frac{\sum_{t=1}^{T} (\gamma_t(j) \cdot I(x_t = o_k))}{\sum_{t=1}^{T} \gamma_t(j)};$$

3: Calculate the probability of observed sequence $p(\boldsymbol{x}_{1:T}|\boldsymbol{\lambda}) = \sum_{i=1}^{S} \alpha_T(i)$.

1. Initialize the parameters by using the completely labeled sequence data, that is, the state sequence and the observed sequence have labeled data;
2. Ignore the Markov dependency of HMM and directly use the standard hybrid model method to estimate the initial parameters, such as using K-means;
3. Multiple random initializations, picking the optimal solution.

11.2.4. *Some Extensions of HMM*

HMM is the most basic model for modeling sequence data, but there are more complex variable dependencies or multiple types of state variables in practice. In this section, we briefly introduce some extended models based on HMM. Interested readers can turn to references for further discussion.

1. Auto-regressive HMM

The standard HMM assumes that the observed variables are independent of each other given a hidden state. In practice, many problems cannot satisfy such assumptions, such as asset price changes in economic research, and observed asset prices are not only affected by potential other economic factors (hidden states), they are also affected by asset prices in the most recent time period.[111] Therefore, we can add a dependency between $\boldsymbol{x}_{t-1}$ and $\boldsymbol{x}_t$. As shown in Figure 11.5, this model is called auto-regressive HMM. For continuous data, the observation model is

$$p(x_t|x_{t-1}, y_t = s_j, \boldsymbol{\theta}) = N\left(x_t; w_j x_{t-1} + \mu_j, \sigma_j\right). \tag{11.23}$$

This observation model is a linear regression model whose parameters are selected according to the hidden state. Of course, we can consider higher order situations as needed, that is the current observed variables are related to the nearest L observations.

$$p(x_t|\boldsymbol{x}_{t-L:t-1}, y_t = s_j, \boldsymbol{\theta}) = N\left(x_t; \sum_{l=1}^{L} w_j x_{t-l} + \mu_j, \sigma_j\right). \tag{11.24}$$

Auto-regressive HMM is widely used in econometric models[111] and speech recognition.[112] We can observe that auto-regressive HMM actually combines two Markov chains. One based on hidden variables, which is used to capture long-range dependencies. The other is based on observed variables, used to capture short-range dependencies.[113] The estimation problem of auto-regressive HMM can be implemented

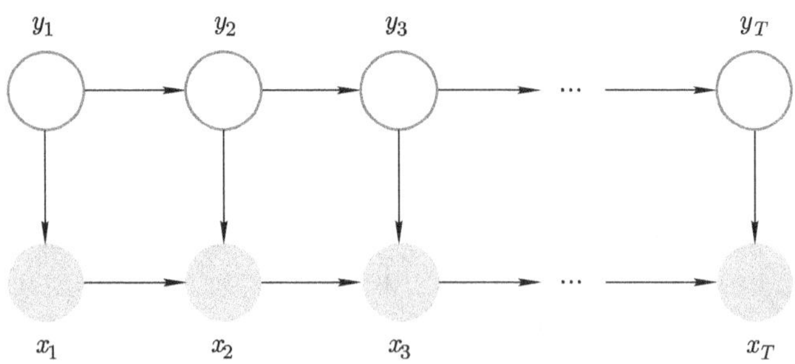

Figure 11.5 Auto-regressive HMM.

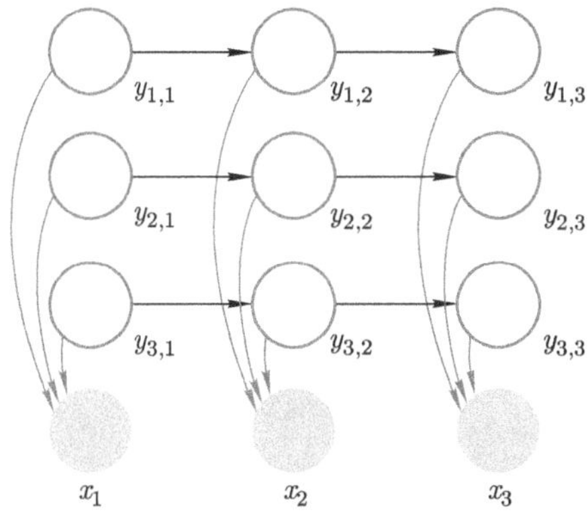

Figure 11.6 Factorial HMM.

by using the forward and backward algorithm. The learning problem can also solved by the EM algorithm. The E-step is the same as standard HMM, while the M-step needs to minimize the following function

$$\sum_t E\left[\frac{1}{\sigma^2(s_t)}(x_t - \boldsymbol{x}_{t-L:t-1}^{\mathrm{T}}\boldsymbol{w}(s_t))^2 + \ln \sigma^2(s_t)\right]. \tag{11.25}$$

2. Factorial HMM

The standard HMM uses a single discrete random variable to represent the hidden state, $y_t \in \{1, 2, \ldots, S\}$. If you want to represent 10 bits of information, you need $S = 2^{10} = 1024$ states. If we consider expressing the hidden state in a distributed way, that is, using the hidden variable $y_{c,t} \in \{0, 1\}$ to represent the c-th bit of the t-th hidden state, we can only use 10 two-dimensional variables to represent the 10 bits of information, as shown in Figure 11.6. This model is called the decomposable HMM.[114] The advantage is that it can capture features of different aspects of the signal. In speech recognition, one of the hidden state chains can represent pronunciation, and the other chain represents words or phrases. The factorial HMM is not only used to solve problems of speech recognition, but also used in energy problems, such as how to infer the electricity consumption of each appliance based on the sequence of total electricity consumption of all appliances in the household.[115,116]

For the estimation, decoding and parameter learning of factorial HMM, interested readers can read the literature in detail.[114] HMM and its extensions and variants are powerful models for modeling sequence data, which are generative models. The given hidden state variables generally have practical meaning, and are highly interpretable, which is helpful for explaining the mechanism of data generation. The popular Python toolkit scikit-learn provides an interface to solve estimation, decoding, and learning problems in HMM.

11.3. Conditional Random Field

Many machine learning tasks involve predicting a large number of features, and there are mutual dependencies between features. Such task is called a structured prediction task. Specifically, our goal is to predict the output vector $\boldsymbol{y} = (y_1, y_2, \ldots, y_T)$, given the observed feature vector $\boldsymbol{x}$. There is a dependency between the output features y_s. For example, image segmentation in computer vision is to divide the pixels of different regions (observed variable $\boldsymbol{x}$) into different categories, such as person and background. There is a strong correlation between the labels of adjacent pixels (y_s). Another example is part-of-text tagging in natural language processing. Each y_s is a label of a word at position s. The input vector $\boldsymbol{x}$ can be decomposed into feature vectors $(\boldsymbol{x}_1, \boldsymbol{x}_2, \ldots, \boldsymbol{x}_T)$, and each feature vector contains multiple aspects of the word at position s, such as suffixes, the topic of the word, etc.

The model that can effectively solve such structural prediction problems is Conditional Random Field (CRF).[117] It is a typical undirected graphical model and a discriminative model. CRF is widely used in many fields, including part-of-speech tagging, machine translation, speech recognition, motion recognition, etc. Next, we will introduce the structure of CRF as well as how to perform model inference and parameter learning in CRF.

11.3.1. *Linear Chain CRF and its General Form*

We first consider the conditional probability distribution $p(\boldsymbol{y}|\boldsymbol{x})$ from the joint distribution $p(\boldsymbol{x}, \boldsymbol{y})$ of HMM. In fact, this conditional distribution is the basic model of CRF, except that a special feature function is chosen. Let's review the joint distribution of HMM.

$$p(\boldsymbol{x}, \boldsymbol{y}) = p(y_1)p(x_1|y_1)\prod_{t=2}^{T} p(x_t|y_t)p(y_t|y_{t-1}).$$

The above form is rewritten into a form that is easy to generalize:

$$p(\boldsymbol{x}, \boldsymbol{y}) = \frac{1}{Z}\prod_{t=1}^{T} \exp\left\{\sum_{s_i,s_j} \theta_{ij}\mathbf{1}_{\{y_t=s_i\}}\mathbf{1}_{\{y_{t-1}=s_j\}} + \sum_{s_i,o} \mu_{oi}\mathbf{1}_{\{y_t=s_i\}}\mathbf{1}_{x_t=o}\right\}, \tag{11.26}$$

where Z is the normalized parameter and the model parameter is $\boldsymbol{\theta} = (\theta_{ij}, \mu_{oi})$. In HMM these parameters take the following form:

$$Z = 1,$$
$$\theta_{ij} = \ln p(y' = s_i|y = s_j),$$
$$\mu_{oi} = \ln p(x = o|y = s_i).$$

By introducing a feature function, Eq. (11.26) can be written in a simpler form:

$$p(\boldsymbol{x}, \boldsymbol{y}) = \frac{1}{Z}\prod_{t=1}^{T} \exp\left\{\sum_{k=1}^{K} \theta_k f_k(y_t, y_{t-1}, x_t)\right\}. \tag{11.27}$$

For each state transition probability, the feature function f_k is: $f_{ij}(y, y', x) = \mathbf{1}_{\{y=s_i\}}\mathbf{1}_{\{y'=s_j\}}$; for each state-observation pair, f_k is $f_o(y, y', x) = \mathbf{1}_{\{y=s_i\}}\mathbf{1}_{\{x=o\}}$.

The conditional probability distribution $p(\boldsymbol{y}|\boldsymbol{x})$ is written according to the joint distribution shown in Eq. (11.27).

$$p(\boldsymbol{y}|\boldsymbol{x}) = \frac{p(\boldsymbol{y}, \boldsymbol{x})}{\sum_{\boldsymbol{y}'} p(\boldsymbol{y}', \boldsymbol{x})} = \frac{\prod_{t=1}^{T} \exp\left\{\sum_{k=1}^{K} \theta_k f_k(y_t, y_{t-1}, x_t)\right\}}{\sum_{\boldsymbol{y}'}\prod_{t=1}^{T} \exp\left\{\sum_{k=1}^{K} \theta_k f_k(y'_t, y'_{t-1}, x_t)\right\}}. \tag{11.28}$$

According to Eq. (11.28), the definition of the linear chain CRF is: Let $\boldsymbol{y}$, $\boldsymbol{x}$ be random vectors, $\boldsymbol{\theta} \in R_k$ be the parameter vector,

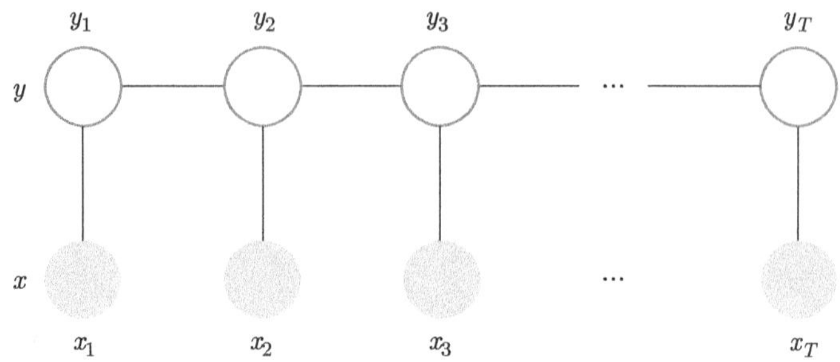

Figure 11.7 Linear chain condition random field.

and $\boldsymbol{f} = \{f_k(y, y', \boldsymbol{x}_t)\}_{k=1}^K$ is a set of feature function. The linear chain CRF models the following conditional probability distribution $p(\boldsymbol{y}|\boldsymbol{x})$.

$$p(\boldsymbol{y}|\boldsymbol{x}) = \frac{1}{Z(\boldsymbol{x})} \prod_{t=1}^{T} \exp\left\{\sum_{k=1}^{K} \theta_k f_k(y_t, y_{t-1}, \boldsymbol{x}_t)\right\}, \tag{11.29}$$

where $Z(\boldsymbol{x})$ is the normalization parameter that depends on the input. Figure 11.7 shows the graphical representation of the linear chain CRF.

CRF is a discriminative undirected graphical model. The feature function $f_k(\cdot)$ can be extended to not only depend on the current observed variable $\boldsymbol{x}_t$, but also on all observed variables $\boldsymbol{x}$, that is $f_k(y_t, y_{t-1}, \boldsymbol{x})$, as shown in Figure 11.8. We can re-write CRF in a more general form, using the potential function of the undirected graph:

$$p(\boldsymbol{y}|\boldsymbol{x}) = \frac{1}{Z(\boldsymbol{x})} \prod_{c} \phi_c(\boldsymbol{y}_c, \boldsymbol{x}_c), \tag{11.30}$$

where c represents the maximal clique, and ϕ_c is a linear combination of some predetermined feature functions:

$$\phi_c(\boldsymbol{y}_c, \boldsymbol{x}_c) = \exp\left\{\sum_{k=1}^{K(C)} \theta_{ck} f_{ck}(\boldsymbol{y}_a, \boldsymbol{x}_a)\right\}. \tag{11.31}$$

The feature function f_{ck} and the weight θ_{ck} are marked with the index c of the clique. How to connect the observed variable $\boldsymbol{x}$ and

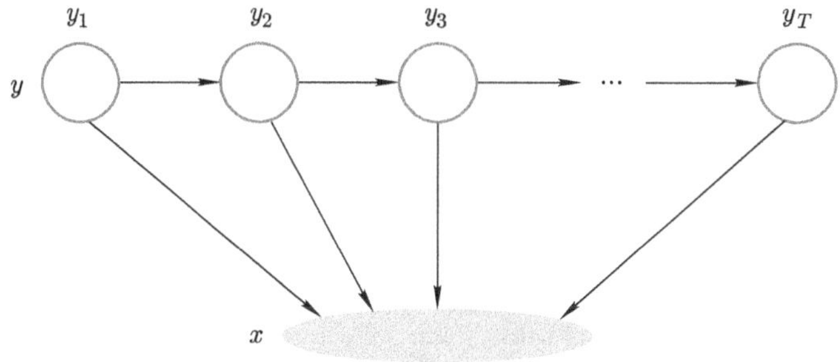

Figure 11.8 General form of conditional random field.

the label variable $\boldsymbol{y}$ is important, which determines how the feature function should be selected, as well as the difficulty of parameter learning and inference.

11.3.2. *Feature Engineering*

In this part, we will give a brief introduction to feature engineering, that is, how to select feature functions. Although these principles are based on natural language processing, they can be generalized and applied to other structured prediction problems. The main trade-off for choosing feature functions is how to use a larger feature set to produce better predictions because the final decision boundary is more flexible. However, large feature set require more memory to access corresponding parameters. If the feature functions are not selected properly, it will lead to over-fitting.

1. Label-observation feature

When the label features are discrete, the feature function f_{ck} generally has the following form:

$$f_{ck}(\boldsymbol{y}_c, \boldsymbol{x}_c) = \mathbf{1}_{\{\boldsymbol{y}_c = \tilde{\boldsymbol{y}}_c\}} q_{ck}(\boldsymbol{x}_c). \tag{11.32}$$

Only when the label is $\tilde{\boldsymbol{y}}_c$, the feature will be non-zero; when this condition is satisfied, the value of the feature function depends only on the input observation variable. We generally call the function $q_{ck}(\boldsymbol{x}_c)$ the observation function. For example, we can determine the value of the observation function based on whether the word x_t is capitalized or whether the word ends with "ing". The label-observation feature generally has two types, the edge-observation feature and the node-observation feature:

$$\begin{cases} f(y_t, y_{t-1}, x_t) = q_m(x_t)\mathbf{1}_{\{y_t=y\}}\mathbf{1}_{\{y_{t-1}=y'\}}; \\ f(y_t, x_t) = q_m(x_t)\mathbf{1}_{\{y_t=y\}}, \end{cases}$$
$$\begin{cases} f(y_t, y_{t-1}, x_t) = \mathbf{1}_{\{y_t=y\}}\mathbf{1}_{\{y_{t-1}=y'\}}; \\ f(y_t, x_t) = q_m(x_t)\mathbf{1}_{\{y_t=y\}}. \end{cases} \tag{11.33}$$

The difference between these two types of features is whether to consider the transformation between labels depends on the current observations.

2. Feature transformation

For *categorical* features, we need to convert them into binary features. This is a standard processing method in text tasks. In computer vision and speech, the observed variables are generally continuous. We can use the Z-score method to normalize them. We can also a convert the real value into a categorical variable, or then convert it into a binary variable.

3. Higher order features

Although we generally denote the feature function as $f(y_t, y_{t-1}, x)$, the feature function does not only depend on the nearest label. Taking the nearby labels into consideration can often improve the prediction performance. For example, the feature "word x_{t+2} is Times and the label y_t is an organization" will be useful for identifying New York Times newspapers.

4. Features from other models

Such features are derived from the results of other simple methods that accomplish the same task. For example, if there is already a rule-based system in a task (for example, the rule "any string of numbers between 1900 and 2050 is year"). For "year", we can use the predictions of this system as an observed function of CRF. A more complicated example is to use the results from another production model as input to the CRF. For example, we can use the feature $f_t(y, \boldsymbol{x}) = p_{\text{HMM}}(y_t = y|\boldsymbol{x})$, where p_{HMM} represents the edge probability of a HMM trained on a similar dataset. One may use the exact same dataset to train the HMM, then use it in the CRF. It may not be a good practice, as we generally expect the HMM to perform well on the corresponding dataset, which will cause the CRF to rely too much on the features provided by HMM and ignore other features. This approach is very helpful in improving the performance of existing prediction systems. The literature[118] provides an example of such genetic recognition.

11.3.3. *Parameter Estimation of CRF*

In general, the essence of parameter estimation is to identify the maximum value of the logarithm maximum likelihood function. Assume that the training set is $D = \{(\boldsymbol{x}^1, \boldsymbol{y}^1), (\boldsymbol{x}^2, \boldsymbol{y}^2), \ldots, (\boldsymbol{x}^N, \boldsymbol{y}^N)\}$. The superscript $1, \ldots, N$ is the sample index which is used to distinguish from the subscript $1, \ldots, T$.

For linear chain CRF, the log likelihood function can be written as

$$L(\boldsymbol{\theta}) = \sum_{i=1}^{N}\sum_{t=1}^{T}\sum_{k=1}^{K} \theta_k f_k(y_t^i, y_{t-1}^i, x_t^i) - \sum_{i=1}^{N} \ln Z(\boldsymbol{x}^i, \boldsymbol{\theta}). \tag{11.34}$$

Let $f_k(\boldsymbol{y}^i, \boldsymbol{x}^i) = \sum_t (y_t^i, y_{t-1}^i, x_t^i)$, take derivative of the log likelihood function with respect to θ_k:

$$\frac{\partial L(\boldsymbol{\theta})}{\partial \theta_k} = \sum_i F_k(\boldsymbol{y}^i, \boldsymbol{x}^i) - \sum_i E_{p(\boldsymbol{y}|\boldsymbol{x}^i, \boldsymbol{\theta})} F_k(\boldsymbol{y}, \boldsymbol{x}^i). \tag{11.35}$$

The gradient is the mathematical expectation of the observed feature and the difference between the mathematical expectation of the feature under the conditional probability model $p(\boldsymbol{y}|\boldsymbol{x}^i, \boldsymbol{\theta})$. The log likelihood maximum estimation may lead to the over-fitting problem. Regularization is usually used to solve this problem. The new objective function is

$$L_R(\boldsymbol{\theta}) = L(\boldsymbol{\theta}) - \frac{\sum_k \theta_k^2}{2\sigma^2}, \tag{11.36}$$

where σ is the regularization parameter. The gradient after regularization becomes

$$\frac{\partial L_R(\boldsymbol{\theta})}{\partial \theta_k} = \frac{\partial L(\boldsymbol{\theta})}{\partial \theta_k} - \frac{\theta_k}{\sigma^2}. \tag{11.37}$$

We use the gradient ascent method to update the parameters. The commonly used two algorithms are GIS and IIS.[117]

11.3.4. *Inference of CRF*

For a given observation sequence $\boldsymbol{x} = (x_1, x_2, \ldots, x_T)^{\mathrm{T}}$, find the label sequence (state sequence) $\boldsymbol{y} = (y_1, y_2, \ldots, y_T)^{\mathrm{T}}$ which can generate the observation sequence with the highest probability. For a chain CRF, we add a starting state y_0 and an ending state y_{T+1}. Define a set of matrices $\{M_t(\boldsymbol{x})|t = 1, 2, \ldots, T+1\}$, where each $M_t(\boldsymbol{x})$ is a matrix of size $|Y| \times |Y|$, and $|Y|$ is the number of all possible values of y. Each element $M_t(y_{t-1}, y_t|\boldsymbol{x})$ is defined as

$$\begin{aligned} &M_t(y_{t-1} = y', y_t = y|\boldsymbol{x}) \\ &= \exp\left(\sum_k \theta_k f_k(y_{t-1}, y_t, x_t)\right) \\ &= \exp\left(\sum_k \theta_k r_k(y_{t-1}, y_t, \boldsymbol{x}) + \sum_k \mu_k s_k(y_t, \boldsymbol{x})\right), \end{aligned} \tag{11.38}$$

where y_{t-1} is the previous token of y_t and $r_k(\cdot)$ and $s_k(\cdot)$ are feature functions. Since $p(\boldsymbol{y}|\boldsymbol{x}, \boldsymbol{\theta})$ is actually a probability of the path from the starting node to the ending node, we can get

$$p(\boldsymbol{y}|\boldsymbol{x}, \boldsymbol{\theta}) = \frac{1}{Z(\boldsymbol{x})} \prod_{t=1}^{T+1} M_t(y_{t-1}, y_t|\boldsymbol{x}), \tag{11.39}$$

where $Z(\boldsymbol{x})$ is the normalization factor and is the (start, stop) element of the product of these matrices:

$$Z(\boldsymbol{x}) = \left(\prod_{t=1}^{T+1} M_t(x)\right)_{(\text{start,stop})},$$

where start $= y_0$ and stop $= y_{T+1}$.

11.4. Summary

One of the great benefits of the probability graphical model is that it provides an intuitive and fast way to define a model for a specific problem, allowing the model builder to determine the dependencies between the features based on the understanding of the problem. HMM and CRF are a representatives of directed graphical and undirected graphical models, respectively. We analyze how HMM and CRF learn and infer in processing sequence data. The difference is the way they represent joint probabilities. Directed graphical model is used to represent conditional probability, while undirected graphical model is used to represent the potential function of largest clique. In this chapter, we only introduce some successful applications of the graphical model on sequence data. Other important models include the topic model (for the directed graphical model) in text analysis, and the restricted Boltzmann machine (RBM) in image modeling.[92,119]

11.5. Case Studies and Exercises

1. Stock market analysis

The Growth Enterprise Market (GEM) is a specialized stock market and an important supplement to the main board market. It is designed to provide a platform for small and medium-sized enterprises (SMEs) and high-growth companies to raise capital and grow their businesses. On October 30, 2009, China's Growth Enterprise Market (GEM) officially went public. On June 1, 2010, the Shenzhen Stock Exchange officially compiled and released the GEM Index to more comprehensively reflect the overall performance of the GEM market.

This exercise uses the GEM index dataset obtained through Python's third-party library, Tus-hare, from March 1, 2014 to March 1, 2017. The dataset contains 729 samples. Each sample includes 8 features, including date, opening price, highest price, closing price, and volume. Please use HMM to analyze the GEM index.

2. Chinese word segmentation

Chinese word segmentation is the task of segmenting Chinese sentences into individual.

The Chinese text used in this case is from the "Action Plan on Promoting Big Data Development" issued by the State Council in 2015. Its contents are summarized as follows:

加强专业人才培养。创新人才培养模式，建立健全多层次、多类型的大数据人才培养体系。鼓励高校设立数据科学和数据工程相关专业，重点培养专业化数据工程师等大数据专业人才。鼓励采取跨校联合培养等方式开展跨学科大数据综合型人才培养，大力培养具有统计分析、计算机技术、经济管理等多学科知识的跨界复合型人才。鼓励高等院校、职业院校和企业合作，加强职业技能人才实践培养，积极培育大数据技术和应用创新型人才。

Please use CRF and HMM based word segmentation tool to transform the above Chinese text into word sequence.

3. Speech recognition

Speech recognition refers to the process of converting audio signals such as sound into corresponding texts through special acoustic models and language models, including acoustic feature extraction, pattern matching, model training, etc. In recent years, speech recognition technology is widely applied, such as car navigation, voice input methods, etc.

The dataset used in this exercise contains 105 audio files of seven English words. These seven words are "apple", "banana", "kiwi", "lime", "orange", "peach" and "pineapple".

Please use the HMM to identify the words read in the audio based on the supplied audio file.

Chapter 12

Text Analysis

Textual data is one of the most important data resources. Nowadays, a great amount of textual data are provided by forums, news, blogs, the Microblog, Wechat, product reviews, comments, e-mail, medical diagnoses and treatment records, questionnaires, and court judgments, etc. Figure 12.1 shows examples of textual data in product reviews, medical diagnoses-and-treatment data, and court judgments.

Besides its independent value, textual data can also helps improve the accuracy of decision making and predicting models when combined with structured data, especially in fields such as news, law, education and etc. For example, the enterprise credit status can be comprehensively evaluated by the textual data combined with Internet public opinions and court judgments; and the accuracy of profile can be improved with the textual data by integrating the transaction flows. Text analysis techniques such as sentiment analysis are widely used in stock market, Internet public opinion, monitoring and commodity service quality assessment, etc.

However, text analysis faces more challenges than traditional structured data analysis. First of all, the expression of natural language is usually ambiguous and needs to be analyzed under the context that is always accompanied with the phenomena such as "polysemy" and "synonyms". For example, there are two sentences "the fuel consumption of this car is very high" and "the cost performance of this new mobile phone is quite high", but the word "high" has completely different meanings. In the words "fast delivery"

Products reviews

The mobile phone is very good, beautiful and very fast. There is a small scratch on the screen of my mobile phone, but I don't usually notice it. I'm too lazy to apply for a replacement.

Medical data

Eyes: no edema of the eyelid, no bleeding spot of the eyelid conjunctiva, no yellow staining of the sclera, transparent cornea, large and equal circles of the pupil, 3-4 mm in diameter, light reflex and collective reflex exist

Court judgment

On July 14, 2015, according to the application of Tianjin XX Real Estate Management Co., Ltd., the Court decided to accept the bankruptcy liquidation case of Tianjin XX Real Estate Management Co., Ltd. It was found that as of July 31, 2015, the debtor had total assets of 10596.56 yuan, total liabilities of 563391.46 yuan, and asset liability ratio of 5616.74%.

Figure 12.1 Text data example.

and "rapid logistics", "fast" and "rapid" share the same meaning. Secondly, to conduct further analysis on unstructured text, it is often necessary to convert the text to a structured vector. The text vector usually has a high dimension, but only one or quite few dimensions take a non-zero value. In addition, with the development of social media, there are a great amount of informal expression, such as language-switching expressions, Internet language, spelling mistakes and abbreviations, which brings difficulty for a machine to understand the meaning.

This chapter briefly introduces some popular text analysis techniques. Section 12.1 introduces the text representation model, how to represent unstructured text data in a structured way, and how to reduce the dimension of text data. Section 12.2 discusses topic models, a technique for mining implicit semantics from text. Section 12.3 introduces sentiment analysis and discusses how to uncover the emotions embodied in textual data.

12.1. Text Representation

The original representation of the textual data is an unstructured string, which cannot be directly handled by the models and algorithms we have described so far. Therefore, we need to convert it into a structured representation. One way is to map the text into a specific feature space, in which the text can be represented by a feature vector to represent its semantic information. Then we can apply the existing data science models to various tasks of text analysis, such as text classification, text clustering, information retrieval, etc.

So, what kind of features can represent the semantic information of the text? An intuitive idea is to choose the words in the text as the features, and a vector representation of the text can be obtained by introducing the weight of the word. However, before representing the text by the vector, we need to convert the text into a sequence of words. This process is called word segmentation. Word segmentation is relatively easy in English, because words and phrases are naturally separated by space. For example, we can easily express the English sentence "In data we trust" as a sequence of words {In, data, we, trust}. Chinese word segmentation is difficult because there is no natural separator-like gap between words in Chinese. For example, for the Chinese sentence "用数据刻画规律", although human can easily divide it into a sequence of words as "用，数据，刻画，规律", it is quite difficult for computers. Therefore, for languages like Chinese, we need to convert the text into a sequence of words through word segmentation before other text processing steps. The HMM model introduced in Section 11.2 and the CRF model introduced in Section 11.3 are the classic models for word segmentation.

In this section we firstly introduce the vector space model, which converts the unstructured text into a high-dimensional vector. Then we introduce several text dimensionality reduction methods to better represent semantic information of text.

12.1.1. *Vector Space Model*

The vector space model was proposed in the 1970s and was applied in the famous information retrieval system SMART. It has become the most classic text representation model.[120]

In vector space model, the text is expressed as a high-dimensional vector. Each dimension of the vector indicates a word, and the value denotes the weight of the word in the text.

Assume that the dictionary is {数据，刻画，中国，规律，描摹，博雅，个体，价值}. Using the vector space model, the sentence "用数据刻画规律，以数据描摹个体，用数据创造价值" is expressed by the vector (3,1,0,1,1,0,1,1). The value of the first dimension is 3, which means that word "数据" appears three times in the sentence. The two zeros mean that word "中国" and "博雅" do not appear in the sentence. In this example, we use word frequency to compute word weight. This model is called the Term Frequency (TF) model. In this section, we first introduce the TF model and the TF-IDF model. Then we introduce the N-gram model that accounts for the order of the words.

1. TF model

A straightforward way of text representation is to express the words based on their frequency in a document, namely the term frequency (TF) model. In the TF model, each dimension of the feature vector corresponds to a word in the dictionary, whose value is its frequency in the document. Therefore, the dimension of the feature vector is the size of the dictionary. Without the loss of generality, the TF model can be described as: given a dictionary $W = \{w_1, w_2, \ldots, w_V\}$, document d can be represented as a feature vector $\boldsymbol{d} = (t_1, t_2, \ldots, t_V)$, where V is the dictionary size, w_i is the i-th word in the dictionary, and t_i represents the frequency of the word w_i in document d.

The TF model records the occurrence of words in the document, which can well describe the important information of the document. The TF model only considers word frequency in a single document, ignoring the role of words in distinguishing information from different documents. For example, functional words such as "的", "了" and "是" will frequently appear in most Chinese documents. However, these words cannot express the semantics of a document. The TF-IDF model is proposed to overcome this shortcoming. Word frequency and the word's document frequency are both considered when computing the weight of a word.

2. TF-IDF model

The TF-IDF model considers document frequency of a word in the entire document collection. Assume that there are n documents in total and $\mathrm{tf}(t, d)$ denotes the frequency of the word t in document d. The document frequency of word t is $\mathrm{df}(t)$, which is the number of documents in which the word t appears. For convenience, we use the inverse document frequency (IDF):

$$\mathrm{idf}(t) = \ln \frac{n+1}{\mathrm{df}(t)+1} + 1. \tag{12.1}$$

The TF-IDF model determines the importance of the word t in the document d by combining the word frequency $\mathrm{tf}(t, d)$ and the inverse document frequency $\mathrm{idf}(t)$. The TF-IDF model computes word weight as follows:

$$\text{tf-idf}(t, d) = \mathrm{tf}(t, d) \cdot \mathrm{idf}(t). \tag{12.2}$$

3. N-gram model

In the TF model and the TF-IDF model, we consider a word as one dimension in the high-dimensional vector space. The implicit assumption of this approach is that words are independent from each other. In other words, we do not consider the order of words in a document. In natural language, however, it is very important to consider the order of words. For this purpose, an improved approach is to use the n consecutive words in the document as one single dimension in the vector space. This representation model is called the N-gram model. When $n = 1$, it is called a unigram. When $n = 2$, it is called a bigram. When $n = 3$, it is called trigram. Although N-gram model is able to consider the information of word order, it will bring an exponential growth of the dimension.

12.1.2. *Text Dimensionality Reduction*

A common method for text dimensionality reduction is keyword selection. For example, when performing user interests analysis, we can just select the words related to "interest" to form the dictionary. We can also remove stop words or filter out the common words without expressing semantics from the document, such as those words "了", "的" and "是".

We will introduce two methods for text dimensionality reduction: latent semantic analysis and probabilistic latent semantic analysis.

1. Latent semantic analysis

The basic idea of Latent Semantic Analysis (LSA) is to map the document to a low dimensional representation.[121] We hope that the low dimensional space can better express semantics. Assume that the document collection is represented as a matrix $\boldsymbol{X}$ of N rows and M columns, where N represents the number of words and M represents the number of documents; and the j-th column of the matrix $\boldsymbol{X}$ represents the document d_j; x_{ij} represents the frequency of occurrence of the i-th word in the document d_j. We then perform Singular Value Decomposition (SVD) on matrix $\boldsymbol{X}$, where $\boldsymbol{U}$ is the left singular matrix, $\boldsymbol{\Sigma}$ is the diagonal matrix formed by the descending singular values, and $\boldsymbol{V}^{\mathrm{T}}$ is the right singular matrix (as shown in Figure 12.2).

$$\boldsymbol{X} = \boldsymbol{U}\boldsymbol{\Sigma}\boldsymbol{V}^{\mathrm{T}}, \tag{12.3}$$

After the singular value decomposition, the first k singular values are selected ($k < l$, l is the rank of matrix $\boldsymbol{X}$), and the corresponding singular vectors in $\boldsymbol{U}$ and $\boldsymbol{V}^{\mathrm{T}}$ are also kept. We denote the corresponding matrix and vector as $\boldsymbol{U}_k$, $\boldsymbol{\Sigma}_k$, and $\boldsymbol{V}_k^{\mathrm{T}}$. In this way, we get an approximate $\boldsymbol{X}_k$ with respect to the original document collection matrix $\boldsymbol{X}$ (shown in Figure 12.3):

$$\boldsymbol{X}_k = \boldsymbol{U}_k\boldsymbol{\Sigma}_k\boldsymbol{V}_k^{\mathrm{T}}. \tag{12.4}$$

In fact, the matrix $\boldsymbol{X}_k$ obtained in the above method is an approximation of the original matrix $\boldsymbol{X}$ with the smallest error in the F-norm. $\boldsymbol{U}_k$ is a matrix of N rows and k columns, and each row represents a word in the k-dimensional semantic space. $\boldsymbol{V}_k^{\mathrm{T}}$ is a matrix of k rows and M columns, and each column represents the document in the k-dimensional semantic space.[a]

We use an example to explain LSA. We extract four sentences from the articles of Xu Zhijie *et al.*[122] and Li Zhiheng

$$\begin{array}{ccccccc} \boldsymbol{X} & = & \boldsymbol{U} & & \boldsymbol{\Sigma} & & \boldsymbol{V}^{\mathrm{T}} \\ \downarrow & & \downarrow & & \downarrow & & \downarrow \\ \begin{bmatrix} x_{11} & \cdots & x_{1M} \\ \vdots & & \vdots \\ x_{N1} & \cdots & x_{NM} \end{bmatrix} & = & \begin{bmatrix} \boldsymbol{u}_1 & \cdots & \boldsymbol{u}_l \end{bmatrix} & \cdot & \begin{bmatrix} \delta_0 & \cdots & 0 \\ \vdots & & \vdots \\ 0 & \cdots & \delta_l \end{bmatrix} & \cdot & \begin{bmatrix} \boldsymbol{v}_1^{\mathrm{T}} \\ \vdots \\ \boldsymbol{v}_l^{\mathrm{T}} \end{bmatrix} \end{array}$$

Figure 12.2 SVD for document collection matrix $\boldsymbol{X}$.

[a]The F-norm of a matrix is the square root of the sum of the squares of all the elements of the matrix.

$$\boldsymbol{X}_k = \boldsymbol{U}_k \quad \boldsymbol{\Sigma}_k \quad \boldsymbol{V}_k^{\mathrm{T}}$$

$$\boldsymbol{X}_k = \begin{bmatrix} \boldsymbol{u}_1 & \cdots & \boldsymbol{u}_k \end{bmatrix} \cdot \begin{bmatrix} \delta_0 & \cdots & 0 \\ \vdots & & \vdots \\ 0 & \cdots & \delta_k \end{bmatrix} \cdot \begin{bmatrix} \boldsymbol{v}_1^{\mathrm{T}} \\ \vdots \\ \boldsymbol{v}_k^{\mathrm{T}} \end{bmatrix}$$

Figure 12.3 Select the singular vectors corresponding to the first k singular values.

et al.[123] respectively. Each sentence is used as a document to form a document collection. The document collection consists of 4 documents on computer vision $\{c_1, c_2, c_3, c_4\}$ and 4 documents on medical big data $\{m_1, m_2, m_3, m_4\}$. The seven sentences are as follows:

- c_1: At the same time, computer vision technology still faces problems such as ambiguous information description, unstable image feature detection, and inefficiency;
- c_2: The application of CV technology is very extensive, such as digital image retrieval management, medical photograph analysis, intelligent security inspection, human-computer interaction, etc.;
- c_3: Digital images and video data contain abundant resources of vision. How to intelligently extract and analyze useful information in photograph has gradually become a research hotspot in recent years;
- c_4: Despite these achievements, the CV system is still too rough compared to the precise and flexible human vision level;
- m_1: With the development of human genomics and high-throughput technologies, the medical literature concerning protein knowledge and related diseases, drugs, and genes has grown exponentially;
- m_2: Semantic output of a MEDLINE document of a specific disease obtained by SemRep, and this semantic output is ranked by a significant information extraction algorithm;
- m_3: Using text mining techniques, it has become possible to discover and extract valuable novel protein knowledge from a large number of biomedical texts;

- m_4: The experimental results have important research significance for understanding disease genes, protein function prediction, and drug-aided design.

For the purpose of simplification, we just consider the underlined words. We use the vector space model to represent documents, and use the TF model to calculate the weight of words. The document collection matrix $\boldsymbol{X}$ is shown in Table 12.1. Then we use LSA model ($k = 2$), and get the approximate matrix $\boldsymbol{X}_2$ of the original matrix $\boldsymbol{X}$, as shown in Table 12.2.

In Table 12.1 and Table 12.2, the semantic similarity between two documents can be obtained by the correlation between the two

Table 12.1 The document collection matrix of scientific articles.

	c_1	c_2	c_3	c_4	m_1	m_2	m_3	m_4
Image	1	1	1	0	0	0	0	0
Human-computer interaction	0	1	0	0	0	0	0	0
Vision	1	0	1	1	0	0	0	0
Photograph	0	1	1	0	0	0	0	0
Research	0	0	1	0	1	0	0	1
Protein	0	0	0	0	1	0	1	1
Gene	0	0	0	0	2	0	0	1
Disease	0	0	0	0	1	1	0	1

Table 12.2 The approximate matrix $\boldsymbol{X}_2$ of the document collection.

	c_1	c_2	c_3	c_4	m_1	m_2	m_3	m_4
Image	0.742	0.829	1.184	0.334	−0.032	−0.038	−0.038	0.018
Human-computer interaction	0.225	0.251	0.353	0.101	−0.042	−0.017	−0.017	−0.019
Vision	0.608	0.679	0.973	0.274	−0.005	−0.027	−0.027	0.031
photograph	0.541	0.604	0.867	0.244	0.003	−0.023	−0.023	0.033
Research	0.266	0.292	0.612	0.122	1.113	0.180	0.180	0.855
Protein	−0.066	−0.078	0.091	−0.027	1.175	0.205	0.205	0.882
Gene	−0.086	−0.103	0.143	−0.036	1.678	0.292	0.292	1.261
Disease	−0.066	−0.078	0.091	−0.027	1.175	0.205	0.205	0.882

columns. Similarly, the semantic similarity between two words can also be obtained by the correlation between two rows. We use the Pearson correlation coefficient, and the semantic similarities between documents before and after latent semantic analysis are shown in Table 12.3 and Table 12.4, respectively. It is observed that although c_2 and c_4 are of the same topic, they are negatively correlated, as shown in Table 12.3. So, the original document representation fails to express the document semantics. When the documents are represented by the LSA model, c_1 and c_2 have a strong correlation, which well reflects the semantic similarity of the documents.

Similarly, we can also compare the correlation between two words in the original representation space and the two-dimensional semantic

Table 12.3 The original similarity between document.

	c_1	c_2	c_3	c_4	m_1	m_2	m_3	m_4
c_1	1							
c_2	0.15	1						
c_3	0.58	0.26	1					
c_4	0.65	−0.29	0.38	1				
m_1	−0.52	−0.7	−0.54	−0.34	1			
m_2	−0.22	−0.29	−0.38	−0.14	0.20	1		
m_3	−0.22	−0.29	−0.38	−0.14	0.20	−0.14	1	
m_4	−0.58	−0.77	−0.50	−0.38	0.90	0.38	0.38	1

Table 12.4 Approximate similarity between documents after SVD.

	c_1	c_2	c_3	c_4	m_1	m_2	m_3	m_4
c_1	1							
c_2	1	1						
c_3	0.99	0.99	1					
c_4	1	1	0.99	1				
m_1	−0.85	−0.85	−0.76	−0.84	1			
m_2	−0.88	−0.88	−0.79	−0.87	1	1		
m_3	−0.88	−0.88	−0.79	−0.87	1	1	1	
m_4	−0.84	−0.84	−0.75	−0.83	1	1	1	1

space. For example, "image" and "photograph" carry similar meanings in computer vision, and their correlations in the original representation and approximate representation are:

$$\begin{aligned} \text{Corr}_1(\text{image}, \text{photograph}) &= 0.75, \\ \text{Corr}_2(\text{image}, \text{photograph}) &= 1. \end{aligned} \tag{12.5}$$

It can be seen that the semantic space obtained by the LSA model shows stronger ability to represent the semantic similarity between words. In other words, LSA can deal with the phenomenon of synonyms (i.e., multi-word synonymy). For example, in computer vision, "image" and "photograph" express the same meaning. There usually exists a lot of noise in the original document representation. LSA can facilitate the filtering of the noise by reducing dimension. Empirical results show that LSA performs better than a traditional vector space model in an information retrieval task.

However, LSA cannot solve the problem of "polysemy" in text mining. In addition, the dimension k needs to be manually selected. Since LSA is based on the vector space model, it still does not consider the information of the word order appearing in the document.

2. Probability latent semantic analysis

The probability latent semantic analysis (pLSA) utilizes probabilistic models to overcome the sparsity problem of LSA.[124] We use $p(d)$ to denote the probability of document d, and $p(w|d)$ to denote the probability that the word w appears in the document d. Assuming there are k latent semantics, we use the symbol z to denote the semantic variable, $p(w|z)$ to denote the probability that a specific word w belongs to the semantic z, and $p(z|d)$ is used to denote the distribution of semantic z in document d. Specifically, by introducing the latent semantics, $p(w|d)$ can be written as follows:

$$p(w|d) = \sum_z p(w|z)p(z|d). \tag{12.6}$$

The pLSA model is a generative model. Given a document d, the semantic z is selected with a certain probability distribution $p(z|d)$, and a specific word is selected to express the semantic z with a certain probability distribution $p(w|z)$, where $p(z|d)$ and $p(w|z)$ are both

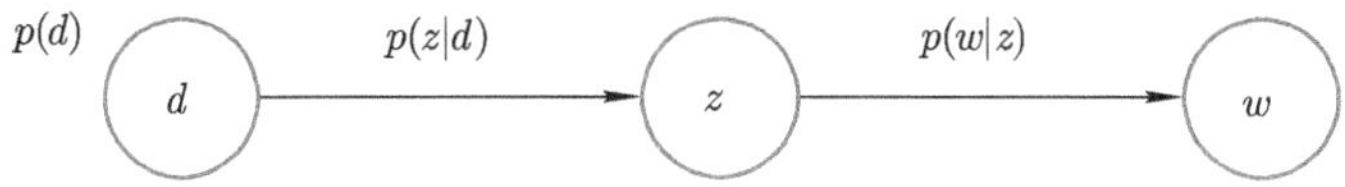

Figure 12.4 The word generation process in the pLSA model.

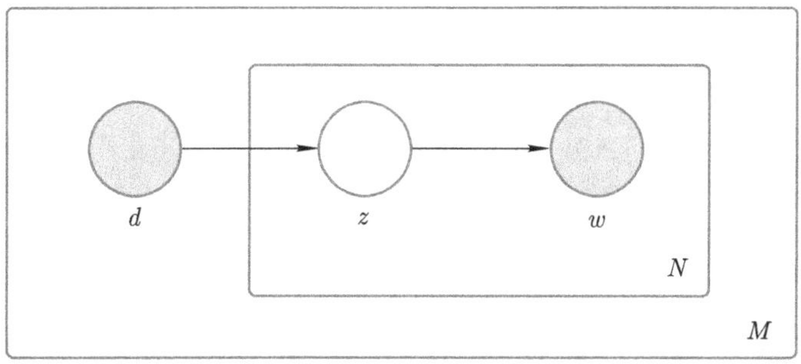

Figure 12.5 The graphical model representation of pLSA.

polynomial distributions.[b] In the pLSA model, the process of word production is shown in Figure 12.4.

Assuming that the document collection contains M documents, the number of words in the document is N. As a directed graph model, the graphical model representation of pLSA is shown in Figure 12.5.

For a document d_i and a word w_j, the joint probability can be written as

$$p(w_j, d_i) = p(d_i) \sum_z p(w_j|z)p(z|d_i). \tag{12.7}$$

$n(w_j, d_i)$ is used to denote the number of occurrences of the word w_j in the document d_i. It is assumed that the generation of a word is independent of each other. In the pLSA model, the likelihood function of the document collection is expressed as:

$$L = \prod_{d_i} \prod_{w_j} p(w_j, d_i)^{n(w_j,d_i)}. \tag{12.8}$$

[b]Please refer to Appendix B for polynomial distribution.

Combined with Eq. (12.7), the log-likelihood function of the document collection in the pLSA model is

$$LL = \sum_{i,j} n(w_j, d_i) \ln p(w_j, d_i)$$
$$= \sum_{i,j} n(w_j, d_i) \ln \left(p(d_i) \sum_z p(w_j|z)p(z|d_i) \right). \tag{12.9}$$

Note that Eq. (12.9) is the objective function with the parameters $p(w_j|z)$ and $p(z|d_i)$. The above optimization problem implies the following constraints $p(w_j|z) \geqslant 0$, $p(z|d_i) \geqslant 0$, $\sum_z p(z|d_i) = 1$, and $\sum_j p(w_j|z) = 1$.

EM algorithm[c] is one of the most popular methods for solving pLSA. The basic idea is to find a simpler lower bound function LL_{lower} w.r.t. the log-likelihood function LL, and optimize LL by continuously optimizing LL_{lower}. Now, let us introduce how to get such a lower bound function LL_{lower}. We use $n(d_i)$ to denote the number of words in the document d_i, which satisfies $n(d_i) = \sum_j n(w_j, d_i)$. Then LL can be rewritten as follows:

$$LL = \sum_{i,j} n(w_j, d_i) \ln p(d_i) + \sum_{i,j} n(w_j, d_i) \ln \sum_z p(w_j|z)p(z|d_i)$$
$$= \sum_{i} n(d_i) \ln p(d_i) + \sum_{i,j} n(w_j, d_i) \ln \sum_z p(w_j|z)p(z|d_i). \tag{12.10}$$

In Eq. (12.10), the first item $\sum_i n(d_i) \ln p(d_i)$ is a constant, then:

$$LL = \sum_{i,j} n(w_j, d_i) \ln \sum_z p(w_j|z)p(z|d_i) + \text{const.} \tag{12.11}$$

Then the lower bound function LL can be obtained by using Jensen's inequality shown in Eq. (C.4).

$$LL_{\text{lower}} = \sum_{i,j} n(w_j, d_i) \sum_z p(w_j|z) \ln p(z|d_i) + \text{const.} \tag{12.12}$$

[c]Chapter 10 gives details about the EM algorithm.

When using EM algorithm to solve the pLSA, the E-step is equivalent to finding the lower bound function LL_{lower}, while the M-step is equivalent to solving the optimal solution of the lower bound function LL_{lower}. Specifically, the iteration formula of the E-step is

$$p(z|d_i, w_j) = \frac{p(w_j|z, d_i)}{p(w_j|d_i)} \frac{p(w_j|z, d_i)p(z|d_i)}{\sum_z p(w_j|z, d_i)p(z|d_i)}. \tag{12.13}$$

The iteration formula for the M-step is

$$\begin{aligned} p(w_j|z) &= \frac{\sum_i n(w_j, d_i)p(z|d_i, w_j)}{\sum_m \sum_i n(w_m, d_i)p(z|d_i, w_m)}, \\ p(z|d_i) &= \frac{\sum_j n(w_j, d_i)p(z|d_i, w_j)}{n(d_i)}, \end{aligned} \tag{12.14}$$

where $n(w_j, d_i)$, $n(d_i)$, and $p(d_i)$ can be directly calculated from the data. E-step and M-step are alternately iterative until convergence.

Compared with LSA pLSA introduces the concept of latent semantics through the hidden variable z. Thus pLSA has a more solid stochastic theoretical basis than LSA. Given a word, its different semantics expressed in different documents can be distinguished through the distribution $p(w|z)$. In this way, the problem of "polysemy" can be solved to some degree. The representation in low-dimensional space of document d is expressed by the distribution $p(z|d)$, while each semantic can be represented by the distribution $p(w|z)$. In Section 12.2, the latent semantic variable z is also called a topic.

Although pLSA is a generative model, it cannot be used to generate a new document. In Figure 12.5, the document variable d appears explicitly in the model, so the pLSA model is not a complete generative model. It is hard to get a low-dimensional representation of a new document.

12.2. Topic Model

With the development of social media such as blogs and microblogs, there are massive textual data generated on the Internet. Such large-scale textual data makes it nearly impossible for human to digest all the information. Therefore, there is an urgent need for new technologies and tools to organize, search and analyze the information. For this purpose, topic model is proposed to discover the topics that can represent the document content. We can annotate the document by the topics. Then we can organize, search and summarize the document according to the topic annotation.

Generally speaking, a topic consists of a set of words describing its meaning. For example, the "diet" topic can be expressed by the words "breakfast", "coffee", "fruit", and "egg". Many topic models have been proposed, and the most typical topic model is the Latent Dirichlet Allocation (LDA) model. LDA was proposed by David Blei, Andrew Ng and Michael Jordan in 2003.[125] In this section, we introduce the basic concept, optimization goal and parameter estimation method of LDA. Then we give a brief introduction to some improved topic models based on LDA.

12.2.1. *LDA Model*

We will first illustrate the structure and the objective of the LDA model as shown in Figure 12.6. It is assumed that the document collection has K topics. Each topic is a multinomial distribution of the dictionary dimension V. For each document d in the document collection, there is a latent distribution of topics to represent the

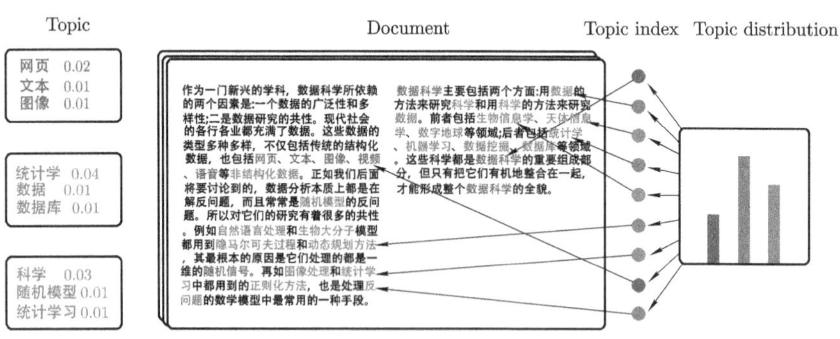

Figure 12.6 The schematic of LDA model.

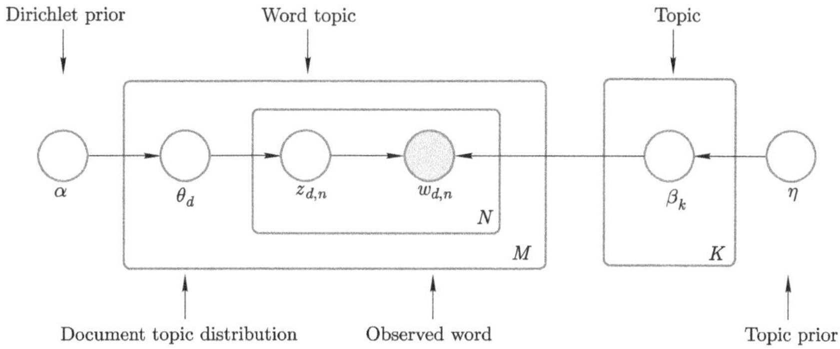

Figure 12.7 The graphical representation of the LDA model.

semantic meaning of the document. Each word in the document is considered to be generated from a certain topic.

Dirichlet distribution[d] is the key of the LDA model, and it is a conjugate prior distribution[e] of the multinomial distribution. We assume that both the word distribution of topic and the topic distribution of document obey multinomial distribution, and these two distributions are generated by sampling from a Dirichlet distribution, which will make it convenient for solving LDA.

Now let us introduce some symbols in the LDA model. Assume that the document collection contains a total of M documents. N denotes the number of words in the document. The document collection has K topics, and $\boldsymbol{\beta}_k$ denotes the k-th topic. $\boldsymbol{\theta}_d$ denotes topic distribution in the document d. $w_{d,n}$ and $z_{d,n}$ denote the n-th word in document d and the topic to which it belongs. α and η are hyperparameters for the topic distribution of document and topic word distribution, respectively. Then the graphical representation of the LDA model is shown in Figure 12.7.

[d]Please refer to Appendix B for Dirichlet distribution.

[e]If the posterior distribution is in the same probability distribution family as the prior probability distribution, the prior and posterior are then called conjugate distributions, and the prior is called a conjugate prior for the likelihood function.

The joint probability of the observed variable $\boldsymbol{w}$ and the latent variable $\boldsymbol{\theta}, \boldsymbol{z}, \boldsymbol{\beta}$ in the LDA model is

$$\begin{aligned} p(\boldsymbol{\theta}, \boldsymbol{z}, \boldsymbol{w}, \boldsymbol{\beta}|\alpha, \eta) &= p(\boldsymbol{\beta}|\eta)p(\boldsymbol{\theta}|\alpha)p(\boldsymbol{z}|\boldsymbol{\theta})p(\boldsymbol{w}|\boldsymbol{z}, \boldsymbol{\beta}) \\ &= \prod_{k=1}^{K} p(\boldsymbol{\beta}_k|\eta) \prod_{d=1}^{D} p(\boldsymbol{\theta}_d|\alpha) \prod_{n=1}^{N_d} \\ &\quad \times p(z_d, n|\boldsymbol{\theta}_d)p(w_{d,n}|z_{d,n}, \boldsymbol{\beta}_{z_{d,n}}), \end{aligned} \tag{12.15}$$

where $p(\boldsymbol{\beta}_k|\eta)$ and $p(\boldsymbol{\theta}_d|\alpha)$ are Dirichlet distributions, and $p(z_d, n|\boldsymbol{\theta}_d)$ and $p(w_{d,n}|z_{d,n}, \boldsymbol{\beta}_{z_{d,n}})$ are multinomial distributions.

12.2.2. *Parameter Estimation*

For the LDA model, there are usually two methods for parameter estimation, variational inference and Gibbs sampling, which are both approximation algorithms. The former one is difficult to derive, but efficient for computation; the latter one is simple to derive, but inefficient for computation.

1. Variational inference

Firstly, we will introduce how to derive the likelihood function of the document collection. In formula (12.15), we integrate the latent variables $\boldsymbol{\theta}, \boldsymbol{z}, \boldsymbol{\beta}$ and take the logarithm to get the log likelihood function L as follows:

$$\begin{aligned} L &= \ln \int_{\boldsymbol{\theta}} \int_{\boldsymbol{z}} \int_{\boldsymbol{\beta}} p(\boldsymbol{\beta}|\eta)p(\boldsymbol{\theta}|\alpha)p(\boldsymbol{z}|\boldsymbol{\theta})p(\boldsymbol{w}|\boldsymbol{z}, \boldsymbol{\beta})\mathrm{d}\boldsymbol{\beta}\mathrm{d}\boldsymbol{z}\mathrm{d}\boldsymbol{\theta} \\ &= \ln \int_{\boldsymbol{\beta}} \prod_{k=1}^{K} \mathrm{Dir}(\boldsymbol{\beta}_k|\eta) \prod_{d=1}^{D} \int_{\boldsymbol{\theta}_d} \mathrm{Dir}(\boldsymbol{\theta}_d|\alpha) \\ &\quad \times \prod_{n=1}^{N_d} \sum_{z_{d,n}} \prod_{v=1}^{V} (\theta_{d,k}\beta_{k,v})^{w_{d,n}^{v}} \mathrm{d}\boldsymbol{\theta}_d \mathrm{d}\boldsymbol{\beta}. \end{aligned} \tag{12.16}$$

In (12.16), since $\boldsymbol{\theta}$ and $\boldsymbol{\beta}$ always appear in pair, the above objective function does not have an analytical solution. We need to use an approximation algorithm to solve this problem. Here we use the EM algorithm. In E-step, we attempt to find a lower bound function that the log-likelihood function L is easy to be optimized, while in M-step we find the optimal solution to this lower bound function. Then, how to find a good lower bound function? We first assume that $q(\boldsymbol{\theta}, \boldsymbol{z}, \boldsymbol{\beta})$ is some distribution on the latent variable $\boldsymbol{\theta}, \boldsymbol{z}, \boldsymbol{\beta}$, and then obtain the lower bound function of L by using Jensen's inequality (C.4).

$$
\begin{aligned}
L &= \ln \int_{\boldsymbol{\theta}} \int_{\boldsymbol{z}} \int_{\boldsymbol{\beta}} p(\boldsymbol{\theta}, \boldsymbol{z}, \boldsymbol{w}, \boldsymbol{\beta}|\alpha, \eta) \mathrm{d}\boldsymbol{\beta} \mathrm{d}\boldsymbol{z} \mathrm{d}\boldsymbol{\theta} \\
&= \ln \int_{\boldsymbol{\theta}} \int_{\boldsymbol{z}} \int_{\boldsymbol{\beta}} \frac{p(\boldsymbol{\theta}, \boldsymbol{z}, \boldsymbol{w}, \boldsymbol{\beta}|\alpha, \eta)}{q(\boldsymbol{\beta}, \boldsymbol{z}, \boldsymbol{\theta}|\boldsymbol{\lambda}, \boldsymbol{\gamma}, \boldsymbol{\phi})} q(\boldsymbol{\beta}, \boldsymbol{z}, \boldsymbol{\theta}|\boldsymbol{\lambda}, \boldsymbol{\gamma}, \boldsymbol{\phi}) \mathrm{d}\boldsymbol{\beta} \mathrm{d}\boldsymbol{z} \mathrm{d}\boldsymbol{\theta} \\
&\geqslant \int_{\boldsymbol{\theta}} \int_{\boldsymbol{z}} \int_{\boldsymbol{\beta}} q(\boldsymbol{\beta}, \boldsymbol{z}, \boldsymbol{\theta}|\boldsymbol{\lambda}, \boldsymbol{\gamma}, \boldsymbol{\phi}) \ln \frac{p(\boldsymbol{\theta}, \boldsymbol{z}, \boldsymbol{w}, \boldsymbol{\beta}|\alpha, \eta)}{q(\boldsymbol{\beta}, \boldsymbol{z}, \boldsymbol{\theta}|\boldsymbol{\lambda}, \boldsymbol{\gamma}, \boldsymbol{\phi})} \mathrm{d}\boldsymbol{\beta} \mathrm{d}\boldsymbol{z} \mathrm{d}\boldsymbol{\theta} \\
&= E_q[\ln p(\boldsymbol{\theta}, \boldsymbol{z}, \boldsymbol{w}, \boldsymbol{\beta}|\alpha, \eta)] - E_q[\ln q(\boldsymbol{\beta}, \boldsymbol{z}, \boldsymbol{\theta}|\boldsymbol{\lambda}, \boldsymbol{\gamma}, \boldsymbol{\phi})] \\
&= L_{\text{lower}}.
\end{aligned}
\tag{12.17}
$$

In order to optimize the lower bound function L_{lower}, the form of the distribution $q(\boldsymbol{\theta}, \boldsymbol{z}, \boldsymbol{\beta})$ should be as simple as possible. In the variational inference of LDA, it is assumed that each variable in the distribution is independent from each other, and i.e., q satisfies the following condition:

$$
\begin{aligned}
q(\boldsymbol{\beta}, \boldsymbol{z}, \boldsymbol{\theta}|\boldsymbol{\lambda}, \boldsymbol{\gamma}, \boldsymbol{\phi}) = &\prod_{k=1}^{K} \mathrm{Dir}(\beta_k|\lambda_k) \\
&\times \prod_{d=1}^{D} \left(\mathrm{Dir}(\theta_d|\gamma_d) \prod_{n=1}^{N_d} \mathrm{Mult}(z_{d,n}|\phi_{d,n}) \right),
\end{aligned}
\tag{12.18}
$$

where $\boldsymbol{\lambda}, \boldsymbol{\gamma}$ and $\boldsymbol{\phi}$ are the corresponding parameters. The graphical representation of the distribution q is shown in Figure 12.8. Now L_{lower} can be expressed by a function of $\boldsymbol{\lambda}, \boldsymbol{\gamma}$, and $\boldsymbol{\phi}$, which is equivalent to the E-step. Then we use the Lagrange method to estimate the parameters $\boldsymbol{\lambda}$, $\boldsymbol{\gamma}$, and $\boldsymbol{\phi}$. $\boldsymbol{\beta}$, $\boldsymbol{\theta}$, and z can be approximated by

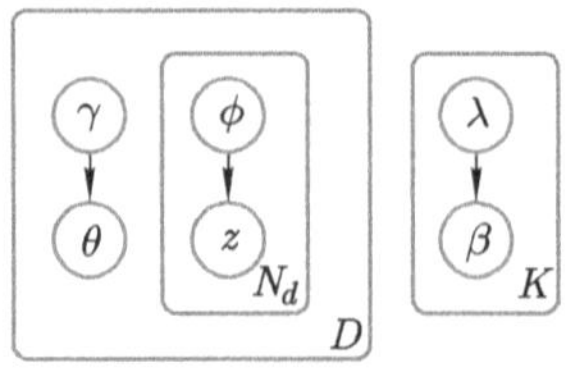

Figure 12.8 Approximate distribution q of latent variables in LDA.

$\boldsymbol{\lambda}$, $\boldsymbol{\gamma}$, and $\boldsymbol{\phi}$ obtained by Algorithm 13, which is equivalent to the M-step.

Algorithm 13 Variational inference for LDA

1: Initialize $\boldsymbol{\lambda}$,$\boldsymbol{\gamma}$, and $\boldsymbol{\phi}$;
2: **while** not converged **do**
3: $\quad \lambda_{k,v} = \eta_v + \sum_{d=1}^{D} \sum_{n=1}^{N_d} w_{d,n}^{v} \phi_{d,n}^{k}$,
4: $\quad \gamma_{d,k} = \alpha_k + \sum_{n=1}^{N_d} \phi_{d,n}^{k}$,
5: $\quad \phi_{d,n}^{k} \propto \exp\left(\Psi(\gamma_{d,k}) + \Psi(\lambda_{d,k}) - \Psi(\sum_{v=1}^{V} \lambda_{k,v})\right)$;
6: return $\boldsymbol{\lambda}$,$\boldsymbol{\gamma}$, and $\boldsymbol{\phi}$.

2. Gibbs sampling

Another method of solving the LDA model is Gibbs sampling. Compared with the variational inference method, Gibbs sampling is simpler and easier to implement. Gibbs sampling only focus on one variable at a time, which means the conditional density of the variable is obtained with other variables fixed. Then we conduct sampling in an iterative manner, and the parameters are estimated according to the sampling values.

To better explain the process, we first introduce two count flags, $\Omega_{d,k}$ and $\Psi_{k,v}$, where $\Omega_{d,k}$ records the number of times that the document d is sampled as topic k, and $\Psi_{k,v}$ records the number of times that the word v is sampled as topic k. For simplicity, we now focus only on the variable $\boldsymbol{z}$. Therefore, we need to integrate out $\boldsymbol{\theta}$ and $\boldsymbol{\beta}$ in Eq. (12.15) to get the joint probability distribution of $\boldsymbol{z}$ and $\boldsymbol{w}$:

$$p(\boldsymbol{z}, \boldsymbol{w}|\alpha, \eta) = \prod_{k=1}^{K} \frac{B(\eta + \Psi_k)}{B(\eta)} \prod_{d=1}^{D} \frac{B(\alpha + \Omega_d)}{B(\alpha)}. \tag{12.19}$$

With the Bayes' theorem

$$p(z_{d,n}|\boldsymbol{z}_{-d.n}, \boldsymbol{w}, \alpha, \eta) = \frac{p(\boldsymbol{z}, \boldsymbol{w}|\alpha, \eta)}{p(\boldsymbol{z}_{-d.n}, \boldsymbol{w_{-d,n}}, w_{d,n}|\alpha, \eta)},$$

Gibbs sampling for z can be obtained:

$$\begin{aligned} p(z_{d,n} = k|\boldsymbol{z}_{-d.n}, \boldsymbol{w}, \alpha, \eta) &= \frac{\eta_v + \Psi_{k,v} - 1}{\sum_{v=1}^{V}(\eta_v + \Psi_{k,v}) - 1} \frac{\alpha_k + \Omega_{d,k} - 1}{\sum_{k=1}^{K}(\alpha_k + \Omega_{d,k}) - 1} \\ &\propto \frac{\eta_v + \Psi_{k,v} - 1}{\sum_{v=1}^{V}(\eta_v + \Psi_{k,v}) - 1}(\alpha_k + \Omega_{d,k} - 1). \end{aligned} \tag{12.20}$$

According to the definition of $\boldsymbol{\theta}$ and $\boldsymbol{\beta}$, we can estimate them as follows

$$\widehat{\theta}_{d,k} = \frac{\Omega_{d,k} + \alpha_k}{\sum_{k=1}^{K} \Omega_{d,k} + \alpha_k}, \tag{12.21}$$

$$\widehat{\beta}_{d,k} = \frac{\Psi_{k,v} + \eta_v}{\sum_{v=1}^{V} \Psi_{k,v} + \eta_v}. \tag{12.22}$$

Algorithm 14 shows the detailed steps of Gibbs sampling for LDA.

Algorithm 14 Gibbs sampling for LDA

Initialize count variables $\Omega_{d,k}$ and $\Psi_{k,v}$ to 0;
for each document d **do**
  for each word w_{dn} in document d **do**
    Sample the topic z_{dn} according to a uniform multinomial distribution,
    Update $\Psi_{k,v}$: $\Psi_{z_{dn}, w_{dn}}$ ++,
    Update $\Omega_{d,k}$: $\Omega_{d,z_{dn}}$ ++;
while not converged **do**
  for each document d **do**
    for each word w_{dn} in document d **do**
      Assume that the topic of the word w_{dn} in the last round of sampling is z_{dn},
      Update $\Psi_{k,v}$: $\Psi_{z_{dn}, w_{dn}}$ −−,
      Update $\Omega_{d,k}$: $\Omega_{d,z_{dn}}$ −−;
      According to Eq. (12.20), sample for topic z'_{dn} of w_{dn},
      Update $\Psi_{k,v}$: $\Psi_{z'_{dn}, w_{dn}}$ ++,
      Update $\Omega_{d,k}$: $\Omega_{d,z'_{dn}}$ ++;
Calculate $\boldsymbol{\theta}$ according to Eq. (12.21);
Calculate $\boldsymbol{\beta}$ according to Eq. (12.22);
return $\boldsymbol{z}$, $\boldsymbol{\theta}$ and $\boldsymbol{\beta}$.

12.2.3. *Summary of the Topic Model*

As a classic topic model, LDA has been widely applied in many areas. Firstly, it can be used to discover latent topics in large-scale document collection. When solving for LDA, $\{\boldsymbol{\beta}_1, \boldsymbol{\beta}_2, \ldots, \boldsymbol{\beta}_k\}$ indicate K topics discovered from the document collection. Each $\boldsymbol{\beta}_k$ is a multi-distribution on the V-dimensional dictionary. Then the topic can be expressed by the words appearing in $\boldsymbol{\beta}_k$ that is sorted by the probabilities in a descending order. Secondly, LDA can be used as a tool for text dimensionality reduction. In the LDA model, the distribution $\boldsymbol{\theta}_d$ is a K-dimensional vector, whose size ranges from tens to hundreds. Compared with the original V-dimensional document representation, $\boldsymbol{\theta}_d$ can be treated as a low-dimension representation of document d.

In this section, we take the hyper-parameters α and β of the LDA model as known quantities. How should we set them? We usually set α and η to be positive numbers less than 1. In some literatures, α and β are treated as parameters that need to be estimated with the EM algorithm.[125]

The LDA model is also used as a basic module for designing more complex topic models. For example, a genetics article is probably relevant to health and disease, but unlikely to be relevant to the ray astronomy. LDA model cannot deal well with such cases. CTM model proposed by Blei and Lafferty has the ability to model the correlation between such topics.[126] CTM considers the topic distribution of document as Logistic distribution instead of the Dirichlet distribution. The number of topics K in LDA model needs to be set manually. However, how to set K is difficult in many tasks. HDP can automatically determine K according to the size of the data.[127] SLDA is an supervised extension of LDA, which can better learn topics and predict document labels.[128]

12.3. Sentiment Analysis

With the rapid development of social media such as forum discussions, blogs, and microblogs, there is an urgent need to analyze these textual data for decision making. In real life, people often want to know the opinions of others when making decisions; businesses want to know what consumers' comments on their products and services; the government needs to understand the public's response to its policies. The development of social media and the application requirements of people have spawned a new field of research: sentiment analysis. The reason why sentiment analysis became a popular research field is mainly due to the following reasons: firstly, there is a wide range of application demand in all walks of life. Sentiment analysis can be applied to many fields, such as social public opinion analysis, product review and recommendation, film and television review, etc.; secondly, there are many unsolved research problems in sentiment analysis; finally, a large amount of user generated content from the Internet provides rich research resources for sentiment analysis. These three reasons promote the development of sentiment analysis.

At present, there are many fields involving the study of sentiment analysis, such as natural language processing, data mining, text mining, and so on. Sentiment analysis is a multidisciplinary and application-oriented research field whose objective is to research and analyze people's opinions, emotions, evaluations, attitudes, sentiments on products, events, and topics. There are many challenging research problems in sentiment analysis, including sentiment classification, aspect-based sentiment analysis, opinion retrieval, and sentiment dictionary construction. In this section we will focus on the sentiment classification and aspect-based sentiment analysis.

12.3.1. *Sentiment Classification*

The objective of sentiment classification is to classify documents into different categories based on the sentiments expressed in the documents. According to different amounts of labels, there are several classification methods listed as follows:

(1) **Binary classification:** Positive and negative;
(2) **Three classification:** Positive, negative, and neutral;
(3) **Multiple classification:** 1 to 5 stars or emotions such as pleasure, anger, sorrow and joy.

The granularity of the classification can be document, sentence, phrase, and word. The document-level sentiment classification is based on a basic assumption: a document contains only one viewpoint towards an entity or feature. This is a very strong hypothesis that is proved to be effective in commodity reviews. To the contrary, online discussions, argument texts, and Weibo often express opinions with different polarities on different aspects or features. For such texts, coarse-grained sentiment classification at document-level is not appropriate. Therefore, under this circumstance, sentiment classification should be made on a fine-grained granularity such as sentences or phrases.

Generally speaking, sentiment classification methods can be divided into two categories: rule-based methods and supervised learning-based methods. Rule-based methods exploit the sentiment polarity of the contextual words and some manually rules (negation rules, shifting rules, etc.) to determine the sentiment orientation of a sentence.[129,130] The supervised learning-based methods treat sentiment classification as a classification problem. The key point is to extract sentiment features as the input for the classification model.

1. Rule-based sentiment classification

The basic idea of the rule-based approach is to use the sentiment polarity of a word in the context to determine the sentiment of the document. This method is divided into two steps: the sentiment orientation score of a word is computed at the first step, and then to get the sentiment polarity of the document through the sentiment orientation scores of the words. How to calculate the sentiment score of a word? A simple approach is to use a human-created sentiment dictionary in which people have manually tagged words and their corresponding sentiment scores. As for some languages, there is no sentiment dictionary, and seed words will be selected for calculating the sentiment scores of the words.

First, we manually select some typical sentiment words and weigh their sentiment scores. For example, in English, we can choose "excellent" and "poor" as positive and negative seed words, and label their sentiment scores as 1 and -1, respectively.

Then, we will calculate the semantic similarity between other words and the seed words, and then we can calculate the sentiment scores of other words by semantic similarity. There are some methods for calculating the similarity, and we will mainly introduce the point-wise mutual information (PMI) method.[129] Assume that we have two words w_1 and w_2, the PMI is computed as follows:

$$\mathrm{PMI}(w_1, w_2) = \ln \frac{p(w_1, w_2)}{p(w_1)p(w_2)}, \tag{12.23}$$

where $p(w_1)$ is the probability of appearance of word w_1, $p(w_2)$ is the probability of appearance of word w_2, and $p(w_1, w_2)$ is the probability of co-occurrence of word w_1 and w_2. If two words are statistically independent, then $p(w_1, w_2) = p(w_1)p(w_2)$. Obviously, PMI measures the statistical correlation between two words, which can be used to represent the semantic similarity between words.

Given a word w, we can calculate its sentiment score $\mathrm{SO}(w)$ by calculating its PMI with the positive seed word "good" and the negative seed word "bad", respectively:

$$\mathrm{SO}(w) = \mathrm{PMI}(w, \text{"good"}) - \mathrm{PMI}(w, \text{"bad"}). \tag{12.24}$$

Finally, we average the sentiment scores of all the words in a document to obtain the sentiment score of the document. Note that the negation expression and polarity shifting should be taken into account.

Rule-based sentiment classification is an unsupervised method that can be applied in any field without data annotation. However, in a specific domain, the classification accuracy of this method is not high. The sentiment of a word will change with changes in domain, the universal sentiment word dictionary or Eq. (12.24) may be inaccurate in calculation in a specific domain. For example, there are two sentences "safety box is heavy" and "the computer's battery is heavy", the word "heavy" indicates the opposite sentiment polarity, positive and negative, respectively.

2. Sentiment classification based on supervised learning

To address the above problem, supervised learning methods have been proposed. Supervised learning methods consider the sentiment classification as a text classification problem, and make use of some traditional classification models.[131] Naive Bayes, maximum entropy, and support vector machine are used to classify English movie reviews into positive and negative. The accuracy of the support vector machine can reach 82.9% by only using n-gram features. Researchers have been gradually improving the accuracy of sentiment classification on the movie review corpus by using different features and classification models. The results are shown in Table 12.5.

We can see that in order to improve the performance of sentiment classification in specific domain, feature selection is very important. Some effective features are listed as follows.

(1) **Term frequency feature:** Word and its frequency in the document.
(2) **Lexical feature:** Part-of-speech features and N-gram phrase features.

Table 12.5 Accuracy comparison of different models on movie review sentiment analysis.

Classification model	Classification accuracy	Features used	Literature
Support vector machine	82.9%	Unigram	[132]
Naive Bayes	86.4%	Unigram + remove objective sentence	[131]
Support vector machine + rules	86.2%	Unigram + bigram	[133]
Support vector machine	89.6%	Unigram	[134]
Support vector machine	90.5%	Unigram, bigrams, trigrams, dependency features, sentiment adjective features	[135]

(3) **Sentiment word feature:** Take words in the sentiment dictionary as separate features, such as the number of positive and negative words.
(4) **Negation feature:** Negation words, modal verbs, reinforcement words, and weakened words.
(5) **Grammatical feature:** Grammatical dependency, etc.

Supervised learning methods mainly use traditional classification models to solve sentiment classification problems. Labeled corpus is required to train a domain-specific classifier. In a specific domain, the supervised learning method usually outperforms the unsupervised learning method. However, it is not easy to obtain labeled data, and quite challenging to transfer a classifier from one domain to another.

12.3.2. *Aspect-based Sentiment Analysis*

Sentiment classification can classify texts into positive and negative ones, but it cannot tell us more fine-grained information. For example, when (time) does the person (opinion holder) express what opinions (opinion) toward which feature (feature) of which entity (target). Let us take a look at a real phone review:

UserID: 27086 Posted on December 20, 2014 "The mobile phone is very good, the service of the merchant is excellent. The gifts are very good and useful, and the logistics is very fast so that it took less than a day for delivery within a thousand miles to reach. But the mobile phone call is not convenient that it can only serve single pass with dual-standby rather than the double passes. This is the problem in the design of the machine. After all, I feel very satisfied with the business. I will come again if I need".

Table 12.6 is the information we extracted from this review. It can be seen that there are many elements extracted are related to opinions. This problem is called feature-based sentiment analysis.

Feature-based sentiment analysis focuses on granularity of target and feature. Opinion mining from the granularity of entity and feature can meet the requirements of many practical applications. For example, given an emergent social event, only understanding the opinion trending in the document level is insufficient, because people often need to understanding the attitude of the public from different

Table 12.6 Examples of opinion extraction from mobile reviews.

Opinion number	Target	Feature	Opinion	Opinion holder	Time
o_1	Cell phone	Cell phone	Well	27086	2014-12-20
o_2	Business	Service	Excellent, thoughtful, timely	27086	2014-12-20
o_3	Cell phone	Gifts	Very good and practical	27086	2014-12-20
o_4	Call phone	Logistics	Ultrafast	27086	2014-12-20
o_5	Cell phone	Dual standby single pass	Not convenient enough	27086	2014-12-20

aspects involved in the event. Similarly, the company wants to know the consumer's evaluation on each aspect or property of its product so as to improve its quality to satisfy the demand of market.

Featured-based sentiment analysis consists of five subtasks, including opinion target extraction, feature extraction, opinion holder identification, opinion mining, and time extraction. These five subtasks are corresponding to the five elements, i.e., opinion target, feature, opinion, opinion holders and time. This section focuses on two core subtasks: feature extraction and the sentiment classification towards the corresponding features.

1. Aspect extraction and grouping

Aspect extraction aims to uncover all the expressions about different features from the documents, and aggregate the expressions into different feature clusters. For example, the electric product "平板电脑 (tablet computers)" mainly includes various features, such as "电池寿命 (battery life)", "音质 (sound quality)" and "显示器 (display)". In different documents, these features may be in different expressions. For example, the feature "显示器 (display)" can be probably expressed as "屏幕 (screen)". Therefore, it is necessary to cluster the expressions with the same meaning into one feature. This process

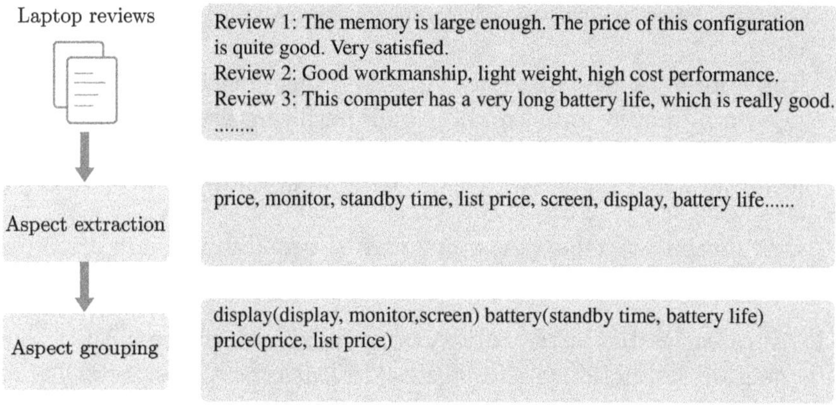

Figure 12.9 Aspect extraction and grouping.

is also called feature classification. Figure 12.9 demonstrates the two main steps in feature extraction and classification.

Technically speaking, aspect extraction can be regarded as an information extraction task. However, in the sentiment analysis task, some special characteristics of language expressions can be used in aspect extraction. For example, the most important characteristics is the opinion expression that describes the opinion holder's opinions towards the target. In an opinionated sentence, the opinion expression and the feature expression often appear at the same time. There are mainly four methods for aspect extraction.

1. The method based on the frequency of noun or noun phrase. This method is simple and effective, which can be easily adapted to other domains. The disadvantage is that it may extract some false candidate features that needs to be further removed by some rules.
2. The method based on the correlation between feature and opinion. This method utilizes the modified relationship between the opinion words and the feature expression in the sentence, but other language tools, e.g., syntactic analysis and dependency parsing are required to identify the modified relationship.
3. Supervised learning method. This method considers aspect extraction as a special case of information extraction. To do this, sequence labeling models (such as HMM and CRF) are used to

extract features, it is usually time-consuming and requires labeling data.

4. Topic model-based method. This is an unsupervised method that does not require labeled data and performs well in some areas (such as product review), but performs poorly in other areas such as social event analysis and public opinion mining.

The method of aspect grouping can be broadly divided into three categories:

(1) **Semantic distance and synonym relationships:** With the help of the dictionary, we aggregate feature expressions that are belonging to a synonym or with close semantic distance into a category;[136]
(2) **Expression similarity:** Some feature expressions with the same feature are similar strings. For example, the first two characters of "screen (显示屏)" and "display (显示器)" are the same. In this case, we can measure the similarity of feature expressions based on the similarity of the strings;[137]
(3) **Clustering and topic models:** Su, Xu, and Guo *et al.* proposed a method of collaborative clustering to aggregate the feature expressions and the sentiment words at the same time. The feature cluster can be formed by aggregating different feature expressions that carry the same meaning; sentiment words cluster can be formed by the similar sentiment words together. The collaborative clustering method cannot only use the similarity information of the feature expressions and that of the sentiment words, but also use the co-occurrence information between the feature expressions and the sentiment words. After collaborative clustering, a feature cluster can be represented by all the relevant feature expressions, while the corresponding sentiment information is determined based on the degree of the association between the feature expressions and the sentiment words.[138]

2. Aspect-based sentiment classification

The objective of the aspect-based sentiment classification is to determine the sentiment orientation or polarity of each extracted feature. The aspect-based sentiment classification task is often based on the assumption that the feature has been extracted, and therefore it is considered to be aware of the feature or the entity. The solution for the aspect-based sentiment classification can be divided into two categories: supervised learning based methods and sentiment dictionary based methods. The disadvantages of supervised learning based methods lie in: (1) the labeled corpus is difficult to be constructed, and it is difficult to adapt to other domains. Since aspect-based sentiment classification is a fine-grained analysis, the classifier trained based on one specific domain would likely perform poorly on the other domain; (2) it is difficult to determine the range of opinion expressions. It is often difficult to determine whether an opinion expression covers its corresponding feature in a sentence. Dictionary based methods can overcome the above problems. In fact, most of the existing aspect-based sentiment classification adopt dictionary-based methods, which make use of sentiment dictionary and intuitive rules to determine the opinion orientation toward a feature. Ding, Liu, and Yu proposed a 4-step method for aspect-based sentiment classification.[139]

(1) **Sentiment word identification:** For those sentences containing features, identify all the sentiment words occurring in the sentence. For the positive sentiment words, assign the sentiment-orientation score as +1; for the negative sentiment words, assign the sentiment-orientation score as −1. For example, given a sentence "The sound quality of the phone is not good, but the battery life is very long", it will be recorded as "The sound quality of the phone is not good [+1], but the battery life is very long [+1]" after the sentiment words are identified. "Good" is a positive sentiment word, and "long" is generally not considered as a sentiment word, but expresses a positive polarity in the context of "battery life".
(2) **Polarity shifter:** Polarity shifters are words or phrases that can shift the polarity of the sentiment orientation. Polarity shifters mainly involve negation words, reinforcement words, and weakened words, etc. After employing the polarity shifters, the sentence becomes as "The sound quality of the phone is not good [−1], but the battery life is very long [+1]" because the negation word "not" appears in front of "good".

(3) **Transition processing:** The sentiment orientation will also be affected by the transition relation in a sentence. Therefore, we need to process the cases of transition relations. There are some words frequently used to indicate the transition relations, including "but", "however", "nevertheless", etc. These transition indicators usually divide the sentence into two parts. A straightforward way for processing transition is: when it is difficult to determine the sentiment orientation of one part, then its polarity can be determined as the opposite of the sentiment orientation of the other part.

(4) **Sentiment aggregating:** This step uses a sentiment clustering function to aggregate the sentiment orientations towards the specific feature from all the sentences to obtain the sentiment orientation of the feature. Suppose there is a feature set $\{a_1, \ldots, a_m\}$ and a set of sentiment words $\{sw_1, \ldots, sw_n\}$ in the sentence s. The sentiment orientations of the words in the sentiment word set have been determined by the above three steps, then the sentiment orientations of the sentence s for the feature is calculated by the following sentiment clustering functions:

$$\text{score}(a_i, s) = \sum_{j=1}^{n} \frac{sw_j.\text{SO}}{\text{dist}(sw_j, a_i)}, \tag{12.25}$$

where $\text{dist}(sw_j, a_i)$ denotes the distance between the sentiment word sw_j and the feature expression a_i in the sentence s, and $sw_j.\text{SO}$ denotes the sentiment orientations of the sentiment word sw_j in the sentence s. The sentiment clustering function in Formula (12.25) can also be replace by other clustering functions. For example, add up or multiply the sentiment orientations of all the sentiment words appearing in the sentence.

3. Target, opinion holder and time extraction

Target, opinion holder, and time extraction can be considered to be the problem of named entity recognition. Named entity recognition has been widely studied in the fields of information retrieval, text mining, data mining, machine learning, and natural language processing. There are two main methods for named entity recognition: rule-based methods and sequence labeling methods. In the early stage, named entity recognition uses manually or automatically constructed rules to identify entities from the text, and existing methods for named entity recognition are mainly based on sequence labeling models, such as HMM and CRF. As to the time extraction, most systems will record the time when the user posts the texts, so it is an easy task for obtaining the time. What needs to be done is to normalize the times indifferent formats. In the area of sentiment analysis, in many application scenarios, the opinion holder is the user who publishes the texts. In some cases, the opinion holder appears in the published texts, and then opinion holder extracted is required. Kim and Hovy confine the opinion holders to person names or organization names, and use named entity marker to identify them. Choi, Breck, and Cardie use the CRF model to extract the opinion holders. To train a CRF model some classic features are employed, including sentiment words, part-of-speech, contextual words, etc. Johansson and Moschitti use the SVM model to extract the opinion holders. Other studies use semantic role labeling (SR) methods to identify the opinion holder.[140–144]

12.3.3. *Summary*

The main purpose of sentiment analysis is to discover sentiments, opinions, emotions, and other subjective information from the textual data. This section describes two types of tasks: sentiment classification and aspect-based sentiment classification. Sentiment classification is one of the most typical tasks in sentiment analysis, which attempts to classify the document into positive, negative, or neutral ones. Aspect-based sentiment classification is to discover fine-grained sentiment information from the perspective of information extraction, including opinion holder, target, feature, opinion, and time. In addition, sentiment analysis also includes some fundamental tasks such as the construction of sentiment dictionary. In sentiment analysis, sentiment dictionary is the most important resource widely used in rule-based sentiment analysis. In supervised learning methods, sentiment word is also used as an important feature. Some popular sentiment dictionary resources are described in Table 12.7.

For readers interested in gaining a comprehensive understanding of sentiment analysis, it is recommended to read a classic survey

Table 12.7 Sentiment dictionary resources.

Sentiment dictionary	Instructions
General inquirer	Manually annotated sentiment dictionary, with each word being labeled as positive or negative
Opinion lexicon	English sentiment dictionary provided by Bing Liu *et al.*, consists of about 6,800 words
MPQA	Sentiment dictionary resources used by the OpinionFinder system
SentiWordNet	A sentiment dictionary extended from WordNet with each word being labeled as positive, negative, or neutral
Emotion lexicon	Each word is labeled as positive or negative, and as one of the eight basic emotions
HowNet	HowNet Chinese sentiment dictionary

paper on sentiment analysis[145] and a systematic and comprehensive book on sentiment analysis and opinion mining.[146]

12.4. Case Studies and Exercises

1. APP classification

There are many types of APPs on the market today, including social, reading, gaming, and ecommerce. We classify the APPs according to their name and specific description. The accurate classification of the APPs can help the building of the APP label system, which will facilitate offering accurate recommendations through APP label system to target customers in marketing.

The dataset used in this exercise is crawled from the Web and contains information about 1000 APPs. Each APP contains two features: one is description, which represents the information of the APP; the other is the target feature category, which represents the category to which the APP belongs. Please train a APP classification model to predict APP categories.

2. Sentiment classification of movie review
Sentiment classification plays a very important role in assisting decision-making in many areas such as public opinion monitoring. Analyzing movie reviews posted by users on the Internet can help us know the quality of movies, and predict the box office. The movie review dataset used in this case contains 1,429 movie reviews, and each review has its sentiment label (1 represents positive, 0 represents negative). Please train a sentiment classifier to automatically determine the sentiment polarity of the movie review.

3. Polarity classification of Chinese online reviews
Online shopping has played an increasingly important role in our daily lives. When we buy a product, our first move is to refer to the corresponding review. Each review has its emotional tendency, whether it is positive, negative or neutral. By analyzing the reviews, the probability of making better decisions will increase.

This case provides a comment data set captured from an ecommerce platform, including 2 features: one is Comment, which represents the text content of the comment; the other is Class, which represents the emotion category to which the comment belongs. Each comment has been marked as positive (1), negative (−1) or neutral (0) according to the emotion category it contains. Please construct a classifier to determine the emotion category of user comments.

4. Spam message filtering

Some advertisers use short message service (SMS) text messages to target potential consumers and send them ads they don't need, the so-called spam messages. Spam messages often disrupt our normal lives. With the knowledge of text analysis, you can construct a classifier that automatically identify spam messages.

The spam data set used in this case has a total of 2 columns: one column records the text content of the SMS, and the other is the label of the text content (spam/nonspam). The dataset contains 5,568 samples. Please identify spam messages based on text content.

Chapter 13

Graph and Network Analysis

In the past few decades, people have become increasingly interested in the "connectivity" formed in the development of modern society. The core of this interest lies in the network. Network is a pattern of the relationship between things. In practical applications, various networks have emerged, such as social networks, the Internet, semantic networks, communication networks, transaction networks and industrial internet.

Figure 13.1 shows an example of a karate club network. In the early 1970s, American sociologist Zachary spent two years observing the social relations among 34 members in a karate club of an American university.[149] He constructed a network of relationships between members based on the internal and external interactions of club members. The network consists of 34 nodes, each of which represents a member. If two members are frequently contacted friends, there is an edge between the corresponding nodes. Due to the dispute between the club coach (node 1) and the supervisor (node 34), the club splits into two separate clubs with their own lead. The members in the two small clubs are represented by pink and green nodes respectively.

This chapter first discusses some basic concepts of graph theory in Section 13.1. We will then introduce centrality in Section 13.2, and discuss how to measure the importance of nodes in the network. Section 13.3 discusses link analysis, including PageRank and HITS algorithm. Section 13.4 discusses community discovery, i.e., how to find community structures from the network. Finally, we introduce knowledge graph in Section 13.5, a frontier topic in network analysis.

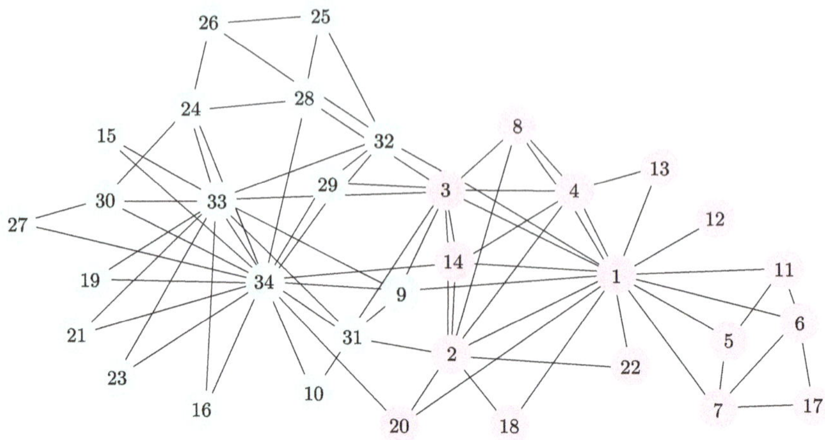

Figure 13.1 The karate club network.

13.1. Basic Concepts

Graph theory, as the basis of the network, was first proposed by the famous Swiss mathematician Euler for the Seven Bridges of Königsberg problem. Today, graph theory has become an important branch of discrete mathematics and is widely used in various fields. The Seven Bridges of Königsberg is one of the famous classical mathematical problems of the 18th century. Euler laid the foundation of graph theory by solving this problem. In a park in Königsberg, there are seven bridges in the Pregel River that connect two islands and riverbanks (as shown in Figure 13.2(a)). Can you start from any of the four lands, crossing each bridge only once and returning to the starting point again?

Euler studied and addressed this problem in 1736. He attributed the problem to the "one stroke" problem shown in Figure 13.2(b), which gives the necessary and sufficient conditions of "one stroke" problem for the connected graph: the number of nodes with odd edges is 0 or 2.

13.1.1. *Basic Definition*

The graph here is a tool that represents several objects and the relationships between these objects in an abstract form. The graph can

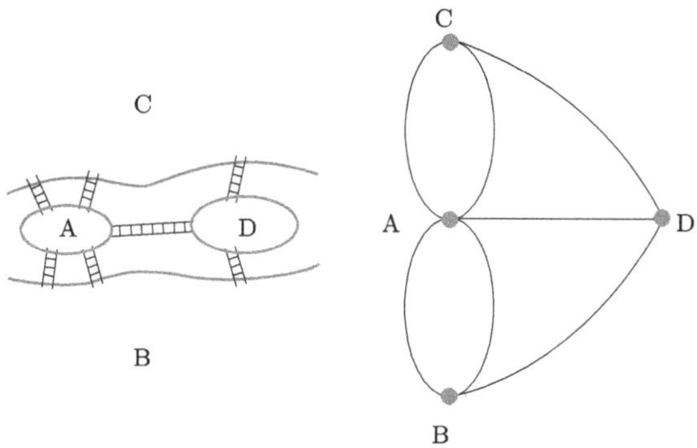

Figure 13.2 The seven bridges of Königsberg problem. (a) The seven bridges of Königsberg. (b) Multigraph of four nodes and seven edges.

be represented as a tuple $G = (V, E)$, where V is a set of nodes, E represents a set of edges. Each edge is composed of two nodes in V. The two nodes are called the endpoints of the edge. The two endpoints of an edge can be the same or different in V. An edge with two same endpoints is called rings.

If the edges are undirected, G is an undirected graph. If the edges are directional, G is a directed graph. In the undirected graph $G = (V, E)$, for any $v \in V$, the sum of the times when v is the endpoint of G is the degree of v, denoted as $\deg(v)$. In the directed graph $G = (V, E)$, for any $v \in V$, the number of times of v as the starting point of G is the out-degree of v, denoted as $\deg^+(v)$. The in-degree of v is the number of times v is the end point in G, which is denoted as $\deg^-(v)$. The degree of node v in the directed graph is the sum of the in-degree and the out-degree, i.e., $\deg(v) = \deg^+(v) + \deg^-(v)$.

Figure 13.3 shows an example of an undirected graph and a directed graph. For node 3 in Figure 13.3(a), its degree is 4. For node 3 in Figure 13.3(b), its in-degree is 3, out-degree is 1, and its degree is 4.

There are many ways to represent graphs based on the actual applications. The two most common representations are adjacency list and adjacency matrix.

Given a graph $G = (V, E)$, its adjacency list is a collection of lists, where the set of neighbors of a node in the graph is described in each list. For an unweighted graph $G = (V, E)$, where $|V| = n$, the

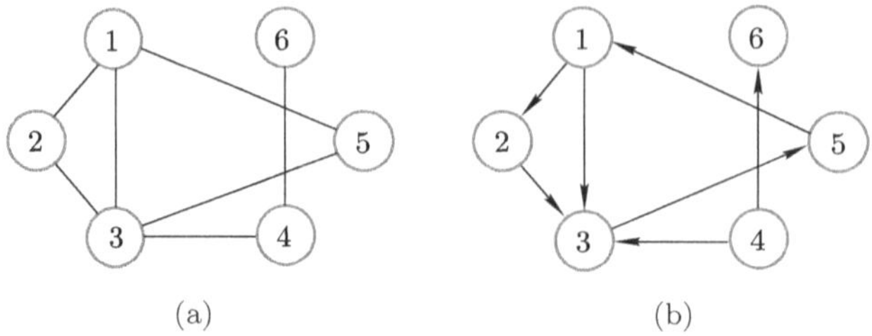

Figure 13.3 Examples of directed graph and undirected graph. (a) Undirected graph. (b) Directed graph.

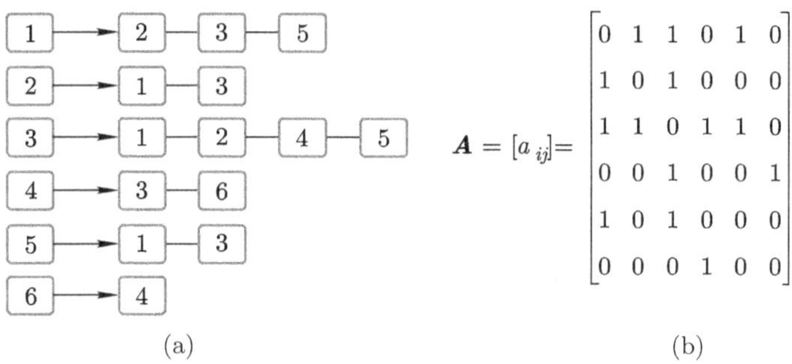

Figure 13.4 Two graph representations. (a) Adjacency list. (b) Adjacency matrix.

nodes in V are $v_1, v_2, \ldots, v_n$. The adjacency matrix of G is denoted as $\boldsymbol{A} = [a_{ij}]$, where $\boldsymbol{A}$ is a $n \times n$ (0-1)-matrix. The elements of $\boldsymbol{A}$ are defined as follows: if there is an edge $e = (v_i, v_j) \in E$, then $a_{ij} = 1$; otherwise, $a_{ij} = 0$. If G is a weighted graph, for each edge $e = (v_i, v_j) \in E$, $a_{ij} = w(e)$, where $w(e)$ is the weight of e. Figure 13.4 shows the adjacency list and adjacency matrix of the graph shown in Figure 13.3(a).

13.1.2. *Commonly Used Graphs*

1. ER random graph

The ER random graph is a random graph generated by a random graph model proposed by the famous Hungarian mathematicians Erdoš and R'enyi in 1959.[150] ER random graph provides good references for studying properties satisfied by any graphs.

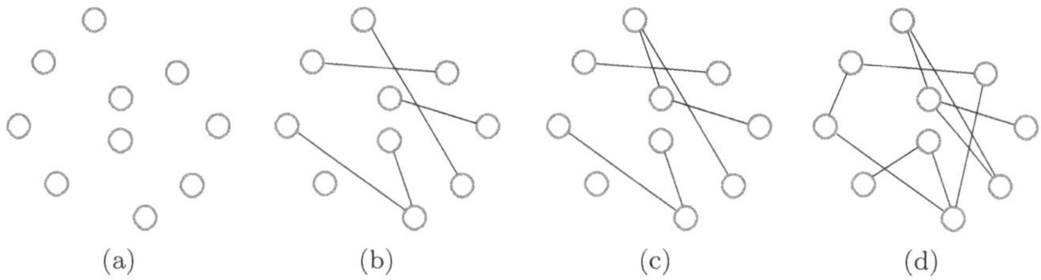

Figure 13.5 Examples of ER random graph. (a) $G(10,0)$. (b) $G(10,0.1)$. (c) $G(10,0.15)$. (d) $G(10,0.25)$.

In this model, a graph is generated as follows. Given n nodes and a probability $p(0 \leqslant p \leqslant 1)$, each node pair is connected with an undirected edge with probability p. The existence of edges is independent to each other. The graph is denoted as $G(n,p)$. Figure 13.5 shows the examples of $G(10,0)$, $G(10,0.1)$, $G(10,0.15)$ and $G(10,0.25)$ respectively.

As the most classic random graph model, ER random graphs are widely used. However, there are two common properties of real-world networks that are not satisfied by ER random graphs. The first one is clustering. Real-world networks often aggregate into several communities, which is difficult to observe in ER random graphs. The second one is centrality. In the ER random graph, the node degree will not be very high, while a real-world network usually contains some nodes with high degrees.

2. Small-world network

To reflect the aggregation of real-world networks, Watts and Strogatz proposed the Watts-Strogatz model (WS model).[151] The graph generated by WS model shows high clustering and small average shortest path length. Since the WS model generates a graph that is more in line with the small world phenomenon than the ER random graph, the graph generated by the WS model is called the small-world network.

The small world phenomenon was derived from a well-known experiment conducted by Harvard University's famous social psychologist Stanley Milgram in the 1960s.[152] In this experiment, Milgram randomly found 160 people in Omaha, Nebraska, USA and asked each of them to send a package to a stock trader in Boston. When forwarding the package, Milgram requires that the participants can just forward the package to people they know. Surprisingly, Milgram found that the first package was sent to Boston in only four days and was only forwarded twice. Overall, these packages were forwarded 2 to 10 times, and the average forwarding time was 5. Milgram concluded that the average distance between any two Americans at that time was 6, and that the average time that one person could contact any other person is 5. This phenomenon has now been gradually proved in many other networks.

3. Scale-free network

In practice, there are some graph whose degree distribution follows the power law. In statistical physics, the phenomenon of following the power law is called the scale-free phenomenon, that is, the scale of the individual in a system is quite different, lacking a characteristic scale. Thus, a network that follows the power law is also called a scale-free network.

The power law distribution can be expressed as $y = cx^{-r}$, where x and y are two variables, and c and r are constants greater than zero. It can be seen that the change of the variable x causes the variable y to change according to the power index of x. If we take the logarithm of both sides of the above formula, we can find that $\ln x$ and $\ln y$ have the linear relationship $\ln y = \ln c - r \ln x$.

A typical power law distribution is the word frequency distribution. In 1932, Zipf from Harvard University studied the frequency of English words and found that the frequency of a word has a simple inverse relationship with constant power of ranking. This relationship is called Zipf's law. Zipf's law states that only a very small number of words are used frequently, while most words are rarely used.

In addition to Zipf's law, power law distribution has been found in many fields such as physics, earth and planetary science, computer science, social sciences, and economics and finance. However, the power law distribution is not followed by ER random graphs or WS random graphs. Hungarian scientists Barabási and Albert proposed the Barabási–Albert model (BA model) to generate scale-free networks.[153]

The BA model generates a network as follows: first, initialize m_0 connected points; then, add nodes one by one, the probability that the added node is connected to an existing node is proportional to the degree of the existing nodes. Formally, for node i, the probability p_i of adding an edge between a newly added node and node i is

$$p_i = \frac{\deg(i)}{\sum_j \deg(j)}, \tag{13.1}$$

where $\deg(i)$ is the degree of node i and $\sum_j \deg(j)$ is the sum of the degrees of all existing nodes.

In general, the BA model can continuously increase the size of the graph. The nodes added by the model are more likely to be connected with existing high-degree nodes. The graph generated by the BA model also satisfies the following properties: the node degree follows the power law distribution $p(k) \sim k^{-3}$; the average path length l follows $l \sim \frac{\ln n}{\ln \ln n}$.

13.2. Geometric Property

To understand the network, scientists have proposed many metrics including centrality, clustering coefficients and modularity.

13.2.1. *Centrality*

Centrality is used to measure the importance of a node in the graph. Centrality is not an attribute of the node itself, but a network structural attribute. Centrality can be used to solve problems in different areas, such as finding the most influential users in social networks, finding critical infrastructure on the Internet or urban networks, discovering super-spreaders in disease networks and so on.

So how to measure the centrality? This section will introduce four different centrality metrics: degree, betweenness, closeness, and eigenvector centrality.

1. Degree centrality

Degree centrality measures the node importance by its number of neighbors. The idea behind degree centrality is quite simple. In the network, a node interacts with other nodes through its neighbors. Therefore, the more neighbors a node has, the more information the

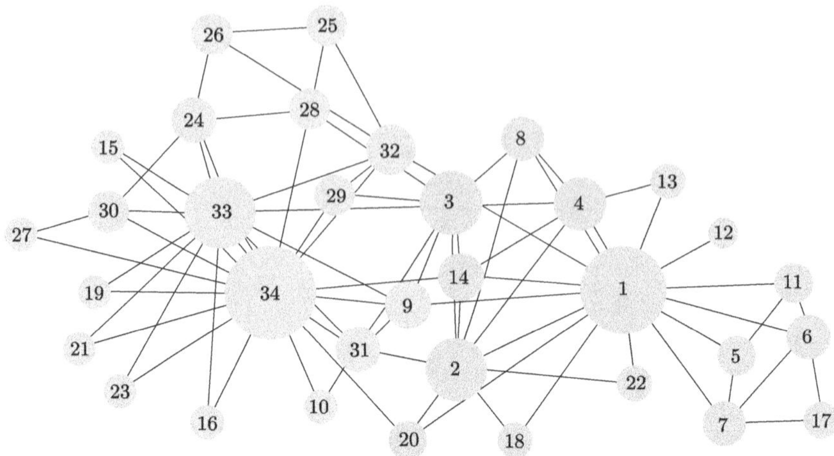

Figure 13.6 The degree centrality of nodes in karate club network.

node can transmit to the outside, and the easier it is to receive information from the outside of the network. Taking social network as an example, when degree centrality is used to represent user's influence, the more friends a user has, the greater influence the user has.

The calculation of degree centrality is also quite simple. Assuming the network is represented as a graph $G = (V, E)$, the degree centrality of node v is

$$C_d(v) = \deg(v). \tag{13.2}$$

The degree centrality often needs to be normalized. In a graph containing $|V|$ nodes, the maximum node degree is $|V| - 1$. The normalized degree centrality is calculated as

$$C_d^{\text{norm}}(v) = \frac{\deg(v)}{|V| - 1}. \tag{13.3}$$

For directed networks, we can further define out-degree centrality and in-degree centrality.

Figure 13.6 shows the distribution of the degree centrality in the karate club network shown in Figure 13.1. The larger the node is, the larger the node's degree centrality is.

The degree centrality only uses local information, that is, the number of neighbor nodes directly connected to the node. The centralities that will be introduced next incorporate global information

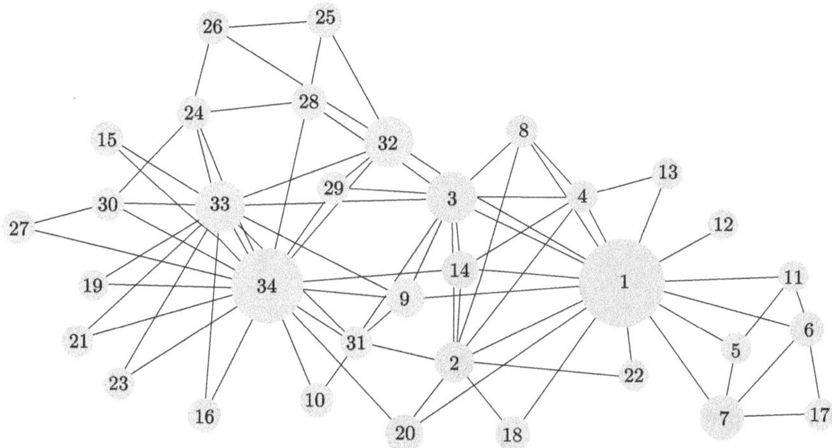

Figure 13.7 The betweenness of nodes in karate club network.

of the node in the graph. For example, betweenness measures the importance of a node by counting the number of shortest paths passing through it. Closeness further examines the length of these shortest paths.

2. Betweenness

It is a natural idea to use the bridge role that a node plays in connecting other nodes to measure its importance. Betweenness is such a centrality, which is first proposed by Freeman to quantify the ability of individuals to control the communication between other people in a social network.[154] For node v in the graph $G = (V, E)$, the betweenness centrality is

$$C_b(v) = \sum_{\{s,t\} \in V \backslash v} \frac{\sigma_{st}(v)}{\sigma_{st}}, \tag{13.4}$$

where σ_{st} represents the number of shortest paths from node s to t. $\sigma_{st}(v)$ represents the number of shortest paths from node s to t passing through node v. Obviously, the betweenness centrality characterizes how many shortest paths pass through node v in G. Figure 13.7 shows the distribution of betweenness centrality of the karate club network. The larger the node is, the larger the betweenness *centrality* is.

Betweenness can also be extended to edge. Suppose the two end-points of edge e are i and j, σ_{st} represents the number of shortest paths between node s and t, and $\sigma_{st}(e)$ represents the number of shortest paths passing through edge e, then the betweenness centrality of edge e is

$$C_b(e) = \sum_{s \neq t \in V \backslash \{i,j\}} \frac{\sigma_{st}(e)}{\sigma_{st}}. \tag{13.5}$$

3. Closeness

In a connected graph, the distance between two nodes can be measured by the length of the shortest path between them. The closeness of a node is defined as

$$C_c(v) = \frac{1}{\sum\limits_{y \in V \backslash v} d(y, v)}. \tag{13.6}$$

The closer a node is to the center of the graph, the smaller the sum of the distances between the node and other nodes in the graph will be, so that the larger the closeness centrality is. When the graph is not a strongly connected graph, the node closeness is defined as

$$C_c(v) = \sum_{y \in V \backslash v} \frac{1}{d(y, v)}. \tag{13.7}$$

Figure 13.8 shows the distribution of the closeness in the karate club network. The larger a node is, the larger the closeness will be.

4. Eigenvector centrality

The eigenvector centrality is mainly used to measure the influence of nodes in the network. Each node in the network is assigned an influence score. If a node is connected to nodes with high scores, its score tends to be high. If a node is connected to low-score nodes, its score tends to be low. PageRank (refer to Section 13.3) is a variant of the eigenvector centrality, which is mainly used to measure the importance of web pages on the Internet. Assuming that $\boldsymbol{A} = [a_{ij}]$

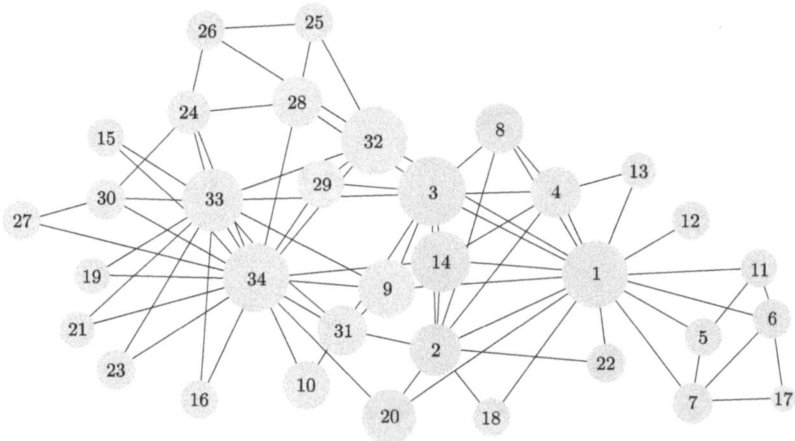

Figure 13.8 The closeness of nodes in karate club network.

is the adjacency matrix of the graph $G = (V, E)$, the eigenvector centrality of node v is defined as

$$x_v = \frac{1}{\lambda} \sum_{t \in V \setminus v} a_{vt} x_t, \tag{13.8}$$

where λ is a positive constant. Eq. (13.8) can be written as a vector form

$$\boldsymbol{A}\boldsymbol{x} = \lambda \boldsymbol{x}. \tag{13.9}$$

We can see that eigenvector centrality is the eigenvector[a] of matrix $\boldsymbol{A}$. If $\boldsymbol{A}$ contains multiple eigenvectors, the eigenvector corresponding to the largest eigenvalue will be used. Figure 13.9 shows the distribution of eigenvector centrality of nodes in the karate club network. The larger the node is, the larger the eigenvector centrality will be.

13.2.2. *Clustering Coefficient*

Clustering coefficient is a measure of the degree to which nodes in a graph tend to cluster together. There are two kind of clustering coefficients: node clustering coefficient and network clustering coefficient.

[a]The definition of eigenvalue and eigenvector of a matrix can be referred to Appendix A.

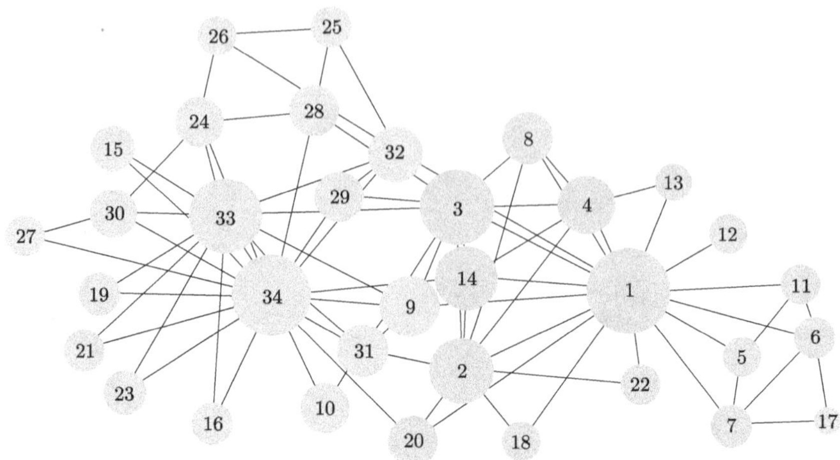

Figure 13.9 The eigenvector centrality of nodes in karate club network.

Given a node i, its clustering coefficient is the ratio of the number of edges between its neighbors to the maximum possible number of edges between these neighbors, i.e.,

$$C_i = \frac{2|E_i|}{k_i(k_i - 1)}, \tag{13.10}$$

where k_i is the degree of node i, and E_i is the number of connected edges between the k_i neighbors of node i.

Given a graph $G = (V, E)$, its network clustering coefficient is the ratio of the number of edges in the graph to the maximum possible number of edges between its nodes, i.e.,

$$C = \frac{2|E|}{|V|(|V| - 1)}. \tag{13.11}$$

Obviously, the node clustering coefficient C_i and the network clustering coefficient C are both greater than 0 and less than 1. When $C = 0$, it means that all nodes in the network are independent nodes, that is, there is no edge in the network. When $C = 1$, it means that the network is a complete graph, namely, there is a edge between any two nodes in the network.

For the network shown in Figure 13.10, the degree of node 3 is 5, and there are 6 edges between its five neighbors, so the node

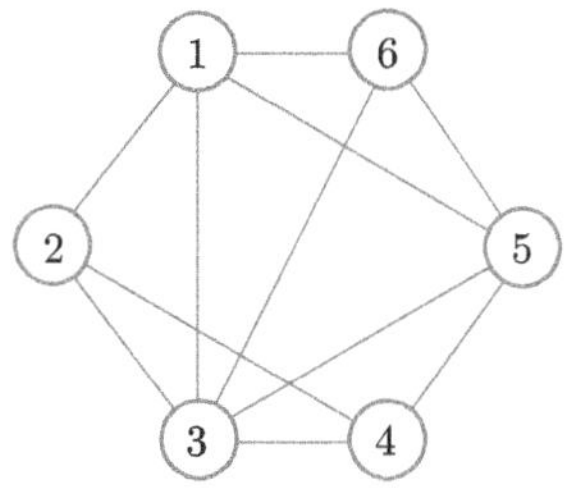

Figure 13.10 Example of clustering coefficient.

clustering coefficient of node 3 is $C_3 = \frac{2\times 6}{5\times 4} = 0.6$. For the network, it contains 6 nodes, 11 edges, so its network clustering coefficient is $C = \frac{2\times 11}{6\times 5} = 0.73$.

13.2.3. *Modularity*

In social networks, a common requirement is to divide nodes into several communities according to their characteristics. Newman and Girvan proposed the concept of modularity in 2004 to evaluate the result of community division.[155] Modularity of a network is the fraction of edges that fall within communities minus the expected fraction at a random network where the degree distribution is consistent with the original network.

To formally describe the concept of modularity, we first discuss the case of only two communities (community 1 and community 2). Given an undirected unweighted graph $G = (V, E)$ and its adjacency matrix $\boldsymbol{A} = [a_{ij}]$, we define a member variable s, when node v belongs to community 1, $s_v = 1$; when node v belongs to community 2, $s_v = -1$.

In order to calculate the expected fraction of edges that fall within communities at a random network where the degree distribution is consistent with the original graph, we divide each edge into two halves, each of which is called a partial edge. The total number of partial edges here is $l_{|V|} = \sum_v \deg(v) = 2|E|$. Then, we randomly reconnect each partial edge to the partial edge of another edge. Thus, in a random network where the degree distribution is consistent with the original graph, the expected number of edges between node u and

v is $\frac{\deg(u)\deg(v)}{2|E|}$. Therefore, the modularity between the two communities is

$$Q = \frac{1}{2|E|}\sum_{u,v}\left(a_{uv} - \frac{\deg(u)\deg(v)}{2|E|}\right)\frac{s_u s_v + 1}{2}. \tag{13.12}$$

Generalizing Eq. (13.12) to the case of multiple communities, the definition of modularity is:

$$Q = \sum_{c\in C}\left[\frac{l_c}{|E|} - \left(\frac{D_c}{2|E|}\right)^2\right], \tag{13.13}$$

where l_c is the number of internal edges in community c, D_c is the sum of node degree in community c. D_c can also be written as $D_c = 2l_c + O_c$, where O_c is the edge between community c and other communities.

13.3. Link Analysis

Link analysis refers to the analysis method based on the edge in the graph. In web search, link analysis is also called hyperlink analysis. Link analysis has been widely used in various fields, such as search engines, website design, network health check and knowledge mining. We will mainly discuss two common link analysis algorithms: PageRank and HITS.

13.3.1. *PageRank*

1. Algorithm introduction

The PageRank algorithm is a page ranking algorithm on the Google search engine designed by Google founders Larry Page and Sergey Brin. It was first published in a paper in 1998.[156]

PageRank is mainly used to calculate the importance of each website on the Internet. First, all the web pages on the Internet are modeled as a graph. The node in this graph is a web page, and a hyperlink on the web page correspond to a directed edge between two nodes.

For web page a, its PageRank value is denoted as $\mathrm{PR}(a)$, which indicates the probability that a user randomly jumps to page a by clicking the hyperlink. For example, a page with a PageRank value of 0.5 indicates a probability of 0.5 for jumping to the page from other pages. PR(a) can also represent the importance of page a.

Figure 13.11 shows an example network and the PageRank value for each web page. It is assumed that when the user browses a web page, the user can jump to another web page by clicking on a hyperlink with a probability of 85%, and randomly jumps to another web page with a probability of 15%. The 85% probability of jumping by clicking a hyperlink is called the damping coefficient. Without the damping coefficient, the PageRank value of all web pages except page a, b and f in Figure 13.11 will be 0. We will introduce the definition of damping coefficient in detail in the following section.

In the network shown in Figure 13.11, the PageRank value is 33.7% for web page b, and 7.6% for web page c. Although there are more hyperlinks pointing to web page c than web page b, the hyperlink pointing to page b comes from page with a high PageRank value, so the PageRank value of page b is higher than the web page c.

PageRank is an iterative algorithm with multiple iterations. At the beginning, the PageRank value of each node is set to $\frac{1}{|V|}$, where $|V|$ is the number of nodes in the graph. In each iteration, each node v passes $\frac{1}{\deg^+(v)}$ of its PageRank value to all of its neighbors. For example, for node c in Figure 13.11, it will pass 2.53% of its PageRank value along each of its three out edges to a, d, and e.

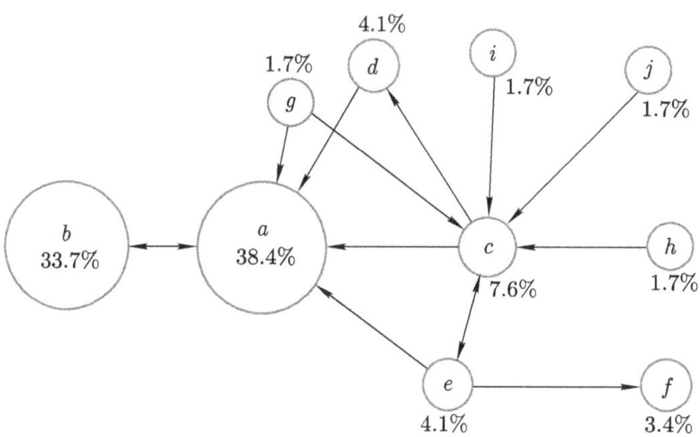

Figure 13.11 PageRank values of a example network.

Then, after the $(t+1)$-th iteration, the PageRank value of each node v is

$$\mathrm{PR}(v, t+1) = \sum_{u \in N^-(v)} \frac{\mathrm{PR}(u, t)}{|N^+(u)|}, \tag{13.14}$$

where $N^-(v)$ is the set of nodes pointing to node v, and $N^+(u)$ is the set of nodes that u points to. In Eq. (13.14), the PageRank algorithm actually assumes that the user is constantly jumping to the next page by clicking on the hyperlink. Thus, the importance of each page is completely related to the number of the pages pointing to it and their importance. In a real scenario, the user may not jump along the hyperlink but will open a separate web page. To this end, the PageRank algorithm defines the damping coefficient to account for this phenomenon.

2. Damping coefficient

In PageRank, β is used to indicate the probability that a node jumps to another node along its out edge. Thus, $(1 - \beta)$ means the probability of randomly picking the next node. Therefore, the PageRank value of a node v is

$$\mathrm{PR}(v, t+1) = \frac{1-\beta}{|V|} + \beta \sum_{u \in N^-(v)} \frac{\mathrm{PR}(u, t)}{|N^+(u)|}. \tag{13.15}$$

Practical experiments prove that 85% is a suitable choice of β.

3. Vectorization of the PageRank algorithm

The PageRank algorithm can also be expressed as a vector form. Given a graph G and its adjacency matrix $\boldsymbol{A} = [a_{ij}]$. The definition of the element a_{ij} in $\boldsymbol{A}$ is:

$$a_{ij} = \begin{cases} \dfrac{1}{|N^+(j)|}, & \text{there exists an edge from } j \text{ to } i; \\ 0, & \text{otherwise,} \end{cases} \tag{13.16}$$

where the column sum of $\boldsymbol{A}$ is 1. The PageRank values in the t-th iteration form a column vector $\boldsymbol{r}^t = (\mathrm{PR}(v_1, t), \mathrm{PR}(v_2, t), \ldots,$

$\mathrm{PR}(v_{|V|}, t))^{\mathrm{T}}$. Thus, the PageRank algorithm can be expressed as follows:

$$\boldsymbol{r}^{t+1} = \beta \boldsymbol{A}\boldsymbol{r}^t + \frac{(1-\beta)}{|V|}\mathbf{1}. \tag{13.17}$$

Although PageRank is the most classic link analysis algorithm, it has some limitations. One limitation is that it only reflects the general popularity of a page, but not the authoritativeness of a page. For example, search engines (Google, Baidu, etc.) and navigation sites (such as "hao123") usually have high PageRank values, but they lack relevance for queries on specific topics. Other link analysis algorithms such as the topic-sensitive PageRank algorithm and the HITS algorithm have been proposed to overcome this limitation.

13.3.2. *Topic-sensitive PageRank*

The topic-sensitive PageRank algorithm is an improvement to PageRank.[157] Its main idea is to define a set of topic-based offset vectors, where each offset vector corresponds to a topic and is used to consider the relevance of the query to the topic in the process of calculating the PageRank value.

In the data preprocessing stage, the topic-sensitive PageRank algorithm first determines the topic categories. In the initial research, Haveliwala used the open directory management system to divide all web pages into 16 categories according to their themes, including art, business, computer, games and so on. In practice we can use other classification methods. Then the topic-sensitive PageRank algorithm updates the PageRank value of each page under different topics. The update formula is as follows:

$$\boldsymbol{r}^{t+1} = \beta \boldsymbol{A}\boldsymbol{r}^t + (1-\beta)\frac{\boldsymbol{s}}{|\boldsymbol{s}|}, \tag{13.18}$$

where $\boldsymbol{s}$ is such a vector: for a topic if page k is in this topic, then the k-th element of $\boldsymbol{s}$ is 1, otherwise 0. Note that each topic has a corresponding $\boldsymbol{s}$, and $|\boldsymbol{s}|$ indicates the total number of 1 in $\boldsymbol{s}$.

When a user submits a search query, the topic-sensitive PageRank algorithm first determines the user's topical tendency to select the appropriate topic vector $\boldsymbol{s}$. In the initial research, Haveliwala used the naive Bayesian method to determine the topic that each query is most likely to belong to. There are many other ways to do this. We can create a theme menu, from which the user can select the item of interest. This method is often used when registering on some social Q&A websites. We can also track user behavior (such as cookie) and analyze the data to determine the user's tendency. The historical information of all queries can also be used. For example, when one user first queries "basketball" and then "Jordan", we can judge the user query is the basketball star Michael Jordan.

13.3.3. *HITS Algorithm*

The HITS algorithm is proposed by Jon Kleinberg.[158] It introduces two scores for each node in the network: hub score and authority score. Web page with high hub value serve as a hub, which contains many hyperlinks that link to other web pages. Web page with high authority value is often linked to by other web pages, especially hub pages. These two values are interdependent and mutually influential. The hub score is the sum of the author scores of all the web pages that the current web page links to. The authority score is the sum of the hub score of the web pages linked to the current page.

Given a node v in the network $G = (V, E)$, assuming that its hub score and authority score are hub (v) and auth (v), which are defined as follows:

$$\text{hub}(v) = \sum_{u \in N^{+}(v)} \text{auth}(u), \tag{13.19}$$

$$\text{auth}(v) = \sum_{u \in N^{-}(v)} \text{hub}(u). \tag{13.20}$$

The HITS algorithm is also an iterative algorithm for a particular query. Initially, given a query, HITS first finds several pages to form a root set. Then, it finds all the pages linked to the root set and some web pages that link to the root set. These two types of web pages form the base set. The web pages in base set and the hyperlinks between them form a graph, which is the focus subgraph for user query.

During the first iteration, HITS performs two update steps: authority score update and hub score update. The authority score update is to calculate the authority score of a node based on the hub scores of the nodes pointing to it. The hub score update is to calculate the hub score of a node based on the authority scores of the nodes it points to. To ensure convergence, HITS will normalize the authority scores and the hub scores according to Eq. (13.21) and Eq. (13.22).

$$\text{hub}^*(v) = \frac{\text{hub}(v)}{\sqrt{\sum_{w \in V} \text{hub}^2(w)}}, \tag{13.21}$$

$$\text{auth}^*(v) = \frac{\text{auth}(v)}{\sqrt{\sum_{w \in V} \text{auth}^2(w)}}. \tag{13.22}$$

13.4. Community Discovery

In practice, the community structure is a common character of the network. The network is composed of many communities. For example, the karate club network shown in Figure 13.1 is a typical network containing two communities.

There is no precise definition of community. The most widely accepted definition was proposed by American scientists Newman and Girvan in 2001.[159] A community means the connection between nodes inside it is very dense, and the connection of nodes in different communities is sparse.

There are many effective algorithms for community discovery. The two most widely used types of algorithms are: hierarchical clustering-based algorithms and *modularity*-based algorithms. This chapter will give a brief introduction to the typical representatives of them.

13.4.1. *Algorithms Based on Hierarchical Clustering*

1. Givan–Newman algorithm

The Givan–Newman (GN) algorithm is a split-type community discovery algorithm.[160] The basic idea is, according to the characteristics of high internal cohesion within the community and low cohesion between communities, gradually remove the edges between communities and obtain relatively cohesive community structure. By gradually removing the edge with highest betweenness, the network is split into several communities.

If an edge bridges two communities, the fraction of the shortest paths between two nodes in different communities will be highest. The betweenness of the edge will be the largest. Thus, we can obtain two communities by removing the edge with the largest betweenness. The GN algorithm works as follows:

(1) Calculate the betweenness centrality of each edge in the network;
(2) Remove the edge with the largest betweenness in the network;
(3) Update the betweenness centrality of all edges in the network;
(4) Repeat steps (2) and (3) until all edges in the network are removed.

It should be noted that the updating of betweenness in step (3) cannot be omitted. If there are multiple edges connecting between two communities, then it cannot be guaranteed that the betweenness of all these edges are high enough.

2. Newman's fast algorithm

As a global search algorithm, GN algorithm has strong practicability. However, there is no quantitative definition of community in the network. Therefore, it cannot directly judge whether the community structure is consistent with the real community structure. Additional information is needed to determine whether the obtained community structure is practical. In addition, the GN algorithm cannot determine the termination condition of the algorithm.

To solve this problem, Newman introduced the concept of modularity (refer to Eq. (13.13)). Based on this concept, the Newman's fast algorithm was proposed.[155] The Newman's fast algorithm works as follows:

1. Treat each node in the network as a community, and the initial value of modularity Q is 0;
2. Computing the increment of modularity ΔQ if merging two communities which are connected by any edge. Then merge the two communities with largest (or least) ΔQ;
3. Repeat step (2) until the network is merged into a single community.

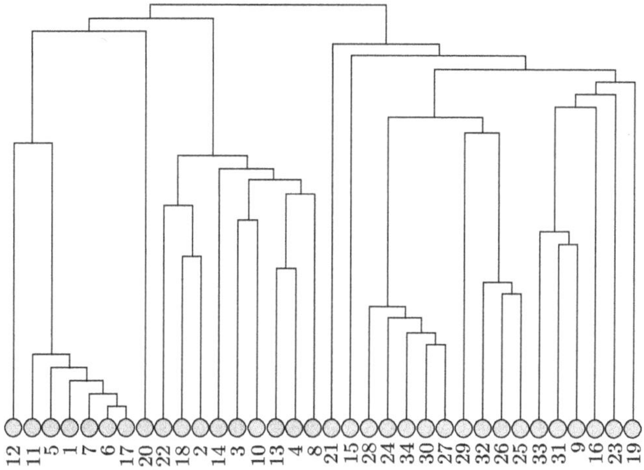

Figure 13.12 The community partition results of the karate club network by Newman's fast algorithm.

The result of community partition can be visually expressed by a dendrogram, in which the optimal partition is the level with highest modularity. Figure 13.12 shows the partition results of the karate club network by Newman's fast algorithm. The circles of different colors represent two real communities. We can see that node 10 is mis-divided.

13.4.2. *Algorithm Based on Modularity Optimization*

As the size of the real-world network increases, the computation time of the above methods becomes increasingly longer. In addition, communities in real-world network are hierarchical, that is, some large communities can be further divided into several small communities. In order to improve the efficiency of community discovery and find communities with hierarchical structure, Vincent Blondel *et al.* proposed the fast unfolding algorithm in 2008.[161] It is an algorithm based on modularity optimization.

Given an undirected weighted graph $G = (V, E)$, the fast unfolding algorithm is divided into two stages. In the first stage, each node in the network is regarded as a community. Then it moves nodes among these communities in sequence. Taking node i in Figure 13.13 as an example (dotted ellipse represents community; green circle represents node), it has three neighbors j_1, j_2 and j_3. The fast unfolding

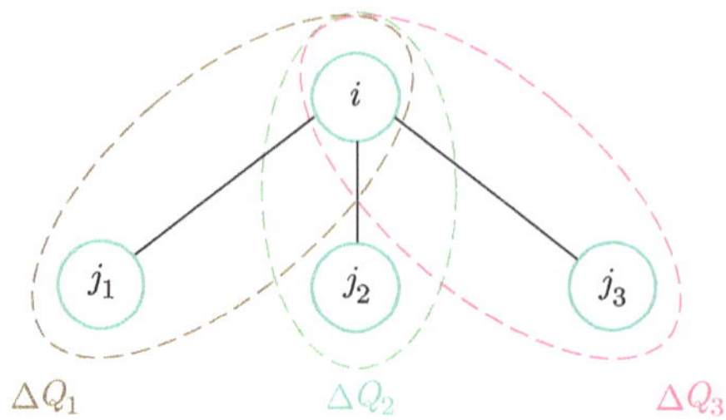

Figure 13.13 Example of fast unfolding algorithm.

algorithm tries to move node i to the communities which j_1, j_2 and j_3 belong to. Node i will be moved to the community with highest modularity increment ΔQ. This iteration is repeated until any node movement can no longer improve the total modularity of the network.

The efficiency of fast unfolding algorithm is reflected in that ΔQ can be calculated efficiently according to the following equation:

$$\Delta Q = \left[\frac{\sum_{\text{in}} + 2k_{i,\text{in}}}{2|E|} - \left(\frac{\sum_{\text{tot}} + k_i}{2|E|}\right)^2\right] - \left[\frac{\sum_{\text{in}}}{2|E|} - \left(\frac{\sum_{\text{tot}}}{2|E|}\right)^2 - \left(\frac{k_i}{2|E|}\right)^2\right], \tag{13.23}$$

where $\sum_{\text{in}}$ is the sum of all edge weights in community C, $\sum_{\text{tot}}$ is the sum of all edges connecting the nodes in community C, $k_{i,\text{in}}$ is the sum of edge weights of all edges starting from node i, and ending in community C, $k_i = \deg(i)$ is the degree of node i.

In the second stage, the fast unfolding algorithm treats the community obtained in the first stage as a new node (one community corresponds to one node). Then it reconstructs a subgraph of these new nodes. The weight between two new nodes is the sum of the weights of edge that corresponds to the two communities. Finally, the fast unfolding algorithm iteratively executes the first stage to discover hierarchical communities.

13.5. Knowledge Graph

In recent years, with the development of artificial intelligence, knowledge graph has attracted extensive attention in academia, industry and the investment community. Knowledge graph has been applied to many fields, including the Internet, industrial design, product management, knowledge publishing, health care, intelligence analysis and so on.

Knowledge graph is a graph representation of entities, entity attributes, and relationships between entities. In traditional database theory, entity relationship (ER) diagram is the most classic conceptual model of entity and entity relationships. We call the ER diagram a conceptual model because it is designed to help people understand the objective world, not a computer-implemented model. The knowledge graph is different from the ER diagram because it not only shows the entity and the entity relationship, but also defines a computer-implemented data model by itself.

In this section, we first introduce the data model of knowledge graph, then introduce the knowledge graph data management method. Finally, we introduce the knowledge graph research in related fields such as natural language processing and machine learning.

13.5.1. *Data Model of Knowledge Graph*

At present, the knowledge graph generally uses the resource description framework (RDF) model in the semantic web to represent data. The core of the semantic web is to enable computers to understand the data in the document and the semantic relationship between data, thereby making the computer smarter to process information. The RDF model mainly includes three object types: resource, predicate and statement. The objects that can be represented by the RDF model are called resources, including all information on networks, virtual concepts, real things, and so on. Resources are uniquely identified using uniform resource identifiers (URIs). Predicates describe the property of resource or the relationship between resources. Each predicate has its own meaning. It is used to define the property value of a resource on a predicate or the relationship with other resources. A statement consists of a subject, a predicate, and an object, usually represented as a triple <subject, predicate, object>, where the subject is a descriptive resource, the predicate represents the property of the subject or represents a relationship between the subject and object. When the predicate represents a property, the object is the property value; otherwise, the object is also a descriptive resource.

We can also integrate the RDF data into graph. The subject and object can be represented as nodes in the RDF graph. The statement (i.e., RDF triple) can be represented as an edge, where the predicate is the label of the edge. Figure 13.14 shows an RDF dataset. This dataset is extracted from the well-known RDF dataset DBpedia which describes some of the famous philosophers and related information in European history.[162]

The Internet consortium (W3C) proposes a structured query language SPARQL for RDF datasets. Like SQL in relational databases, SPARQL is also a descriptive structured query language. This means that users only need to follow SPARQL grammar rules to describe the information they want to query. There is no need to explicitly specify the implementation steps. We can also represent SPARQL as a query graph. At this point, a SPARQL query is essentially a subgraph matching problem.

13.5.2. *Knowledge Graph Data Management Methods*

A core topic of knowledge graph data management is how to efficiently store and query RDF datasets. In general, there are two completely different ideas. The first idea is to use existing database management systems (such as relational database systems) to store knowledge graph data. The SPARQL queries for RDF knowledge graph are converted into queries in the database management systems, such as SQL queries for relational databases. Then answer the queries using existing relational database products or related technologies. The core problem is how to construct a relational table to store RDF knowledge graph data and make execution of the SQL queries efficient. The second idea is to directly develop RDF-oriented data storage and query system. The database is optimized from the bottom considering the characteristics of RDF knowledge graph management. We will look into the second idea now.

Subject	Property	Object
Aristotle	influencedBy	Plato
Aristotle	mainInterest	Ethics
Aristotle	name	"Aristotle"
Aristotle	placeOfDeath	Chalcis
Aristotle	wikiPageUsesTemplate	Template:Planetmath
Boethius	influencedBy	Aristotle
Boethius	mainInterest	Religion
Boethius	name	"Boethius"
Boethius	placeOfDeath	Pavia
Boethius	viaf	100218964
Friedrich_Nietzsche	influencedBy	Aristotle
Friedrich_Nietzsche	mainInterest	Ethics
Friedrich_Nietzsche	name	"Friedrich Nietzsche"
Friedrich_Nietzsche	placeOfDeath	Weimar
Max_Horkheimer	influencedBy	Friedrich_Nietzsche
Max_Horkheimer	influencedBy	Karl_Marx
Max_Horkheimer	mainInterest	Social_theory
Max_Horkheimer	mainInterest	Counter-Enlightenment
Max_Horkheimer	name	"Max Horkheimer"
Max_Horkheimer	placeOfDeath	Nuremberg
Max_Horkheimer	wikiPageUsesTemplate	Template:Persondata
Chalcis	imageSkyline	Chalkida.JPG
Chalcis	country	Greece
Chalcis	postalCode	34100
Pavia	country	Italy
Pavia	postalCode	27100
Weimar	country	Germany
Weimar	postalCode	99401-99441
Weimar	wappen	Wappen Weimar.svg
Nuremberg	country	UnitedStates
Nuremberg	postalCode	90000-90491

(a)

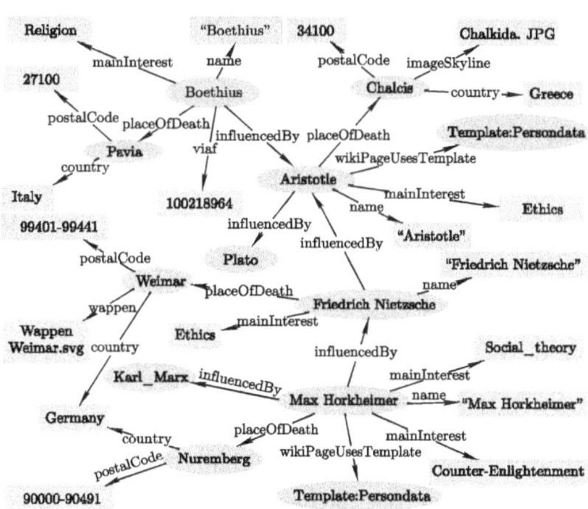

(b)

Figure 13.14 Examples of RDF dataset. (a) RDF tuple; (b) RDF graph.

By treating the RDF triples as labeled edges, the RDF knowledge graph data fits naturally into the graph structure. Therefore, we can look at the RDF data from the perspective of the RDF graph model. The RDF data storage problem is solved by storing the RDF graph. The graph model conforms to the semantic level of the RDF model, which can maximize the semantic information of the RDF data and facilitate the query of the semantic information. By storing RDF data as graph, we can use the mature graph algorithms and graph database to design the storage scheme and query algorithms of RDF data. However, there are difficulties using graph model to design RDF storage and query. Compared with graph model, the edges on the RDF graph have labels and may become query targets. Typical graph algorithms tend to be time-intensive and require careful design to reduce the time complexity of the query.

A typical system for managing and querying knowledge graph data from the perspective of graph model is gStore. gStore is an open source graph database system developed by the data management laboratory of Wangxuan Institute of Computer Technology of Peking University for RDF knowledge graph. Unlike traditional knowledge graph data management method based on relational databases, gStore is native graph-based and it maintains the graph structure of the original RDF knowledge graph. gStore converts SPARQL queries for RDF into subgraph matching for RDF graphs. Index based on graph structure is used to accelerate the query execution. For more introduction of the system principles, see Zou Lei *et al.*[163,164]

Another graph-based system is Trinity RDF.[165] It is a distributed memory-based graph data engine developed by Microsoft Asia Research Institute. Due to the "locality" of graph operations, the system manages RDF graph data in the form of memory cloud.

13.5.3. *Research on Knowledge Graph in Different Disciplines*

Knowledge graph is a multidisciplinary research field of database, natural language processing, knowledge engineering and machine learning. In the above sections, we introduce the knowledge graph from the perspective of database and graph theory. Below we will briefly introduce the research hotspots of knowledge graph from the perspective of natural language processing, knowledge engineering and machine learning.

The study of knowledge graph in the field of natural language processing is mainly conducted from two aspects. One is information extraction. At present, most of the data on the Internet are still unstructured textual data. How to extract RDF triples from unstructured textual data is still a challenging task.[166] Another active research topic is the semantic parser, which converts the natural language query input by the user into a structured query problem oriented to the knowledge graph.[167]

There are two main hot research problems in the field of knowledge engineering. One is the construction of large-scale ontology and knowledge base. For example, DBpedia[162] and YAGO[168] both build large-scale knowledge graph dataset by acquiring knowledge from Wikipedia. The construction of knowledge graph for closed domain is widely used in industry. The second is the study of reasoning on knowledge graph. Different from the closed-world assumption of traditional databases, the knowledge graph uses the open-world assumption. Under the open-world assumption, we do not assume that the stored data is complete. The statement that is not explicitly stored in the system but can be reasoned is still considered to be the correct data.

In recent years, the field of machine learning has also set off a wave of research on knowledge graph. Popular topics include representation learning for knowledge graph, where the TransE model is one of the most representative works.[169] The TransE model maps the subject, object, and predicate of each triple of the knowledge graph into vectors. For any triple in the knowledge graph G, assume that the vectors of its subject, predicate, and object are represented as $\boldsymbol{s}$, $\boldsymbol{p}$, and $\boldsymbol{o}$. The optimization goal can be described from the following two perspectives. For the existence of the triples $(\boldsymbol{s}, \boldsymbol{p}, \boldsymbol{o})$ in the knowledge graph, try to make $\boldsymbol{s} + \boldsymbol{p}$ and $\boldsymbol{o}$ close; for the triples $(\boldsymbol{s}, \boldsymbol{p}, \boldsymbol{o})$ that do not exist in the knowledge graph, try to make $\boldsymbol{s} + \boldsymbol{p}$ and $\boldsymbol{o}$ not close. Besides the TransE model, researchers also proposed a number of improved representation learning models. These models have improved the accuracy of many tasks such as predicate prediction and knowledge completion of knowledge graph.

13.6. Case Studies and Exercises

1. Community discovery of the karate club network

The karate club network shown in Figure 13.1 is a classic social network proposed by American sociologist Zachary in the early 1970s after nearly three years of research.[149] The network contains two real communities. Many researchers use this network to test the effectiveness of the community discovery algorithms. Try to use the community discovery algorithm described in this chapter to divide the network into communities (the number of communities is 2) and compare it with the real communities. Readers can use the Gephi or igraph to perform network visualization and community discovery.

2. European letter network analysis

This exercise is based on a European letter network dataset. The dataset contains a network with 1000 nodes and 14116 directed edges (letter sent from A to B). There are also multiple features, such as Id, Label (the node label), Attribute (indicates the gender, 1 for male and 2 for female), City, Latitude and Longitude. Please use the Gephi tool to visualize the network, divide the network into communities and map the network to the map based on the latitude and longitude information.

3. Twitter friends network analysis and community discovery

Twitter, Facebook, Google+, and Weibo provide APIs or Apps to acquire user social network data, such as Facebook's Netvizz and myFnetwork, which provide raw data for social network analysis.

This exercise uses a social network dataset for @wiredUK extracted via the Twitter social networking API. The dataset contains 254 nodes and 3834 edges. Use the Gephi tool to visualize the social network and perform community discovery.

Chapter 14

Deep Learning

Artificial intelligence (AI) has become a field with many practical applications and active research attracting wide attentions from both academy and industry communities. We expect to simulate humans to automatically handle different types of tasks through artificial intelligence systems, such as understanding natural language, speech, image, video, and auxiliary medical diagnosis, etc. It is usually difficult to describe these tasks in a formal way, but humans can solve them with intuition and experience.

Deep learning, which has been developed in the past decade, provides a promising solution for these tasks. As described by Goodfellow *et al.*: "This solution is to allow computers to learn from experience and understand the world in terms of a hierarchical of concepts, with each concept defined through its relation to simpler concepts. By gathering knowledge, this approach avoids the need for human operators to formally specify all the knowledge that the computer needs. The hierarchy of concepts enables the computer to learn complicated concepts by building them out of simpler ones".[172] If we draw a graph showing the concepts from simple ones to complicated ones, we will get a "deep" layer graph, which is also the reason why deep learning gets its name.

In recent years, deep learning has made milestones in many fields, in both academy and industry communities. For example, ResNet[173] has exceeded the average level of human beings in the object detection task on large-scale image dataset ImageNet;[173] DeepMind's robot AlphaGo[174] based on deep learning defeated South Korean Go master Lee Se-dol with 4:1 and Chinese Go master Jie Ke with 3:0, respectively; The core of Google's machine translation system also gradually relies on deep learning. The breakthrough of deep learning in these areas demonstrates its key role in the future development of artificial intelligence.

An important nature of deep learning is the ability to learn the complex features or representations of data from the process of building a hierarchical network. For example, we consider the task of recognizing people or animals from images. The input data is the pixel array of the image. As we can imagine, it is very complicated to construct a function that can map a set of pixels to an image class. It is almost impossible to directly learn such a mapping without making any assumptions or restrictions on the function construction. If it is approximated by a high-order polynomial, the number of combinations between features will increase exponentially due to the high input dimension. To solve the core problem of representation learning, deep learning allows computers to build a complex concept through simple concepts, and model multi-scale features on different data hierarchies. Figure 14.1 demonstrates how does deep learning solve representation learning and perform image recognition.

The framework of deep learning tactfully proposes a solution to solve the representation learning problem. It gradually approximates the final complex function through the continuous composition of multiple simple nonlinear mappings. Note that each simple nonlinear mapping is represented by one layer of the deep networks. The network firstly receives the observation data at the input layer,

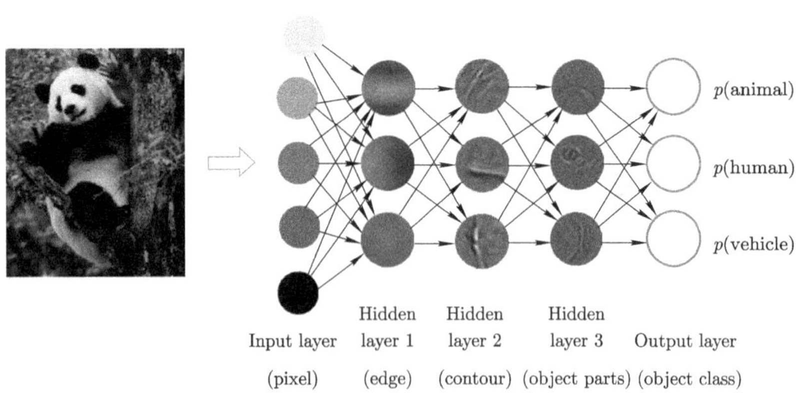

Figure 14.1 Demonstration of deep learning for image recognition.

and then continuously extracts more and more abstract features through multiple hidden layers. In the image recognition problem shown in Figure 14.1, the first hidden layer extracts the image edges. The second hidden layer describes the corners and outlines based on the extracted image edges. The third hidden layer describes the entire features of the input image by combining the corners and outlines. Finally, the output layer predicts the probabilities of each category for the extracted object.

It is worth mentioning that deep learning is quite different from the traditional machine learning methods, such as logistic regression and naive Bayes, due to its powerful representation ability. The performance of traditional machine learning methods depends on the feature engineering. If we directly take the pixel value of the original image as input, it is difficult to obtain the correct recognition result. It is because there is almost no correlation between the pixel information and the class information of the image without feature extraction. Classical methods in computer vision extract features from the image by designing various feature extraction algorithms (such as HoG[175] and SIFT[176]), and then feed these features into the classifier. Here, feature extraction and image classification are two independent processes. The deep learning framework provides a unified solution that combines feature extraction and classification together. This also reflects the remarkable end-to-end learning characteristic of deep learning: the input is the observed raw data, the output is the classification result we want, with the network structure in the middle.

One of the characteristics of deep learning is to "design an appropriate network structure" for different problems. In this chapter we will introduce several typical network structures in different fields such as computer vision, natural language and speech.

Another characteristic of deep learning is that it often relies on "large-scale training data". Complex network structures are often required in complex tasks, which also means that there are many network parameters (hundreds of thousands, millions, or even hundreds of millions) to learn from training data. If we do not have enough training samples, it will lead to overfitting. However, even with sufficient training samples, deep learning systems also require powerful computing resources for training. GPU is usually used to accelerate the training process.

In this chapter, we will firstly introduce multi-layer perceptron and the backpropagation algorithm. Then we will introduce the optimization methods commonly used in deep learning. Finally, we will discuss several deep learning networks, including convolutional neural networks, recurrent neural networks and long-short term memory networks. The development of deep learning is very fast, and its applications are becoming increasingly widespread. This chapter covers only the basic knowledge of deep learning, and the readers interested in this area can refer to the work of Goodfellow *et al.*[172] and related academic conference papers.

14.1. Multi-Layer Perceptron

Multi-layer perceptron (MLP) is a kind of deep feedforward network, which is the most basic and typical deep learning model. Perceptron is usually used as function approximators to approximate a function $f(\cdot)$. For example, for a classification problem, a function $y = f(\boldsymbol{x})$ is used to map the input $\boldsymbol{x}$ to the output category y. MLP performs a function mapping $y = f(\boldsymbol{x}; \boldsymbol{\theta})$, where $\boldsymbol{\theta}$ is parameters that can be learnt from the training data, to obtain the best approximation function.

MLP is composed of many different functions. The network is a directed acyclic graph that describes the composite form of the functions. For example, given three functions $f^{(1)}$, $f^{(2)}$, $f^{(3)}$ to construct the composite function $f(\boldsymbol{x}) = f^{(3)}(f^{(2)}(f^{(1)}(\boldsymbol{x})))$. In a multi-layer perceptron, $\boldsymbol{x}$ is the input layer of the network, and $f^{(i)}$ is called the i-th hidden layer of the network, and the last layer is usually called the output layer. This chain structure is most commonly used in neural networks. Figure 14.2 shows the structure of a multi-layer perceptron with two hidden layers.

The input $\boldsymbol{x} = (x_1, x_2, x_3)^{\mathrm{T}}$ is fed into the first hidden layer through a linear transformation $\boldsymbol{W}_1^{\mathrm{T}}\boldsymbol{x} + \boldsymbol{b}_1$ and a nonlinear function $\phi(\cdot)$:

$$\boldsymbol{f}^{(1)}(\boldsymbol{x}) = \phi(\boldsymbol{W}_1^{\mathrm{T}}\boldsymbol{x} + \boldsymbol{b}_1), \tag{14.1}$$

where the weight matrix $\boldsymbol{W}_1$ and bias $\boldsymbol{b}_1$ are network parameters. $\phi(\cdot)$ is called activation function, which is a nonlinear function.

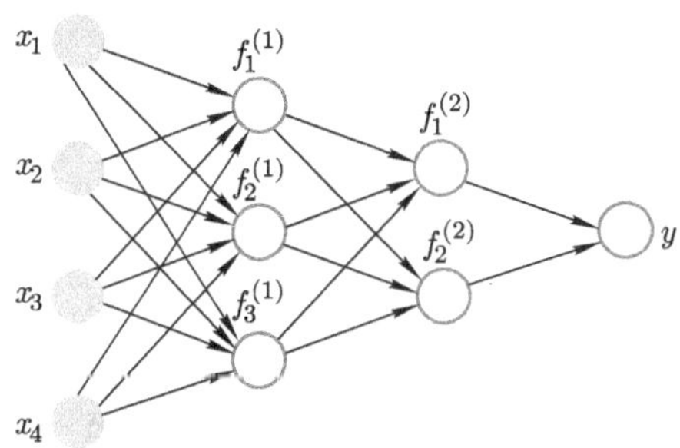

Figure 14.2 Multi-layer perceptron.

The second hidden layer $\boldsymbol{f}^{(2)}$ is obtained by making a set of linear transformations and nonlinear transformations on the output of the first hidden layer $\boldsymbol{f}^{(1)}$, that is $\boldsymbol{f}^{(2)}(\boldsymbol{x}) = \phi(\boldsymbol{W}_2^{\mathrm{T}}\boldsymbol{f}^{(1)} + \boldsymbol{b}_2)$. Then the output can be obtained through a linear transformation on $\boldsymbol{f}^{(2)}$:

$$y = \boldsymbol{W}_3^T\boldsymbol{f}^{(2)} + b_3. \tag{14.2}$$

Similarly, $\{\boldsymbol{W}_3, b_3\}$ are network parameters. The final output is obtained through a three-layer function, which also demonstrates the hierarchical structure of the deep learning model. Such networks constructed by the hierarchical functions are referred to as artificial neural networks, because they were initially inspired by the information transferring between neurons in neuroscience. However, it should be emphasized that the deep learning model does not intend to model the brain, but to solve some specific AI tasks.

In the networks, each hidden layer contains multiple nodes (neurons), which can be represented by a vector, such as $\boldsymbol{f}^{(1)}$ and $\boldsymbol{f}^{(2)}$. The number of nodes in each hidden layer determines the width of the model, and the number of hidden layers determines the depth of the model. The mapping between adjacent layers corresponds to a vector-to-vector function. The layer can also be considered as the parallel operations by multiple units, and each operation defines a vector-to-scalar function. The deep learning models are nonlinear, and usually only one single nonlinear function (e.g., activation function) is used.

Many factors need to be considered in designing a neural network, including the selection of activation function, the determination of network depth and width, the determination of output layer form and loss function, etc. How to learn network parameters is a major

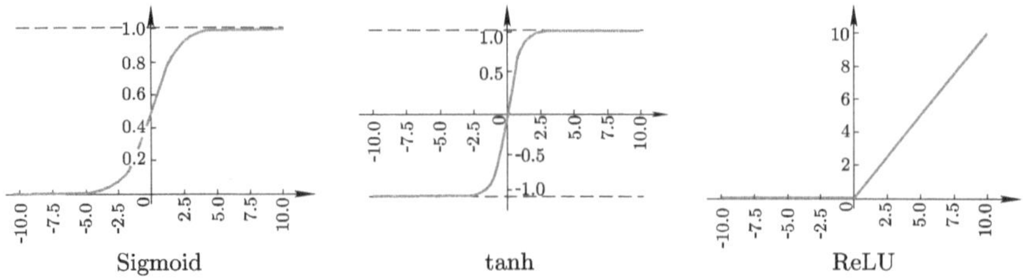

Figure 14.3 Three commonly used activation functions.

consideration when training a neural network. Next, we will introduce these factors.

14.1.1. *Activation Function*

Commonly used activation functions are sigmoid function, tanh function, and Rectified Linear Unit (ReLU),[177] as shown in Figure 14.3. The ReLU function is the default activation function of most deep feedforward networks. The ReLU function consists of two linear function segments. Similar to linear models, deep networks with ReLU can be optimized by using the gradient methods. Therefore, we often use ReLU function with wide and deep network to build a global function approximator.

Several variants of the ReLU function are used in different tasks, such as the maxout unit,[178] the leaky ReLU function,[179] and the noisy ReLU function.[180]

Before the ReLU function, most neural networks use the S-shaped activation function, such as the sigmoid function

$$\sigma(z) = \frac{1}{1 + \mathrm{e}^{-z}}, \tag{14.3}$$

or tanh function

$$\tanh(z) = 2\sigma(2z) - 1. \tag{14.4}$$

The sigmoid function can also be used as a Bernoulli output to deal with the binary classification problem. The sigmoid function is saturated in most regions, that is, when the absolute value of z becomes very large, $\sigma'(z)$ is very close to 0. $\sigma(z)$ has a very good gradient property only when z is close to 0. This characteristic makes the gradient-based learning very difficult. Therefore, the sigmoid function is not recommended as the activation function of deep feedforward networks. However, in the recurrent networks, probabilistic models, and auto-encoders, the sigmoid function is still widely used, because ReLU is unsuitable due to their special requirements.

14.1.2. *Network Structure Design*

Designing the network structure is very important in the use of neural networks. For example, we should choose the number of hidden layers (network depth), the number of nodes in each layer (network width), and how these nodes should be connected, etc. Research has shown by a general universal approximation theorem that, a single-hidden layer feedforward neural network could approximate arbitrary function mapping from input to output with any accuracy on a given training set.[181] However, such "shallow" networks are not commonly used, because deep networks with fewer nodes per layer can greatly reduce the number of parameters and obtain better generalization results.[182,183] For a specific task, we need to constantly test, observe the generalization results on the validation set, and then determine the network structure.

14.1.3. *Output Layer*

Suppose perceptron provides a set of hidden features, denoted as $\boldsymbol{h} = f(\boldsymbol{x}; \boldsymbol{\theta})$, the output layer performs additional transformations on these features to get the output we are interested in. Some commonly used output units are:

1. **Linear output unit:** If a continuous output is required, the simplest output unit is to linearly transform the feature $\boldsymbol{h}$ obtained by the hidden layer to get $\hat{\boldsymbol{y}} = \boldsymbol{W}^{\mathrm{T}}\boldsymbol{h}+\boldsymbol{b}$. This output can be used as the mean of the conditional normal distribution

$$p(\boldsymbol{y}|\boldsymbol{x}) = N(\boldsymbol{y}; \hat{\boldsymbol{y}}, \boldsymbol{I}). \tag{14.5}$$

In this case, maximizing the log likelihood is equivalent to minimizing the mean square error.

2. **Sigmoid output unit:** For a binary classification problem, the neural network only needs to output $p(y = 1|\boldsymbol{x})$. We first use the linear transformed value $z = \boldsymbol{w}^{\mathrm{T}}\boldsymbol{x} + b$ to construct an unnormalized probability distribution $\tilde{p}(y)$, and then normalize to get the new probability distribution. Thus, we can construct the unnormalized distribution as $\ln \tilde{p}(y) = yz$, that is, $\tilde{p}(y) = \mathrm{e}^{yz}$. It will be normalized as:

$$\tilde{p}(y) = \frac{\mathrm{e}^{yz}}{\sum_{y'=0,1} \mathrm{e}^{y'z}} = \sigma((2y-1)z), \tag{14.6}$$

   where $\sigma(\cdot)$ is the Sigmoid function, and the objective function is

$$L(\boldsymbol{\theta}) = -\ln p(y|\boldsymbol{x}) = -\ln \sigma((2y-1)z). \tag{14.7}$$

3. **Softmax unit:** For a multi-class classification problem, we usually use the softmax function. The softmax function can be regarded as a multivariate extension of the Sigmoid function. Similar to the Sigmoid output unit, an unnormalized log probability will be firstly produced, $\boldsymbol{z} = \boldsymbol{W}^{\mathrm{T}}\boldsymbol{h} + \boldsymbol{b}$, and it will be exponentiated and normalized by the softmax function to get the output $\hat{\boldsymbol{y}}$:

$$\mathrm{softmax}(\boldsymbol{z})_i = \frac{\exp(z_i)}{\sum_j \exp(z_j)}. \tag{14.8}$$

   Each component of $\hat{\boldsymbol{y}}$ represents the probability that the input sample belongs to the corresponding class.

14.1.4. *Loss Function*

Similar to other models, the loss function of neural networks is generally defined by the maximum likelihood principle, such as negative log likelihood

$$L(\boldsymbol{\theta}) = -\underset{\boldsymbol{x},\boldsymbol{y}\sim\hat{p}_{\mathrm{data}}}{E}[\ln p_{\mathrm{model}}(\boldsymbol{y}|\boldsymbol{x})], \tag{14.9}$$

where $\hat{p}_{\mathrm{data}}$ represents the distribution of the observed data and p_{model} is the distribution of the model. For example, if we use the

Gaussian distribution model $p_{\text{model}}(\boldsymbol{y}|\boldsymbol{x}) = N(\boldsymbol{y}; f(\boldsymbol{x}, \boldsymbol{\theta}), \boldsymbol{I})$, we can get the mean square error (MSE) loss function

$$L(\boldsymbol{\theta}) = \frac{1}{2} \underset{\boldsymbol{x},\boldsymbol{y} \sim \hat{p}_{\text{data}}}{E} \|\boldsymbol{y} - f(\boldsymbol{x}; \boldsymbol{\theta})\|_2^2 = \frac{1}{2n} \sum_{i=1}^{n} \|\boldsymbol{y}_i - f(\boldsymbol{x}_i; \boldsymbol{\theta})\|_2^2. \quad (14.10)$$

When the output layer is Sigmoid or softmax unit, the loss function is cross entropy. Usually regularization terms are added to the loss function. The objective function, which is the function that is to be minimized, is constructed as the sum of loss function and regularization terms.

After the loss function is determined, the neural network can be trained by gradient descent algorithms. How to train neural networks with gradient descent algorithms will be discussed in Section 14.2.

14.1.5. *Backpropagation Algorithm*

Most neural networks use gradient descent algorithms to learn parameters. The core of gradient-based methods is to calculate the gradient of the objective function w.r.t. parameters. In deep networks, however, that is not a simple problem. Backpropagation (backprop, BP) provides a simple and effective gradient calculation method for neural networks.[184] It applies the chain rule of derivative of composite function in calculus.

To better understand the backpropagation algorithm, we first introduce the computational graph and the chain rule.

1. Computational graph

Computational graph can help us to describe backpropagation algorithm precisely. In a computational graph, each node represents a variable, such as scalar, vector, and matrix, while the directed edge represents the operation, which is a function of single-variable or multi-variable. By using computational graph, we can visually demonstrate a complex computational process of a composite function. The computational graph of equation $\boldsymbol{a} = \max\{\boldsymbol{0}, \boldsymbol{W}^{\mathrm{T}} + \boldsymbol{b}\}$ is shown in Figure 14.4, where "matmul" indicates matrix multiplication.

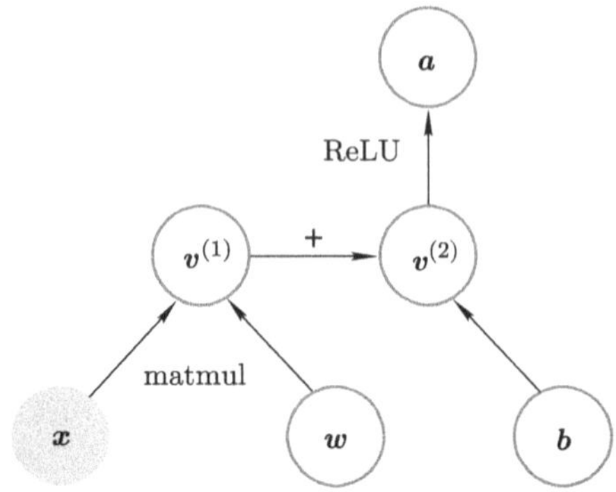

Figure 14.4 Computational graph.

2. Chain rule

By applying the chain rule recursively according to the hierarchical structure of the network, backpropagation can calculate the gradients of the parameters in each layer efficiently.

Assuming that $y = f(x)$, $z = g(y)$, the chain rule of derivation can be described as follows:

$$\frac{\mathrm{d}z}{\mathrm{d}x} = \frac{\mathrm{d}z}{\mathrm{d}y}\frac{\mathrm{d}y}{\mathrm{d}x}. \tag{14.11}$$

In neural networks, the variables are usually vectors and tensors. We first consider the case of vectors, $\boldsymbol{y} = f(\boldsymbol{x})$ and $z = g(\boldsymbol{y})$, then

$$\frac{\partial z}{\partial x_i} = \sum_j \frac{\partial z}{\partial y_j}\frac{\partial y_j}{\partial x_i}. \tag{14.12}$$

In addition to scalars and vectors, the chain rule can be generalized to tensors of any dimension. Now, we will show how to use the chain rule to calculate derivation of parameters in a multi-layer perceptron. Before performing backpropagation to calculate the gradients, we need to perform forward propagation to provide some necessary intermediate variables. Assuming the depth of the network is l, the parameters are $\{\boldsymbol{W}^{(i)}, \boldsymbol{b}^{(i)}\}_{i=1}^{l}$. For training sample $(\boldsymbol{x}, \boldsymbol{y})$, $\boldsymbol{x}$ is input and $\hat{\boldsymbol{y}}$ is the network output. By comparing the sample label

$\boldsymbol{y}$ with the network output $\hat{\boldsymbol{y}}$, the loss can be calculated by $L(\hat{\boldsymbol{y}}, \boldsymbol{y})$. The forward propagation process is described as follows:

1. Initialization: $\boldsymbol{h}^{(0)} = \boldsymbol{x}$;
2. Forward propagation along the computational graph. From layer $i = 1$ to l, calculate the pre-activation value $\boldsymbol{a}^{(i)} = \boldsymbol{b}^{(i)} + \boldsymbol{W}^{(\mathrm{T})}\boldsymbol{h}^{(i-1)}$ and the hidden layer output $\boldsymbol{h}^{(i)} = f(\boldsymbol{a}^{(i)})$, where $f(\cdot)$ is the activation function;
3. The final output and loss function values are $\hat{\boldsymbol{y}} = \boldsymbol{h}^{(l)}$ and $L = L(\hat{\boldsymbol{y}}, \boldsymbol{y})$, respectively.

The gradient of the loss function against the parameters of each layer can be calculated by backpropagation:

1. Calculate the gradient of the output layer:

$$\nabla_{\hat{\boldsymbol{y}}} L = \nabla_{\boldsymbol{h}^{(l)}} L = \nabla_{\hat{\boldsymbol{y}}} L(\hat{\boldsymbol{y}}, \boldsymbol{y}); \tag{14.13}$$

2. For each layer, $i = l-1, l-2, \ldots, 1$:
   (a) Calculate the gradient of the pre-activation $\boldsymbol{a}^{(i)}$,

$$\nabla_{\boldsymbol{a}^{(i)}} L = \nabla_{\boldsymbol{h}^{(i)}} L \odot f'(\boldsymbol{a}^{(i)}), \tag{14.14}$$

   where $\odot$ is the operation of multiplying the corresponding elements of two vectors. For simplicity, we will denote the above gradient as $\boldsymbol{g}$;
   (b) Calculate the gradients of weight matrix and bias in the current layer

$$\nabla_{\boldsymbol{W}^{(i)}} L = \boldsymbol{g}\boldsymbol{h}^{(i-1)\mathrm{T}}, \tag{14.15}$$

$$\nabla_{\boldsymbol{b}^{(i)}} L = \boldsymbol{g}; \tag{14.16}$$

   (c) Pass the gradients to the previous hidden layer and output

$$\nabla_{\boldsymbol{h}^{(i-1)}} L = \boldsymbol{W}^{(i)\mathrm{T}}\boldsymbol{g}. \tag{14.17}$$

The BP algorithm calculates the gradient of the loss function against the pre-activation value $\boldsymbol{a}^{(i)}$ for each layer i. Starting from the output layer to the first hidden layer, the parameters gradients of each layer (including weights and bias) can be obtained according to these gradients of pre-activation values. After obtaining the parameters gradients of each layer, the parameters will be updated.

In addition to calculating the gradients in strict accordance with the chain rule, numerical differentiation provides another way. This method is widely used in many deep learning software architectures, such as TensorFlow,[185] Theano[186] and Mxnet *et al.*[187] Numerical differentiation uses the definition of derivatives to approximate the gradient

$$\nabla_\theta f \approx \frac{f(\theta + h/2) - f(\theta - h/2)}{h}, \tag{14.18}$$

where h is a small positive number.

14.2. Optimization of Deep Learning Model

The parameter size of deep learning model is relatively huge, and the parameter learning often requires a lot of computational resources and time. Therefore, it is important to design efficient optimization algorithms to speed up the network training and expand the application scope of the deep learning model. This section will focus on several classic optimization algorithms in deep learning.

The aim of optimization in deep learning model is to minimize the following loss function:

$$J(\boldsymbol{\theta}) = \underset{\boldsymbol{x},y \sim \hat{p}_{\text{data}}}{E} [L(f(\boldsymbol{x};\boldsymbol{\theta}), y)] = \frac{1}{n}\sum_{i=1}^{n} L(f(\boldsymbol{x}_i;\boldsymbol{\theta}), y_i), \tag{14.19}$$

where n is the sample size, $L(f(\boldsymbol{x}_i;\boldsymbol{\theta}), y_i)$ is the loss error on the i-th sample. To avoid confusion, we use $J(\boldsymbol{\theta})$ to denote the expectation of the loss function on the sample. The typical learning algorithm is the gradient descent method for minimizing the loss function $J(\boldsymbol{\theta})$ (referring to Appendix C). After the parameter initialization, the gradient descent method iteratively updates the parameters by using the following formula:

$$\boldsymbol{\theta}^{(t+1)} = \boldsymbol{\theta}^{(t)} - \eta_t \nabla_{\boldsymbol{\theta}} J(\boldsymbol{\theta}^{(t)}), \tag{14.20}$$

where t is the iteration step, η_t is the learning rate, and the gradient can be calculated by the following formula:

$$\nabla_{\boldsymbol{\theta}} J(\boldsymbol{\theta}^{(t)}) = \frac{1}{n}\sum_{i=1}^{n} \nabla_{\boldsymbol{\theta}} L(f(\boldsymbol{x}_i;\boldsymbol{\theta}), y_i). \tag{14.21}$$

In each iteration, the gradients of n samples should be calculated to get their mean. With the increase of the sample size ($n > 10^6$), the computational cost will become very high. To solve this problem, the calculation of the gradient should be simplified. Usually, we will approximate rather than accurately calculate $\nabla_{\boldsymbol{\theta}} J(\boldsymbol{\theta})$.

Stochastic approximation is a widely used method. In each iteration, a subset with size m (typically tens to hundreds) is randomly sampled from the dataset. This subset is called mini-batch. The gradient $\nabla_{\boldsymbol{\theta}} \tilde{J}(\boldsymbol{\theta})$ of the subset is used to approximate $\nabla_{\boldsymbol{\theta}} J(\boldsymbol{\theta})$. $\nabla_{\boldsymbol{\theta}} \tilde{J}(\boldsymbol{\theta})$ is calculated by

$$\nabla_{\boldsymbol{\theta}} \tilde{J}(\boldsymbol{\theta}) = \frac{1}{m} \sum_{j=1}^{m} L(f(\boldsymbol{x}_j; \boldsymbol{\theta}_t), y_j). \tag{14.22}$$

It can be proved that $\nabla_{\boldsymbol{\theta}} \tilde{J}(\boldsymbol{\theta})$ is an unbiased estimation of $\nabla_{\boldsymbol{\theta}} J(\boldsymbol{\theta})$, that is $\nabla_{\boldsymbol{\theta}} J(\boldsymbol{\theta}) = E[\nabla_{\boldsymbol{\theta}} \tilde{J}(\boldsymbol{\theta})]$. Then in each iteration, the parameters are updated by

$$\boldsymbol{\theta}^{(t+1)} = \boldsymbol{\theta}^{(t)} - \eta_t \nabla_{\boldsymbol{\theta}} \tilde{J}(\boldsymbol{\theta}^{(t)}), \tag{14.23}$$

where η_t is the learning rate. This learning method relies on stochastic gradients, so it is called stochastic gradient descent (SGD). SGD is one of the most classic optimization methods in deep learning, and many other algorithms are designed based on the idea of SGD.

How to set the learning rate is a key in SGD algorithm. Bottou gives a sufficient condition which theoretically guarantees the convergence of SGD:

$$\sum_{t=1}^{\infty} \eta_t = \infty, \quad \sum_{t=1}^{\infty} \eta_t^2 < \infty. \tag{14.24}$$

This condition tells us that it is necessary to gradually reduce the learning rate as the number of iterations increases.[188] Specifically, the learning rate can be determined by the change of the objective function value with the number of iterations. If it is observed that there is no significant change over time, the learning rate can be reduced to $\frac{1}{10}$ to $\frac{1}{2}$ of the previous step.

Since mini-batch is used to approximate the gradient, the computation amount in each iteration can be predicted. For large-scale datasets, SGD may converge to a satisfactory result before the entire training set is processed.

Below we introduce some typical variants of SGD.

14.2.1. *Momentum Method*

The main idea of the momentum method is to consider the information of the past gradient when the parameter is updated.[189] The rules of parameter update for the momentum method are as follows:

$$\boldsymbol{v}^{(t+1)} = \alpha \boldsymbol{v}^{(t)} + \eta \nabla_{\boldsymbol{\theta}} \tilde{J}(\boldsymbol{\theta}^{(t)}), \tag{14.25}$$

$$\boldsymbol{\theta}^{(t+1)} = \boldsymbol{\theta}^{(t)} - \boldsymbol{v}^{(t)}, \tag{14.26}$$

where the hyper-parameter $\alpha \in [0, 1)$ determines the decay rate of the contribution of the past gradient. From the above formula, we can see that the larger α is, the greater the influence of the past gradient will be on the current update direction. In practice, α is generally set as 0.5, 0.9 or 0.99. Similar to the learning rate, α also needs to be adjusted as the number of iterations increases. Generally, α is initialized with a small value, and then gradually increases with the number of iterations increases. It is more important to adjust α than η.

The momentum method overcomes two shortcomings of SGD: the poor conditioning of Hessian matrix, and the instability brought by the variance of the stochastic gradient. It has one more hyper-parameter α than SGD, which increases the difficulty of parameter tuning. If adjusted properly, the momentum method can achieve better results than SGD.

14.2.2. *Nesterov Momentum Method*

Sutskever *et al.* proposed a variant of the momentum method,[190] which combines the well-known Nesterov accelerated gradient algorithm in convex optimization[191] with SGD. The update strategies are as follows:

$$\boldsymbol{v}^{(t+1)} = \alpha \boldsymbol{v}^{(t)} + \eta \nabla_{\boldsymbol{\theta}} \tilde{J}(\boldsymbol{\theta}^{(t)} - \alpha \boldsymbol{v}^{(t)}), \tag{14.27}$$

$$\boldsymbol{\theta}^{(t+1)} = \boldsymbol{\theta}^{(t)} - \boldsymbol{v}^{(t)}. \tag{14.28}$$

The difference between the Nesterov momentum and the standard momentum methods lies in the gradient calculation. In the Nesterov momentum method, the gradient is calculated after updating the parameters (i.e., after adding the current speed), while in the standard momentum method the gradient is calculated before updating the parameters. Therefore, the Nesterov momentum method can be understood as the exploration and correction of the updated direction $\boldsymbol{v}$ before updating the parameters.

14.2.3. *Optimization of Adaptive Learning Rate*

In the training process of neural networks, the learning rate has a significant impact on the performance of the model, but it is difficult to set. Meanwhile, due to the characteristics of the Hessian matrix, the objective function is very sensitive to different parameters. If all parameters adopt the same learning rate, it is obvious that the efficiency will be quite low. Therefore, we expect that the optimization algorithm can automatically set the learning rate for different parameters throughout the iteration process to improve the efficiency. In this section, we will introduce several optimization algorithms which can adaptively set the learning rate.

1. AdaGrad

In the AdaGrad algorithm, the learning rate of each parameter is set separately. Specifically, it uses the square sum of past gradients to automatically adjust the learning rate.[191]

The stochastic gradient of the parameter $\boldsymbol{\theta}$ in the time step t is $\boldsymbol{g}^{(t)} = \nabla_{\boldsymbol{\theta}} \tilde{J}(\boldsymbol{\theta}^{(t)})$. For the i-th component of the parameter $\boldsymbol{\theta}$, let us denote the square sum of the past gradients till time step t as $r_i^{(t)}$, i.e.,

$$r_i^{(t)} = \sum_{j=0}^{t} g_i^{(j)} \cdot g_i^{(j)}. \tag{14.29}$$

Assume the global learning rate is η, the update formula for the AdaGrad algorithm is as follows:

$$\theta_i^{(t+1)} = \theta_i^{(t)} - \frac{\eta}{\sqrt{r_i^{(t)} + \delta}} g_i^{(t)}, \tag{14.30}$$

where δ is an integer for maintaining the stability of the numerical calculation, which can be generally set to 10^{-8}. It can be seen that the learning rate is relatively small in the direction where the square sum of the past gradients is big. In the direction where the square sum of the past gradients is small, the learning rate is relatively big. Thus, AdaGrad makes bigger updates in the more gently sloped directions in the parameter space.

Since AdaGrad sets different learning rates for different components of the parameters, it is not easy to stay at the saddle point of the objective function. Although AdaGrad can automatically adjust the learning rate, the global learning rate η needs to be set manually. Besides, since the square sum of the gradients is calculated at the beginning of each time step, AdaGrad may easily cause fast decay of the learning rate. As a result, it will make the learning rate too small, and the parameter updates approach zero, thus stopping the training process early.

2. RMSprop

RMSprop was proposed by Geoffrey Hinton.[237] Similar to AdaGrad, RMSprop automatically adjusts the learning rate by the past gradient information. In AdaGrad, the learning rate is adjusted only based on the square sum of the past gradients, while in RMSprop, the mean of the past gradients and the square sum of the current gradients are considered simultaneously.

The stochastic gradient of parameter $\boldsymbol{\theta}$ in time step t is $\boldsymbol{g}^{(t)} = \nabla_{\boldsymbol{\theta}} \tilde{J}(\boldsymbol{\theta}^{(t)})$. For the i-th component of $\boldsymbol{\theta}$, we use $r_i^{(t)}$ to denote the adjustment factor of learning rate at time step t, which is calculated as

$$r_i^{(t)} = \rho \frac{1}{t-1} \sum_{j=0}^{t-1} g_i^{(j)} \cdot g_i^{(j)} + (1-\rho) g_i^{(t)} \cdot g_i^{(t)}, \tag{14.31}$$

where $\rho \in [0,1]$ is a hyper-parameter to adjust the relative importance of the mean of the past gradients and the square sum of the current gradients.

Based on the adjustment factor $r_i^{(t)}$, the parameter update formula for RMSprop is

$$\theta_i^{(t+1)} = \theta_i^{(t)} - \frac{\eta}{\sqrt{r_i^{(t)} + \delta}} g_i^{(t)}, \tag{14.32}$$

where η is the global learning rate.

In RMSprop, the learning rate of the current time step is mainly influenced by the gradient of the latest step, which greatly reduces the dependence on gradients of the earlier steps. In this way, it avoids the case where the learning rate is reduced too early and too much in some directions. RMSprop has one more hyper-parameter ρ than AdaGrad. In RMSprop, it is suggested to set ρ as 0.9 and the global learning rate η as 0.001.

3. AdaDelta

The AdaDelta algorithm further uses the updated information of the parameters in each time step to adjust the learning rate.[193] Assume that $r_i^{(t)}$ is defined as equation (14.31), and the updated value of parameter in time step t is $\Delta\theta_i^{(t)}$:

$$\Delta\theta_i^{(t)} = -\frac{\eta}{\sqrt{r_i^{(t)} + \delta}} g_i^{(t)}, \tag{14.33}$$

then the parameter update factor in step t is:

$$s_i^{(t)} = \gamma \frac{1}{t-1} \sum_{j=1}^{t-1} \Delta\theta_i^{(j)} \cdot \Delta\theta_i^{(j)} + (1-\gamma)\Delta\theta_i^{(t)} \cdot \Delta\theta_i^{(i)}, \tag{14.34}$$

where γ is a hyper-parameter.

The parameter update formula for AdaDelta is

$$\theta_i^{(t+1)} = \theta_i^{(t)} - \frac{\sqrt{s_i^{(t-1)} + \delta}}{\sqrt{r_i^{(t-1)} + \delta}} g_i^{(t)}. \tag{14.35}$$

It can be seen that in AdaDelta, there is no need to set learning rate manually. However, two extra parameters ρ and γ should be set when calculating $\boldsymbol{r}$ and $\boldsymbol{s}$.

4. Adam

Adam is a recently proposed algorithm that can adaptively adjust the learning rate.[194] Adam uses the current gradient and the past gradients to determine the parameter update direction. Assuming that parameter update direction in time step $t-1$ is $\boldsymbol{s}^{(t-1)}$, then

$\boldsymbol{s}^{(0)} = \mathbf{0}$ and

$$\boldsymbol{s}^{(t)} = \rho_1 \boldsymbol{s}^{(t-1)} + (1 - \rho_1)\boldsymbol{g}^{(t)}. \tag{14.36}$$

At the same time, the square sum of past gradients $\boldsymbol{r}^{(t-1)}$ is also used to adjust the learning rate, with $\boldsymbol{r}^{(0)} = \mathbf{0}$ and

$$\hat{r}_i^{(t)} = \rho_2 r_i^{(t-1)} + (1 - \rho_2) g_i^{(t)} \cdot g_i^{(t)}, \tag{14.37}$$

where $\boldsymbol{g}$ is the stochastic gradient. $\rho_1, \rho_2 \in [0, 1]$ and their default values are 0.9 and 0.999. Besides, bias correction can also be performed for updating $\boldsymbol{s}^{(t)}$ and $\boldsymbol{r}^{(t)}$

$$\hat{\boldsymbol{s}}^{(t)} = \frac{\boldsymbol{s}^{(t)}}{1 - \rho_1^t}, \tag{14.38}$$

$$\hat{\boldsymbol{r}}^{(t)} = \frac{\boldsymbol{r}^{(t)}}{1 - \rho_2^t}. \tag{14.39}$$

The parameters will be updated after the above steps:

$$\theta_i^{(t+1)} = \theta_i^{(t)} - \frac{\eta}{\sqrt{\hat{r} + \delta}} \hat{s}_i^{(t)}. \tag{14.40}$$

Adam is relatively insensitive to the hyper-parameters, and it can achieve good performance in practice.

14.2.4. *Batch Normalization*

Batch Normalization (BN) is one of the most efficient strategies for optimizing deep learning models in recent years.[195] It has become the standard configuration of the current network structures and great success has been made in many tasks. BN is actually a reparameterization strategy rather than an optimization algorithm for deep model training.

Deep models involve the composition of multiple functions (or layers). Gradient method usually updates parameters of all layers simultaneously. However, the parameters update of current layer is based on the assumption that parameters in other layers remain unchanged, which is not taken into consideration by current study. The parameters updates of i-th layer will lead to the changes of inputs in the

$(i+1)$-th layer in both the distribution and the scale. It is difficult for an algorithm to select a proper learning rate, because the parameter update of each layer will have a great impact on the other layers.

To solve this problem, BN performs Gaussian normalization (also known as Z-score normalization) to the activation value of each layer based on a mini-batch. It can reduce the changes of output distributions. BN first calculates the mean and variance of the mini-batch samples as follows:

$$\boldsymbol{\mu}_B = \frac{1}{m}\sum_{i=1}^{m} \boldsymbol{x}_i, \tag{14.41}$$

$$\boldsymbol{\sigma}_B^2 = \frac{1}{m}\sum_{i=1}^{m} (\boldsymbol{x}_i - \boldsymbol{\mu}_B)^2, \tag{14.42}$$

where the dimension of vector $\boldsymbol{\sigma}_B$ is the same as the input sample. $\boldsymbol{\sigma}_B$ only consider the variance of $\boldsymbol{x}$ in each dimension, regardless of the covariance between the dimensions. Then the input on the mini-batch will be normalized as

$$\hat{\boldsymbol{x}} = \frac{\boldsymbol{x}_i - \boldsymbol{\mu}_B}{\sqrt{\boldsymbol{\sigma}_B^2 + \delta}}, \tag{14.43}$$

where δ is a small positive number to increase the stability of numerical calculation. Since the mean and variance of mini-batch cannot represent the statistics on the entire dataset, BN introduces two parameters γ and β to adjust the normalized output:

$$\boldsymbol{y}_i = \gamma \hat{\boldsymbol{x}}_i + \beta. \tag{14.44}$$

Note that γ and β are also learnt through backpropagation and gradient descent algorithm. Such parameterization can also avoid the lack of expression ability to the network after the normalization. For detailed discussion of BN, please refer to Loffe and Szegedy's paper.[194]

14.2.5. *Summary*

In this section, we discussed several typical optimization algorithms and strategies for neural networks. Schaul *et al.*[197] demonstrated the comparison between different optimization algorithms on different learning tasks. The results showed that adaptive algorithms, such as RMSprop and AdaDelta[196] were quite robust. For the mathematical analysis of general SGD methods and the design of adaptive algorithms, please refer to the literature.[197]

14.3. Convolutional Neural Networks

Convolutional neural network (CNN) is a class of neural networks applied to grid data (such as time series data and image data).[198] CNN has been widely used in many fields such as computer vision, speech recognition, video analysis and natural language processing. Especially in computer vision, with strong performance, CNN has become the mainstream solution for most tasks.

A typical CNN usually consists of one or more convolutional layers and one or more top-level fully connected layers. The fully connected layer has the same structure as the multi-layer perceptron. The convolution layer is a special network layer, usually with three steps, as shown in Figure 14.5. Firstly, the convolution layer takes multiple parallel convolution operations on the input to produce multiple feature maps. Secondly, in the detection layer, each feature is transformed by a nonlinear activation function (such as ReLU). Finally, further adjustments are made by the pooling layer to get the final output.

In this section, we will first introduce two important operations in convolutional neural networks: convolution and pooling. Then we will introduce several widely used convolutional neural network structures.

14.3.1. *Convolution*

In neural networks, the convolution operation is a weighted calculation of the lattice structure of the input. For example, if the

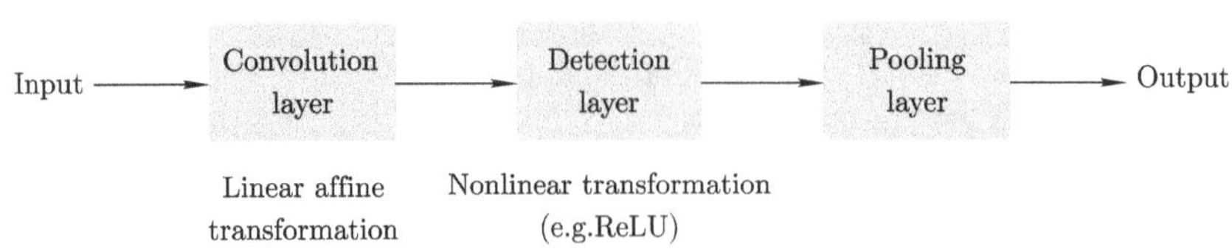

Figure 14.5 Common steps in the convolution layer.

two-dimensional image $\boldsymbol{I}$ is taken as input, we use the weight matrix $\boldsymbol{K}$ for weighted averaging:

$$\boldsymbol{S}(i,j) = (\boldsymbol{I} * \boldsymbol{K})(i,j) = \sum_m \sum_n \boldsymbol{I}(m,n)\boldsymbol{K}(i-m, j-n). \quad (14.45)$$

The weight matrix $\boldsymbol{K}$ is also called kernel function, which is the parameter of the convolutional neural network. The output $\boldsymbol{S}$ is called feature map. The convolution operation can be comprehended as: a small window $\boldsymbol{K}$ is used to slide on the image by pixel to calculate the weight of the corresponding pixel, as shown in Figure 14.6. Convolution is commutative, so it can be denoted equivalently:

$$\boldsymbol{S}(i,j) = (\boldsymbol{K} * \boldsymbol{I})(i,j) = \sum_m \sum_n \boldsymbol{I}(i-m, j-n)\boldsymbol{K}(m,n). \quad (14.46)$$

Convolution is one of the important means for CNN to deal with grid data successfully.

Why convolution?

Convolution have three important characteristics to improve model performance: sparse connectivity, parameter sharing, and equivariant representation.

1. Sparse connectivity

The adjacent two layers of the traditional neural networks are fully connected, that is, every output unit interacts with every input unit. Convolutional neural network uses a kernel that is much smaller than the input dimension to convolute, making the connectivity between the input and output units sparse. Sparse connectivity means that

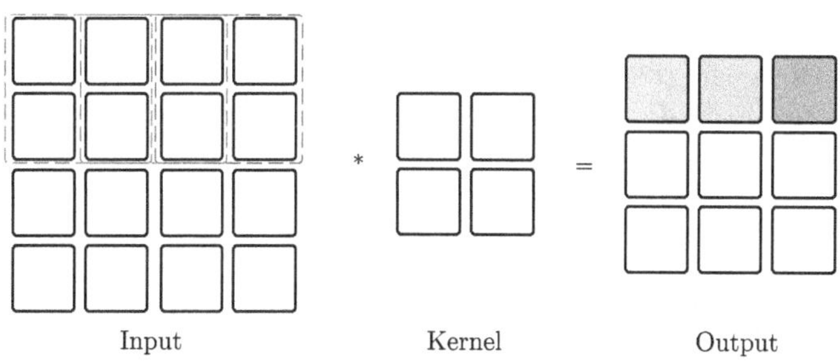

Figure 14.6 Convolution operation on two-dimensional image, with kernel size 2×2.

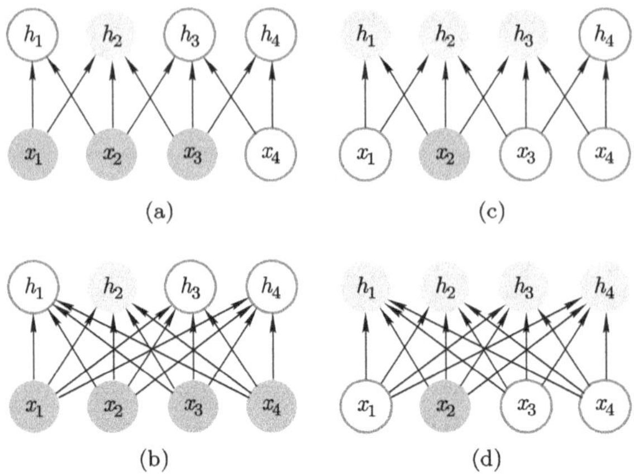

Figure 14.7 Comparison of sparse networks and fully connected networks.

the number of parameters and the amount of calculation are greatly reduced. Figure 14.7 shows the difference between convolution operation and fully connected network. The expression ability of CNN will not be affected by the adoption of sparse connectivity.

As shown in Figure 14.7, (a) and (c) depict the connection under the convolution operation. The kernel size is 3, and (b) and (d) are fully connected networks. Comparing (a) and (b), under convolution operation, the output unit h_2 is only related to the three input units, and each output unit of the fully connected network is related to all the input units. Comparing (c) and (d), in the convolution operation, the input x_2 only affects three (the kernel size) output units, while in the fully connected network, x_2 affects every output unit.

In a deep convolutional neural network, units in deeper layers indirectly connect to most inputs, as shown in Figure 14.8. The network can model complicated relationships efficiently with a few parameters. Units in the deep layer are indirectly connected to most or even all units of the shallow layer. For example, s_2 has indirect connections to all units in the bottom layer through h_1, h_2, h_3.

2. Parameter sharing

Parameter sharing means that multiple functions or operations share the same set of parameters in the model. In multi-layer perceptron, the weight parameters of each layer are used only once, and the

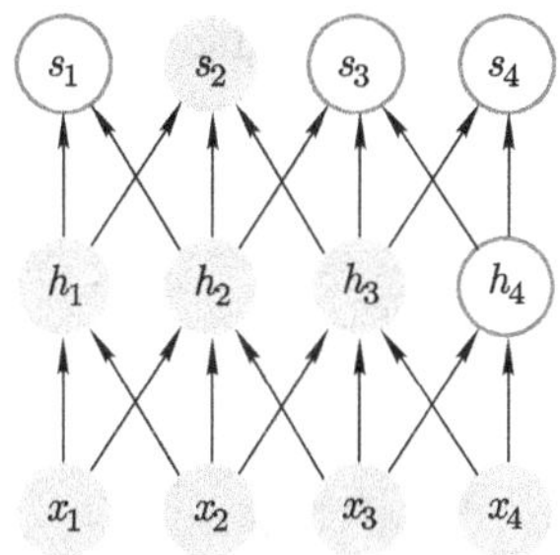

Figure 14.8 Connection of shallow layer units and deep layer units.

number of parameters gradually increases with the number of layers. In CNN, we use the same kernel to perform the convolution operation to get the output in each unit of the input. With parameter sharing, CNN only needs to learn one global parameter set, rather than one parameter set for each position. Parameter sharing can also reduces the number of parameters to k (the kernel size). k is typically much smaller than the input size m. For image data, the input is often tens of thousands of pixels, while the kernel size is in hundreds. To enhance the model expression ability, we can use different kernels in the same layer to conduct feature mapping.

3. Equivariant representation

After the convolution operation, parameter sharing makes the neural network layer equivariant to translation. Assume that $\boldsymbol{I}$ is the input image, and g is the translation transformation of the image. If we first apply translation transformation g to $\boldsymbol{I}$, and then apply convolution, the result will be the same as when we first apply convolution to $\boldsymbol{I}$, and then apply translation transformation g.

The convolution of the image produces a two-dimensional map to obtain a particular feature map of the input image. Note that convolution is not equivalent to some other transformations, such as scaling or rotation.

14.3.2. *Pooling*

Pooling is another local operation to update feature mapping. For image data, we can use a statistic within a rectangular neighborhood at each grid point to replace the output of the feature map at that point. For example, max pooling outputs the maximum value within a rectangular neighborhood. Other pooling operations include average pooling, ℓ_2-norm pooling, and weighted average pooling based on the distance to central pixel.

Pooling operation plays an important role in extracting multi-scale information. It makes the network similar to the cognitive function of human brain. Local features are extracted from shallow layers and relatively global features are extracted from deep layers. Pooling (especially max pooling) can also enhance the robustness of feature extraction. When there exist anomalies or fluctuations in the input features, max pooling can make the representations keep almost unchanged.

Another important function of pooling is to handle inputs of varying sizes. Take image classification as an example, the size of input images may be inconsistent, while the number of input units to the final classification layer needs to be fixed. We can adjust the size of the pooling area to get a fixed number of input units to the classification layer. Pooling has the effect of down sampling, which can reduce output dimension and improve computational efficiency. As Figure 14.9 shows, seven input units are down sampling to three output units through max pooling.

As to how to choose pooling functions, readers can refer to the work by Boureau *et al.*[199] We can also use dynamic pooling method to produce a different pooling region set for each image, such as running a clustering algorithm through the position of a feature.[200] Another method is to learn an individual pooling structure and then apply it to all images.[201]

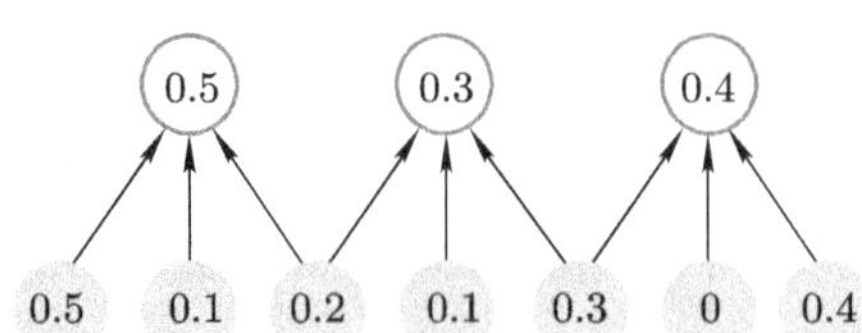

Figure 14.9 Max pooling has the effect of down sampling in CNNs.

14.3.3. *Common Convolutional Neural Networks Structures*

Convolutional neural networks are widely used in image recognition tasks. In these tasks, in addition to convolution and pooling operations, several fully connected layers are built before the output layer. Figure 14.10 shows the structure of the well-known *LeNet5* convolutional neural network in the early days, which has been successfully applied to the handwritten digit recognition problem.[202] LeNet5 has a profound impact on the development of convolutional neural networks.

As shown in Figure 14.10, the input of LeNet5 are images of handwritten digits (from 0–9). There are six kernels with the size of 5×5, followed by 2×2 down sampling (pooling) operation in the first convolution layer. Then the second convolution adopts 16 kernels followed by a 2×2 down sampling operation. Then there are two fully connected layers (i.e., the structure of the multi-layer perceptron). The last layer is a RBF output layer.

LeNet5 passes the input image through two convolutions: nonlinear transformation (using Sigmoid or tanh function) and pooling operation, after two fully concatenated layers (i.e., multi-layer perceptron), the final output layer is the Euclidean radial basis function (RBF) (usually using a Gaussian function). Given an input, the loss function should minimize the distance between the RBF parameter vector (i.e., the desired classification of the pattern) and the configuration of the F6 layer in Figure 14.10. The original LeNet5 output layer can also be replaced with the softmax layer.

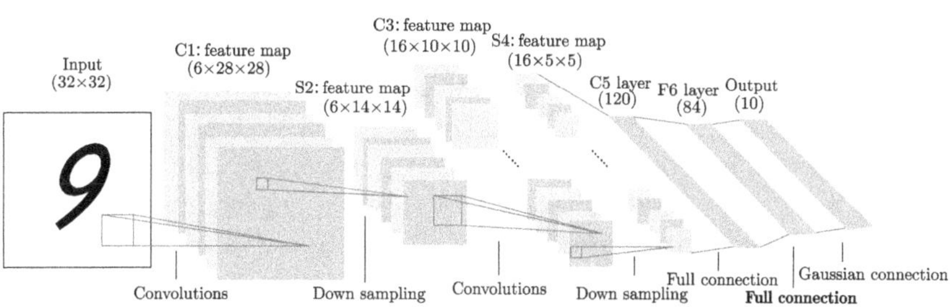

Figure 14.10 The network structure of LeNet5.

For a long time, little progress has been made in the research on neural networks, and CNNs were not popular. In 2012, AlexNet surpassed traditional classification methods by a large margin (such as SVM, random forest, etc.) in the image classification competition on the ImageNet dataset.[203] Since then CNN has attracted the attention to the deep learning model from the academic and industry communities, and gradually set off a wave of research and application of deep learning.

AlexNet made significant improvement based on LeNet5, and the major changes lie in the following aspects. ReLU is adopted as the activation function, instead of Sigmoid or tanh function in traditional neural networks. During the training processing, dropout technique is used to selectively ignore some neurons as a way to avoid overfitting and max pooling is used.[204] These innovations in AlexNet have made a major impact on the structure design and training of future networks, including VGG Network (with billions of parameters),[205] Network-in-network,[206] GoogleNet, Inception V2, Inception V3, Inception V4,[207] and ResNet[173] who won the ImageNet Competition 2016 and outperformed human beings in image recognition.

For the training of convolutional neural networks, the gradient of the parameters can be calculated by the backpropagation algorithm, and then the optimization methods in Section 14.2 can be used for training.

14.4. Recurrent Neural Network

If we use traditional neural networks to model a time series $(\boldsymbol{x}^{(1)}, \ldots, \boldsymbol{x}^{(\tau)})$, a separate set of parameters is required at each time step t. This method makes it difficult to share statistical features between different positions or sequences of variable lengths. It also greatly limits the model generalization ability. For example, it cannot handle sequences with length that does not appear in the training set.

Recurrent neural network (RNN) is a type of neural network designed to model time series and sequences of variable length.[183] The key idea of RNN is to share parameters in different modules of the model, so the model can be applied and generalized to samples of variable lengths. In RNN, each position of the input sequence is usually corresponding with a hidden state. The hidden state of current position is a function of its input value and the hidden state of the previous position. Parameter sharing is achieved through a recursive operation that adopt the functions with the same parameters in different time steps.

In this section, we will first introduce the computational graph of RNN. Then we will introduce several typical recurrent network structures and the gradient calculation in RNN. Finally, we will focus on an improved recurrent network: long short-term memory (LSTM) networks.

14.4.1. *Computational Graph of RNN*

In most recurrent neural networks, hidden units are used to represent the state of the network and data. For sequence $\boldsymbol{x} = (\boldsymbol{x}^{(1)}, \ldots, \boldsymbol{x}^{(T)})$, the hidden layer of RNN is often designed as a recursive form:

$$\boldsymbol{h}^{(t)} = f(\boldsymbol{h}^{(t-1)}, \boldsymbol{x}^{(t)}; \boldsymbol{\theta}), \tag{14.47}$$

where f is a nonlinear transformation and $\boldsymbol{\theta}$ is the parameter. It can be seen that at different time steps, the hidden units share the same form and parameters. Figure 14.11 shows the computational graph of a recurrent network and its corresponding unfolded form. The hidden state $\boldsymbol{h}^{(t)}$ at time step t is related to the input sequence before time step t.

In this RNN, the input sequence $\boldsymbol{x}$ is transformed to the hidden layer $\boldsymbol{h}$ by the nonlinear function f. The current hidden unit $\boldsymbol{h}^{(t)}$ is not only influenced by $\boldsymbol{x}^{(t)}$ but also by the previous hidden unit $\boldsymbol{h}^{(t-1)}$. Figure 14.11(a) is a schematic computational graph, and

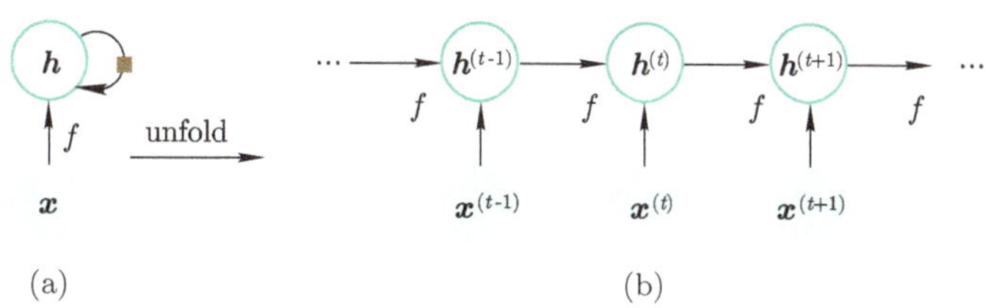

Figure 14.11 The computational graph of a recurrent network. (a) Schematic computational graph and (b) unfolded computational graph.

brown arrow indicates the delay of a single time step. Figure 14.11(b) is an unfolded computational graph, which shows the computation process of each hidden unit.

A typical RNN needs extra structures, such as an output layer that reads the hidden layer $\boldsymbol{h}$ for prediction. When RNN is used to predict future based on the past sequence information, $\boldsymbol{h}^{(t)}$ then represents the features of the past sequence of inputs up to time step t. Such a feature representation needs to be able to map a sequence with arbitrary length $(\boldsymbol{x}^{(1)}, \boldsymbol{x}^{(2)}, \ldots)$ to a hidden unit $\boldsymbol{h}^{(t)}$ with fixed length. This is a kind of lossy feature extraction and representation method. The network can automatically extract important features based on the objective function and discard insignificant features. For example, if a RNN is used to predict the next word in a sentence, the network only needs to extract features that can help predict the rest part of a sentence, rather than represent all the information of the input before time step t. However, other tasks may require the ability to extract the feature of the sentence to a maximum extent to enable the network to recover the original information. Machine translation is such a typical task for RNN, and we will describe this sequence-to-sequence model in the next section.

14.4.2. *Network Structure of RNN*

With the concept of unfolded computational graph and the idea of parameter sharing in mind, we will introduce the structure of recurrent neural network. Figure 14.12 gives several typical structures.

There is an output at each time step, and the shared parameter $\boldsymbol{W}$ has a recursive connection between the hidden layer units. The objective function L computes the distance between each output $\boldsymbol{o}$ and the true value $\boldsymbol{y}$. If a softmax layer is connected to $\boldsymbol{o}$, the objective function will calculate the difference between $\hat{\boldsymbol{y}} = softmax(\boldsymbol{o})$ and the true value $\boldsymbol{y}$. Assume that $\boldsymbol{U}$ is the weight matrix for the input to the hidden layer unit, and $\boldsymbol{W}$ is the weight matrix of the recurrent connection between the adjacent hidden layer, and $\boldsymbol{V}$ is the weight matrix between the hidden layer to the output. As shown in Figure 14.12(a), this network structure can be used to simulate a general Turing machine.

There is an output at each time step, but there is no connection between adjacent hidden layer units. The network structure is shown

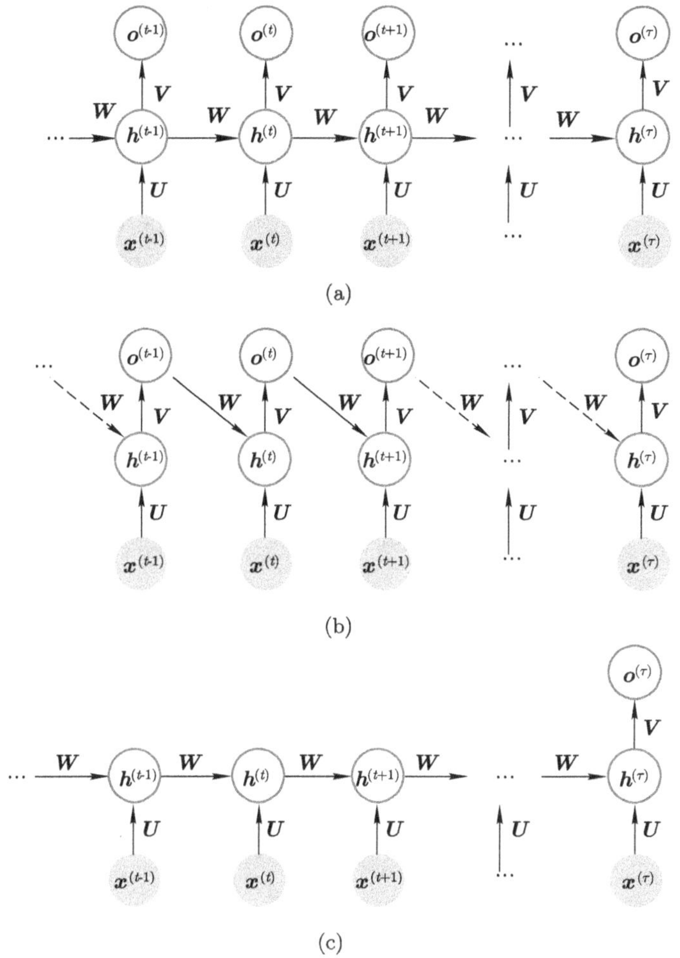

Figure 14.12 Several typical structures of RNN.

in Figure 14.12(b). Since there is lack of direct connection between hidden layer units in this network architecture, its expression ability is quite different from structure (a). However, the advantage of this RNN is that it can achieve decoupling (that is, there is no coupling between hidden layer units), which make it possible to train each time step in parallel.

There are connections between the hidden layer units, but only a single output at the end of the sequence. As shown in Figure 14.12(c), the network structure can extract the information of the input sequence into a fixed-length output, which is equivalent to an encoding process. This output can be used in the subsequent decoding processes, such as sequence-to-sequence models.

Now we will specifically introduce how to utilize the network architecture shown in Figure 14.12(a) to get the output and hidden layer units through forward propagation. We usually use tanh as the activation function and assume the output is discrete, such as word or character. The output $\boldsymbol{o}$ can be computed by the non-normalized probability of possible value of each discrete variable, and then use the softmax function to obtain the output vector with the normalized probability $\hat{\boldsymbol{y}}$. The forward propagation process from the initial state $\boldsymbol{h}^{(0)}$ of RNN is as follows. From $t = 1$ to $t = \tau$:

$$\begin{aligned} \boldsymbol{a}^{(t)} &= \boldsymbol{b} + \boldsymbol{W}\boldsymbol{h}^{(t-1)} + \boldsymbol{U}\boldsymbol{x}^{(t)}; \\ \boldsymbol{h}^{(t)} &= \tanh(\boldsymbol{a}^{(t)}); \\ \boldsymbol{o}^{(t)} &= \boldsymbol{c} + \boldsymbol{V}\boldsymbol{h}^{(t)}; \\ \hat{\boldsymbol{y}} &= softmax(\boldsymbol{o}^{(t)}), \end{aligned} \tag{14.48}$$

where $\{\boldsymbol{b}, \boldsymbol{c}, \boldsymbol{U}, \boldsymbol{V}, \boldsymbol{W}\}$ is the parameter set.

Since the input sequence is mapped to an output sequence of the same length, the total loss is the sum of the loss at all time steps. That is, if $L^{(t)}$ is the negative log likelihood of step t, $L^{(t)} = -\ln p_{\text{model}}\big(y^{(t)}|\{\boldsymbol{x}^{(1)}, \ldots, \boldsymbol{x}^{(t)}\}\big)$, then

$$\begin{aligned} & L\left(\{\boldsymbol{x}^{(1)}, \ldots, \boldsymbol{x}^{(\tau)}\}, \{y^{(1)}, \ldots, y^{(\tau)}\}\right) \\ & \quad = \sum_t L^{(t)} = -\sum_t \ln p_{\text{model}}\left(y^{(t)}|\{\boldsymbol{x}^{(1)}, \ldots, \boldsymbol{x}^{(t)}\}\right), \end{aligned} \tag{14.49}$$

where $y^{(t)}$ is the item with the correct label of $\hat{\boldsymbol{y}}^{(t)}$.

We will then describe a RNN-based modeling architecture called encoder-decoder sequence-to-sequence structure. By using this architecture, the input sequences with different lengths can be mapped to the sequences with different lengths. This network architecture has high flexibility, and it is widely used in speech recognition, machine translation and question answering systems.

Figure 14.13 shows a typical example of this architecture for machine translation. The encoder reads the English input sequence

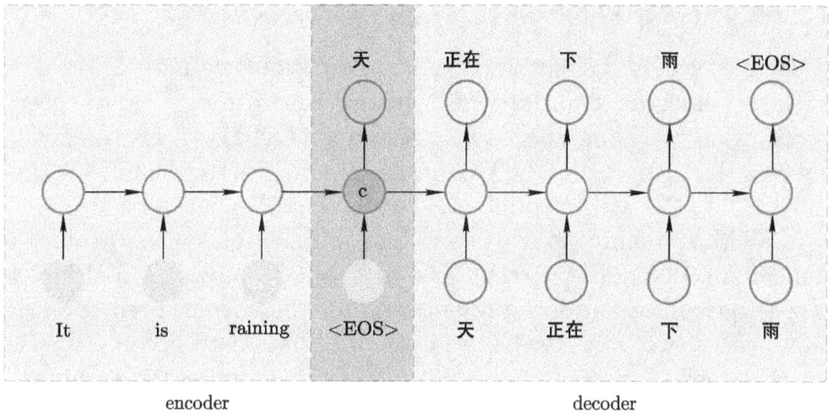

Figure 14.13 Sequence-to-sequence machine translation based on encoder-decoder architecture.

"It is raining", and outputs the representation of the sequence (also called context) c through RNN, which is generally a simple function of the hidden layer state. The symbol "⟨EOS⟩" in the figure is a symbol indicating the termination of the sequence. c is the representation of the decoder with respect to the input sequence, and the output of the translated Chinese sequence is "天正在下雨" through another RNN. During the processing, an output is generated at each time step, then the output is regarded as the input of the next time step to help generate the next output, and so on. It can be seen that the advantage of the architecture is that the length of input and output sequences can be arbitrary.

14.4.3. *Gradient Computation of RNN*

In RNN, gradient computation through backpropagation is often called backpropagation through time (BPTT). The gradient computation of RNN is very computationally expensive, because it needs a forward propagation and a backpropagation process for each step, and the time cost is $O(\tau)$. This propagation process is difficult to implement in parallel because forward propagation follows an inherent time order. In forward propagation, each state must be stored to be reused in backpropagation, so the memory cost is also $O(\tau)$. After obtaining the gradients by BPTT, they can be further used to train the RNN.

14.4.4. *Long Short-term Memory Networks*

RNN has achieved great success in many sequence modeling tasks, such as machine translation,[208] image generation,[209] and speech recognition.[210] Long short-term memory (LSTM)[211] is the key to success. In many tasks, LSTM performs much better than the standard RNN. We will introduce LSTM in the following section.

LSTM is mainly used to solve the problem of long-term dependencies in RNN. Recall that one of the core purposes of RNN is to use past information representation for the current prediction. In some cases, we only need the latest information to accomplish the current task. For example, in natural language processing, the previous words of a sentence are used to predict the next word. If you want to predict the last word in the sentence "the birds are flying in the ____", we only need to predict the last word as "sky" with a larger probability based on the information of the current sentence. Since the position of the word you want to predict is very close to the required information, the standard RNN can make an accurate prediction. But there are also some cases where we need more contextual information to make accurate predictions. For example, if you want to predict the last word in the sentence "He has been living in the USA.... He speaks fluent English", based on the contextual information of the nearest words, you can speculate that the last word may be a language. However, if we want to predict exactly which language is, you need the contextual information of "USA", which is far from the position.

When the standard RNN is used for long-term information representation, the multiplication of the same weight matrix will decrease or increase the gradient exponentially (depending on the eigenvalue of the matrix). If the magnitude of the long-term effect is too small, the long-term dependency of the signal will be easily masked by small fluctuations caused by short-term correlation. As a result, it takes a long time to learn the long-term dependencies. If the gradient magnitude of the long-term is too large, the gradient explosion problem will emerge, which will bring great instability to the gradient-based training method. In the standard RNN, the long-term information is poorly performed with the increase of the distance. For more detailed analysis, please refer to Benigo *et al.*[212,213]

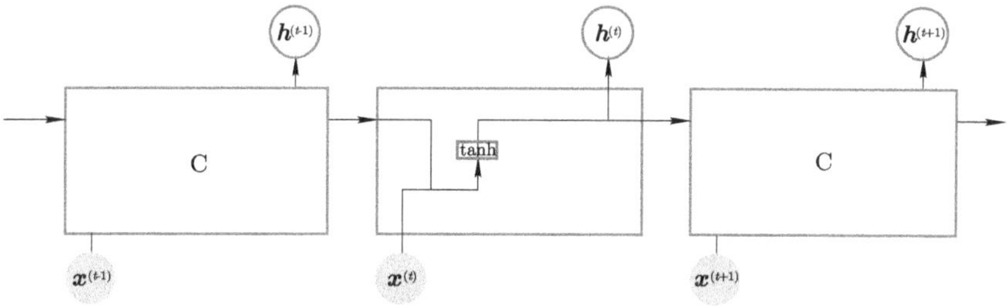

(a)

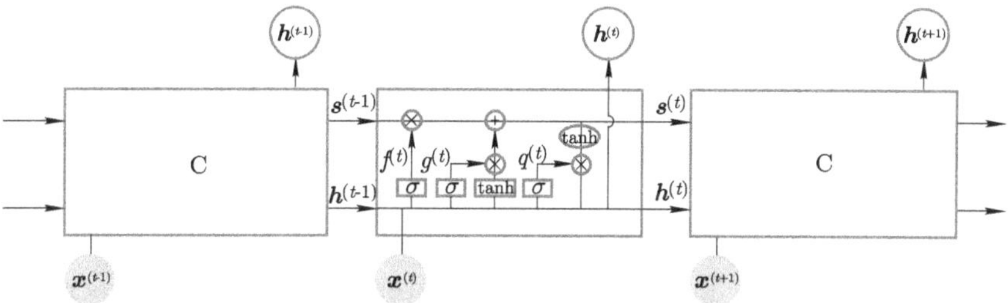

(b)

Figure 14.14 Comparison of cell structure between LSTM and standard RNN. (a) Cell structure of RNN and (b) cell structure of LSTM.

LSTM solves the problem of long-term dependencies. RNN uses recurrent network modules (also known as cells) to form long chains. In RNN, the structure of the cell is very simple (for example, only use the tanh activation function), as shown in Figure 14.14(a).

LSTM changed the internal structure of the cell. As shown in Figure 14.14(b), four nonlinear neural network layers are designed in each cell. LSTM also introduces the cell state $\boldsymbol{s}$. In LSTM, both hidden state $\boldsymbol{h}$ and cell state $\boldsymbol{s}$ are inputs to the next cell. Moreover, with the gate structure, LSTM has the ability to add or remove information to the cell state. A gate is a way of selectively passing information, which consists of an Sigmoid layer and an element-wise multiplication operation. The Sigmoid layer outputs a value between $0 \sim 1$, indicating the amount of information to pass through. 0 denotes that no information is allowed to pass, and 1 denotes that all the information passes. LSTM contains three such gates to protect and control the cell state. The information flow of each cell is updated as follows:

$$\boldsymbol{f}^{(t)} = \sigma\Big(\boldsymbol{b}^f + \boldsymbol{U}^f\boldsymbol{x}^{(t)} + \boldsymbol{W}^f\boldsymbol{h}^{(t-1)}\Big); \tag{14.50}$$

$$\boldsymbol{g}^{(t)} = \sigma\Big(\boldsymbol{b}^g + \boldsymbol{U}^g\boldsymbol{x}^{(t)} + \boldsymbol{W}^g\boldsymbol{h}^{(t-1)}\Big); \tag{14.51}$$

$$\boldsymbol{q}^{(t)} = \sigma\Big(\boldsymbol{b}^q + \boldsymbol{U}^q\boldsymbol{x}^{(t)} + \boldsymbol{W}^q\boldsymbol{h}^{(t-1)}\Big); \tag{14.52}$$

$$\boldsymbol{s}^{(t)} = \boldsymbol{f}^{(t)} \bigodot \boldsymbol{s}^{(t-1)} + \boldsymbol{g}^{(t)} \bigodot \tanh \left(\boldsymbol{b} + \sum_j \boldsymbol{U}\boldsymbol{x}^{(t)} + \boldsymbol{W}\boldsymbol{h}^{(t-1)}\right); \tag{14.53}$$

$$\boldsymbol{h}^{(t)} = \tanh\left(\boldsymbol{s}^{(t)}\right) \bigodot \boldsymbol{q}^{(t)}. \tag{14.54}$$

The forget gate is denoted by the Eq. (14.50), which is to decide how to discard information. $\{\boldsymbol{b}^f, \boldsymbol{U}^f, \boldsymbol{W}^f\}$ are the biases, input weights, and recurrent weights of the forget gate respectively. The external input gate is denoted by the Eq. (14.51) which is to determine the updated information with parameters $\{\boldsymbol{b}^g, \boldsymbol{U}^g, \boldsymbol{W}^g\}$; the output gate is denoted by formula Eq. (14.52) with its corresponding parameters. The cell state $\boldsymbol{s}^{(t)}$ and output $\boldsymbol{h}^{(t)}$ of LSTM are obtained by Eq. (14.53) and Eq. (14.54), respectively. It can be seen that the output is based on current cell state. Filter mechanism is also taken into consideration. Firstly, an output gate of the Sigmoid layer is constructed to determine which parts of the cell are outputs. Then the cell state will be passed through the tanh function (so that the output value is between -1 and 1), to be multiplied by the output gate, and output what we want.

During the development of LSTM, there have been many variants trying to improve the performance of LSTM for different tasks. For example, Gated Recurrent Unit (GRU) is one of the most commonly variants used in machine translation.[214]

This section is only a brief introduction to RNN. There are many extensions of RNN, such as deep RNN and bi-directional RNN[216] that increase the number of hidden layers in RNN. For more information, please refer to the works of Graves[217] and Goodfellow *et al.*[172]

14.5. Summary

This chapter introduced deep learning, which is one of the most popular techniques in current research and application of artificial intelligence. Specifically, we introduced the intuitive understanding of deep learning, three commonly used network architectures and the optimization algorithms of deep learning models.

Due to the rapid development of deep learning, many aspects are not covered in this chapter. For example, unsupervised learning models based on deep learning, autoencoders,[218] Deep Boltzmann Machines (DBM),[219] Generative Adversarial Nets (GAN)[220] and deep reinforcement learning.[174,221]

There are many well-designed open-source frameworks and software packages available for deep learning research and development, including TensorFlow,[185] Torch,[222] Caffe,[223] Theano,[186] Mxnet[187] and Keras.

Although deep learning has become the mainstream for solving many artificial intelligence tasks, there are still many problems. We list the most critical problems for readers to explore:

1. Why is deep learning so effective on image and speech tasks?

We can learn from the intuitive way or cognitive science that it can gradually learn the effective representation of information by building model in a hierarchical manner. However, there has long been a lack of theoretical interpretation ability to explain the effectiveness of deep learning, as many models are almost "black boxes".

2. What are the geometric features of the objective function of deep learning model?

Since the variable dimension of the objective function is extremely high and highly non-convex, it brings a great challenge for model optimization, and it makes difficult to analyze its geometric features.

3. How to choose a network structure?

At present, we can only try to choose the network structure according to the tasks and the performance of the model.

4. How to choose the hyper-parameters in the model training process?

The training process of deep learning model is often sensitive to hyper-parameters. There is still no theory to support how to adjust the learning rate, but mostly rely on tuning experience.

Chapter 15

Distributed Computing

With the development of information technology, our society is generating large-scale data all the time. The data come from a variety of sources, such as Internet data, including social networks, search data, and e-commerce platform data. There are 30,000 hours of music playback, 430,000 Wikipedia page visits, 4 million Google searches, and 10 million WeChat messages in one minute on the Internet. In addition to Internet data, large-scale data come from the fields of finance, carrier and health care. For example, in the financial sector, there are bank data, stock market data and insurance data. In the carrier sector, there are user communication behavior data, SMS and call data, mobile location data, and DPI data, etc. In the health care sector, there are medical records, medical image data, etc.

In the previous chapters of this book, we discussed data processing and analysis. It is generally assumed that data can be stored entirely in the memory of one computer. In fact, the above large-scale data can not be directly stored in the disk of one computer, not to mention the memory. In this large-scale data scenario, data storage, management, and processing are all facing great challenges, not to mention deep-level tasks such as data analysis and modeling.

This chapter first introduces the current popular distributed system Hadoop, including the distributed storage system HDFS and the distributed data processing framework MapReduce. We then describe how to perform data analysis in this new framework, and to implement the models commonly found in data science. After that, we introduce the Spark system which is developed in recent years and suitable for distributed data analysis. Finally, a brief introduction to other commonly used distributed systems is presented. The distributed system itself is complex to implement and requires a lot of knowledge in the field of computer engineering. This chapter is not meant to explain these implementation details, but rather introduces the basic principles and ideas to assist data analysts in choosing tools and systems.

15.1. Hadoop: Distributed Storage and Processing

Hadoop is originates from the open source search engine project Nutch, which is part of the open source text search project Lucene itself. In 2003 and 2004, Google published two papers introducing its large-scale data storage system Google file system (GFS)[226] and MapReduce[226] to the world. In 2004, the Nutch team imitated GFS to implement its own distributed file system Nutch distributed file system (NDFS) and MapReduce, of which NDFS can be regarded as the predecessor of HDFS. In 2006, NDFS and MapReduce in the Nutch project began to be independent and became a subproject of the Lucene project called Hadoop. In 2008, Hadoop officially became a top-level project of the Apache Software Foundation. Today, Hadoop has been adopted by many companies in the industry as standard tools for large-scale data storage and processing. Hadoop includes a series of related projects, Figure 15.1 shows the system architecture of Hadoop 1.x.

In this section, we mainly introduce Hadoop's two core components: HDFS and MapReduce. Among them, HDFS implements data storage, and MapReduce implements data processing. Most remaining components in Hadoop are based on these two components.

15.1.1. *HDFS: Distributed Data Storage*

Hadoop distributed file system (HDFS) is a distributed file system running in a cluster of ordinary computers. Compared with the traditional distributed file system, its main difference is that it has high fault tolerance and strong horizontal scalability. In a cluster environment of multiple ordinary computers, the failure of a single node computer is very common. Once a node fails, it means that the data stored on that node will be unavailable or even lost. HDFS uses the

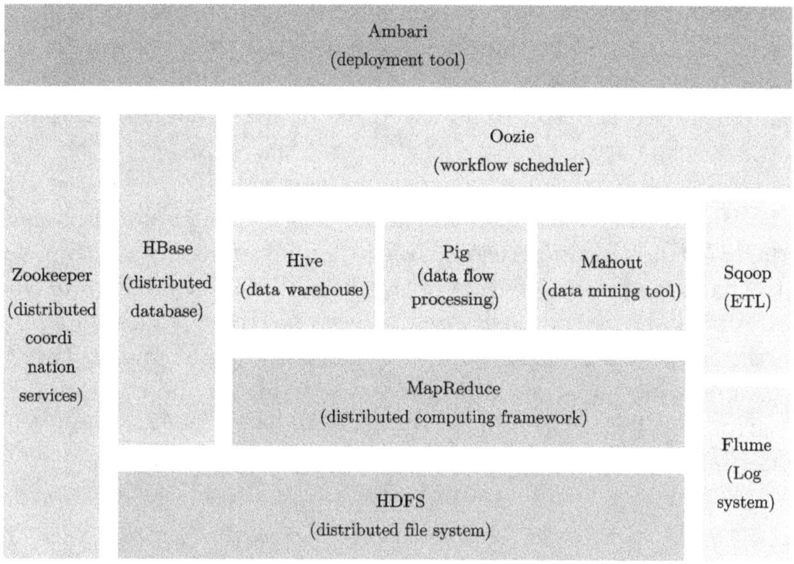

Figure 15.1 System architecture of Hadoop version 1.x.

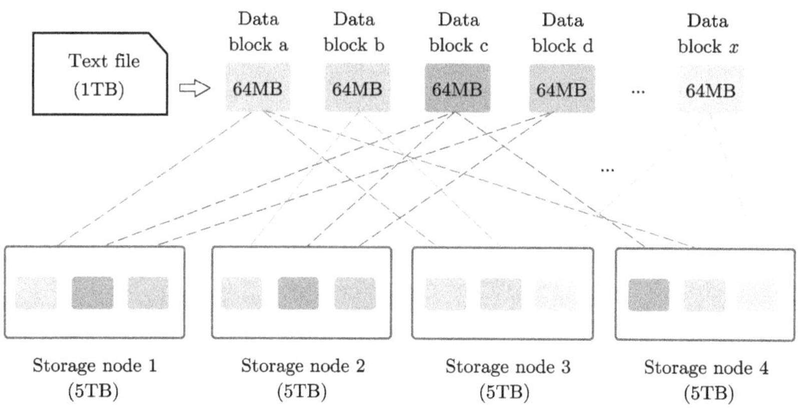

Figure 15.2 A large data file is divided into small data blocks to be stored on different nodes with HDFS.

data backup mechanism to solve this problem. Usually a large data file is divided into many small data blocks with the size of 64MB, each data block is copied into multiple copies (usually 3 copies) and stored on different nodes, as shown in Figure 15.2.

When a node fails, HDFS can get data from other nodes. In the Figure 15.2, assuming that node 1 fails, data block a can also be obtained from node 3 or node 4. Although multiple copies of data mean more storage space, as the price of hard drives continues to drop and capacity continues to increase, the strategy of sacrificing storage for data reliability is generally worthwhile.

In HDFS, the node responsible for storing data blocks is called *DataNode*, and the node responsible for maintaining information such as the backup storage location of the data blocks is called *NameNode*. Figure 15.3 shows the basic architecture of HDFS. When reading a file, the HDFS client obtains file storage location information through the NameNode and then downloads the file from the corresponding DataNode. When writing files, the HDFS client stores different data blocks into different DataNodes through the NameNode.

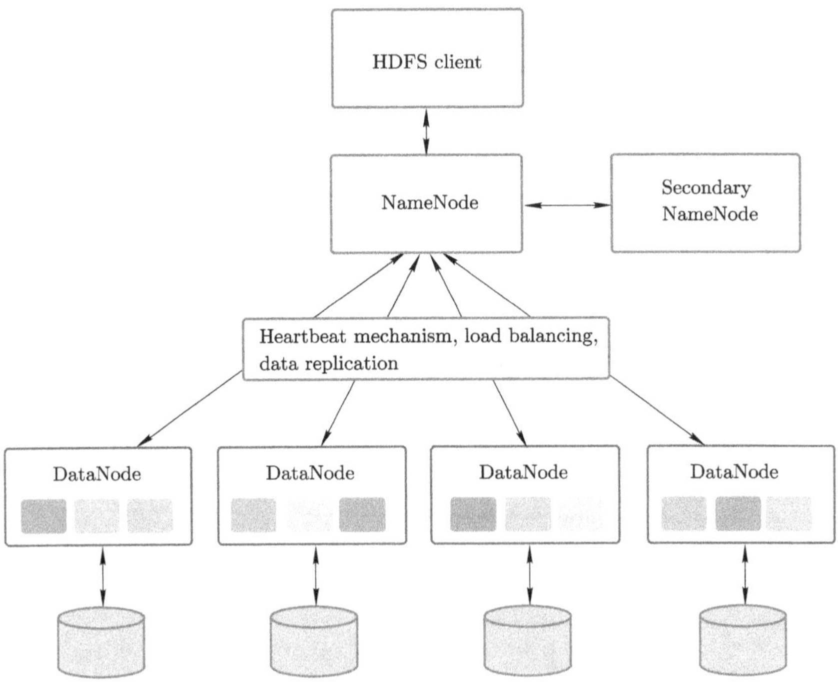

Figure 15.3 Basic architecture of HDFS.

Usually, there is only one NameNode, and there are many DataNodes. Obviously the NameNode will become the bottleneck of the entire distributed file system. Therefore, a *Secondary NameNode* is also set in the HDFS cluster. The Secondary NameNode will synchronize the information of the NameNode. When the NameNode fails, the Secondary NameNode will immediately provide a file read and write service in the place of the NameNode.

HDFS is suitable for large file storage and processing, and can process GB, TB or even PB level data. At the same time, it has a strong horizontal scalability, by increasing the number of nodes to expand the amount and performance of data processing, supporting a size of more than 10,000 nodes. HDFS has high fault tolerance through automatic multiple copies of data blocks. HDFS is very suitable for streaming file access, supporting application scenarios of one-time writing and multiple reading of files.

HDFS is not suitable for some application scenarios. Since it is designed to handle big data files, HDFS is not suitable for handling large-scale small files access. This is because large-scale small files will take up a lot of memory on the NameNode, causing too much load on the NameNode. At the same time, HDFS's support for concurrent writing and random modifications of files is not enough. In HDFS, a data file can only have one written operation at the same time. For the modification of the file, only the file append operation is supported, and the file content cannot be modified flexibly.

15.1.2. *MapReduce: Distributed Data Processing*

In order to use a distributed platform for efficient processing and large-scale data analysis, data analysts need to make some changes to the data representation and processing flow. In MapReduce, data are abstracted into key-value pairs, and the entire data processing procedure revolves around how to handle and convert such key-value pairs. The data processing flow is broken down into MapReduce jobs. A data analysis task may include only one MapReduce job or multiple MapReduce jobs.

MapReduce jobs include several stages: input, map, shuffle, reduce and output. Below, we will introduce each stage through a data

Table 15.1 Example of word frequency statistics in a document set.

	Amy, 2
Jane Amy Lisa	Ella, 3
Ella Ella Lisa	Jane, 2
Jane Ella Amy	Lisa, 2

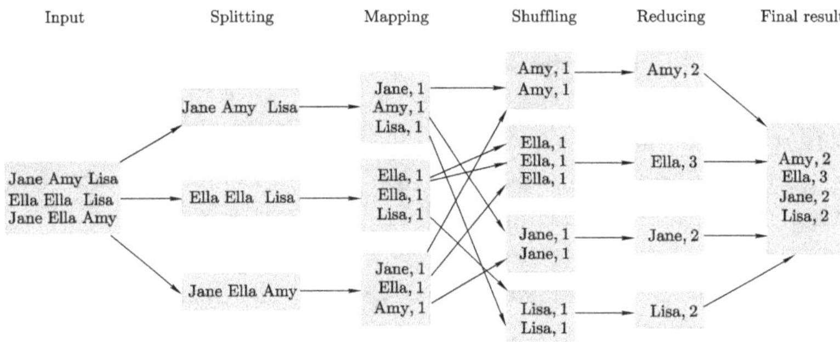

Figure 15.4 Complete process of the MapReduce job.

processing example. Suppose we need to deal with a large document set now, we want to count the frequency of each word in the document set. Suppose the document set is stored in HDFS. Each document is a line in a text file. The format of each line is a sequence of words separated by blank symbols. The data example is shown in the Table 15.1.

The complete process of the MapReduce job for the word count task is shown in Figure 15.4, and the description of each stage is as follows:

1. **Input stage:** A complete document set is stored in HDFS. For ease of explanation, assume that the three rows in Table 15.1 are stored in three DataNodes respectively in HDFS. Assume that each row is a key-value pair, where the key is the row number and the value is the textual content. For example, "Jane Amy Lisa" is represented as the key-value pair ⟨0, "Jane Amy Lisa"⟩;
2. **Map stage:** The map stage first performs word segmentation on the value part of each key-value pair passed in the input stage.

For each occurrence of the word w, the output is key-value pair $\langle w, 1\rangle$. For example, if you input $\langle 0$, "Jane Amy Lisa"$\rangle$, the map stage outputs three key-value pairs: $\langle$"Jane", 1$\rangle$, $\langle$"Amy", 1$\rangle$ and $\langle$"Lisa", 1$\rangle$;

3. **Shuffle stage:** For each key-value pair output in the map stage a hash function on the key will be used to group the pairs with same key to the same node in the cluster. For example, the pairs with the word "Amy" as the key will be assigned to the same node;
4. **Reduce stage:** The reduce node processes the key-value pairs corresponding to the same key. In this example, the reduce stage computes the sum of all the values corresponding to the same key. For example, for the "Amy" key, there are two key-value pairs: $\langle$"Amy", 1$\rangle$ and $\langle$"Amy", 1$\rangle$. The output of the reduce stage is the key-value pair $\langle$"Amy", 2$\rangle$;
5. **Output stage:** Combine all reduce results and output the final results. Usually, the output will also be stored in HDFS for subsequent data processing.

In the above stages, there are two stages that need to be concerned and implemented by the data analyst: the map stage and the reduce stage. These two stages correspond to the implementation of map and reduce functions. Both map and reduce functions use key-value pairs as input and output. The actual content of the key and value can be defined by the data analyst. The abstract definitions of the map function and the reduce function are as follows:

$$\begin{aligned} \text{map}(k_1, v_1) &\quad\rightarrow\quad \text{list}(k_2, v_2), \\ \text{reduce}(k_2, \text{list}(v_2)) &\quad\rightarrow\quad \text{newlist}(k_3, v_3). \end{aligned} \tag{15.1}$$

The map function receives the key-value pair $\langle k_1, v_1\rangle$ as input, and outputs a list of new key-value pair. The reduce function takes all the key-value pairs with the same key as input, and outputs the corresponding result.

Since the data are stored on multiple nodes of the cluster, after the data analyst implements the two core functions of map and reduce, the MapReduce platform distributes the map and reduce code to the node where the real data block is located, and in each node, completes the data processing task. In MapReduce, the node responsible for task management and coordination is called *JobTracker*, and

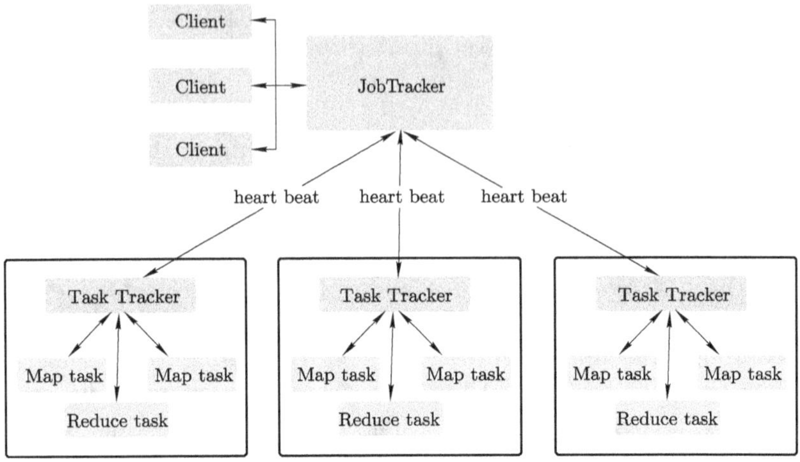

Figure 15.5 MapReduce architecture.

the node that actually performs data processing operations is called *TaskTracker*. The MapReduce architecture is shown in Figure 15.5. In a large-scale cluster, a single node failure is unavoidable. To support fault tolerance, the TaskTracker periodically sends status information to the JobTracker at specific time intervals (for example, every 10 minutes). The JobTracker determines whether the TaskTracker is alive based on the status information. Once a TaskTracker has not sent status information for more than 10 minutes, it will determine that the node has failed. The JobTracker no longer assigns new tasks to the node, and the tasks being performed on that node are assigned to a new node. In MapReduce, the above mechanism for fault tolerance by sending status information is called *heart beat mechanism*.

MapReduce provides a method for processing large-scale data, and it is especially suitable for batch processing of data. To use MapReduce for data analysis and modeling tasks, data analysts need to redesign the implementation logic of the algorithm. We need to split the existing data analysis tasks into MapReduce jobs consisting of map and reduce functions. In the next section, we will discuss how to implement common models in data science in MapReduce to handle large-scale data.

15.2. MapReduce Implementation of Common Models

Data processing and analysis will be significantly different from local environment, when data are stored in HDFS. Specifically, to take advantage of the large-scale data processing of the MapReduce platform, data analysts need to redesign the data processing flow and re-implement the data analysis model. In this section, we use linear regression, support vector machine, K-means, and PageRank as examples to discuss how to use MapReduce to implement the data analysis model discussed in the previous chapters of this book.

15.2.1. *Statistical Query Model*

The result of the statistical query model (SQM) is the expected estimate of the function $f(\boldsymbol{x}, y)$ (weighted average of the functions obtained for different local datasets). Many data analysis models can be solved by SQM.[228] The learning algorithm is implemented by calling the expectation of the statistical function $f(\boldsymbol{x}, y)$ on the dataset. The expected value of the query function can be calculated in parallel by MapReduce. Suppose the dataset is $D = \{(\boldsymbol{x}_1, y_1), (\boldsymbol{x}_2, y_2), \ldots, (\boldsymbol{x}_n, y_n)\}$. The dataset is stored of a Hadoop cluster of k storage nodes. The dataset can then be represented as $D = \{D_1, D_2, \ldots, D_k\}$, where the number of samples in the k-th data subset is denoted as n_k. Assuming the query function is $f(\boldsymbol{x}, y)$, one can use MapReduce to easily calculate:

1. **Map function:** Summing the data in the subset D_j: $f_j = \sum_{i=1}^{n_j} f(\boldsymbol{x}_i, y_i)$;
2. **Reduce function:** The local sum is aggregated to global sum: $\sum_{j=1}^{k} f_j$.

Now, suppose that the model learning depends on two functions $f(\boldsymbol{x}, y)$ and $g(\boldsymbol{x}, y)$. Then its MapReduce implementation is shown in Figure 15.6. Many models in data science can be generalized into statistical query models, such as linear regression, logistic regression, naive Bayes, support vector machine, decision tree, K-means, PageRank, the EM algorithm, and principal component analysis. Therefore, these models can be implemented using MapReduce.

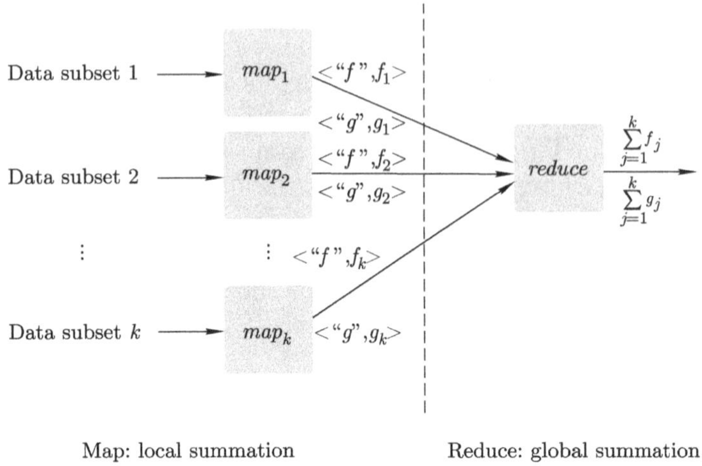

Figure 15.6 MapRuduce implementation example for statistical query.

15.2.2. *MapReduce Implementation of Linear Regression*

Suppose the dataset is $D = \{(\boldsymbol{x}_1, y_1), (\boldsymbol{x}_2, y_2), \ldots, (\boldsymbol{x}_n, y_n)\}$. The dataset is stored a Hadoop cluster with k nodes as $D = \{D_1, D_2, \ldots, D_k\}$, where the number of samples in the k-th subset is denoted as n_k.

The dataset is represented as a feature matrix $\boldsymbol{X}$ of n rows and d columns, and the target feature is represented as n-dimensional vector $\boldsymbol{y}$. According to Eq. (3.10), the parameter $\boldsymbol{w}$ can be computed as $\hat{\boldsymbol{w}} = (\boldsymbol{X}^{\mathrm{T}}\boldsymbol{X})^{-1}\boldsymbol{X}^{\mathrm{T}}\boldsymbol{y}$. To obtain the MapReduce implementation, we decompose the above computation into two parts:

$$\hat{\boldsymbol{w}} = \boldsymbol{A}^{-1}\boldsymbol{b}, \tag{15.2}$$

where $\boldsymbol{A} = \boldsymbol{X}^{\mathrm{T}}\boldsymbol{X}$, $\boldsymbol{b} = \boldsymbol{X}^{\mathrm{T}}\boldsymbol{y}$. Further, rewrite $\boldsymbol{A}$ and $\boldsymbol{b}$ as follows:

$$\begin{aligned} \boldsymbol{A} &= \sum_{i=1}^{n} \boldsymbol{x}_i\boldsymbol{x}_i^{\mathrm{T}} = \sum_{j=1}^{k}\sum_{i\in D_j} \boldsymbol{x}_i\boldsymbol{x}_i^{\mathrm{T}}, \\ \boldsymbol{b} &= \sum_{i=1}^{n} y_i\boldsymbol{x}_i = \sum_{j=1}^{k}\sum_{i\in D_j} y_i\boldsymbol{x}_i. \end{aligned} \tag{15.3}$$

We can calculate $\boldsymbol{A}$ and $\boldsymbol{b}$ by a MapReduce job. The map function implements the internal summation calculation, which calculates $\boldsymbol{A}_j = \sum_{i \in D_j} \boldsymbol{x}_i \boldsymbol{x}_i^{\mathrm{T}}$ and $\boldsymbol{b}_j = \sum_{i \in D_j} y_i \boldsymbol{x}_i$. The reduce function implements a global summation calculation, which computes $\boldsymbol{A} = \sum_{j=1}^{k} \boldsymbol{A}_j$ and $\boldsymbol{b} = \sum_{j=1}^{k} \boldsymbol{b}_j$.

1. **Map function:** Input the data subset D_j, calculate $\boldsymbol{A}_j$ and $\boldsymbol{b}_j$ respectively, and output two key-value pairs: $\langle$"A", $\boldsymbol{A}_j\rangle$ and $\langle$"b", $\boldsymbol{b}_j\rangle$;
2. **Reduce function:** Sum all the data with the key "A", sum all the data with the key "b", and calculate $\boldsymbol{A}$ and $\boldsymbol{b}$.

After completing the calculation of $\boldsymbol{A}$ and $\boldsymbol{b}$ by MapReduce job, we can use Eq. (15.2) to calculate the estimate of linear regression coefficient $\hat{\boldsymbol{w}}$.

15.2.3. *MapReduce Implementation of Support Vector Machine*

Suppose the training set is $D = \{(\boldsymbol{x}_1, y_1), (\boldsymbol{x}_2, y_2), \ldots, (\boldsymbol{x}_n, y_n)\}$, where the target feature y has a value range of $\{+1, -1\}$. Consider the original problem of the soft margin support vector machine, as shown in Eq. (4.61). We can turn it into an equivalent unconstrained optimization problem with the optimization objective

$$L(\boldsymbol{w}, b) = \frac{1}{2}\|\boldsymbol{w}\|_2^2 + C\sum_{i=1}^{n} \max\{0, 1 - y_i(\boldsymbol{w}^{\mathrm{T}} \boldsymbol{x}_i + b)\}. \tag{15.4}$$

Suppose we use the gradient descent method to solve the problem. By observing Eq. (15.4), we found that only the samples that satisfy $y_i(\boldsymbol{w}^{\mathrm{T}} \boldsymbol{x}_i + b) \leqslant 1$ contribute to the calculation of the gradient. These samples are actually the support vectors. In a configuration of $\boldsymbol{w}$ and b, we denote the corresponding support vector set as $S = \{(\boldsymbol{x}_s, y_s) | y_s(\boldsymbol{w}^{\mathrm{T}} \boldsymbol{x}_s + b) \leqslant 1\}$. Then the gradient is calculated as

$$\frac{\partial L(\boldsymbol{w}, b)}{\partial \boldsymbol{w}} = \boldsymbol{w} + C \sum_{(\boldsymbol{x}_i, y_i) \in S} (-y_i \boldsymbol{x}_i), \tag{15.5}$$

$$\frac{\partial L(\boldsymbol{w}, b)}{\partial b} = C \sum_{(\boldsymbol{x}_i, y_i) \in S} (-y_i). \tag{15.6}$$

In each iteration, we design a MapReduce job to compute $\frac{\partial L(\boldsymbol{w},b)}{\partial \boldsymbol{w}}$ and $\frac{\partial L(\boldsymbol{w},b)}{\partial b}$. The corresponding map and reduce functions are as follows:

1. **Map function:** The input is a data subset. For each sample $(\boldsymbol{x}_i, y_i)$, check if $y_i(\boldsymbol{w}^{\mathrm{T}}\boldsymbol{x}_i + b) \leqslant 1$ is satisfied. If it is satisfied, output two key-value pairs $\langle k_1, -y_i\boldsymbol{x}_i\rangle$ and $\langle k_2, -y_i\rangle$;
2. **Reduce function:** Calculate the sum of all values with the key k_1 to get $\sum_{(\boldsymbol{x}_i,y_i)\in S}(-y_i\boldsymbol{x}_i)$; Calculate the sum of all values with the key k_2 to get $\sum_{(\boldsymbol{x}_i,y_i)\in S}(-y_i)$.

After completing the calculation of $\frac{\partial L(\boldsymbol{w},b)}{\partial \boldsymbol{w}}$ and $\frac{\partial L(\boldsymbol{w},b)}{\partial b}$ with a MapReduce job, the parameters can be updated as:

$$\begin{aligned} \boldsymbol{w}^{(t+1)} &= \boldsymbol{w}^{(t)} - \eta \frac{\partial L(\boldsymbol{w}^{(t)}, b^{(t)})}{\partial \boldsymbol{w}^{(t)}}, \\ b^{(t+1)} &= b^{(t)} - \eta \frac{\partial L(\boldsymbol{w}^{(t)}, b^{(t)})}{\partial b^{(t)}}, \end{aligned} \tag{15.7}$$

where η is the learning rate. Suppose we need to perform T iterations, we need T MapReduce jobs to run serially.

15.2.4. *MapReduce Implementation of K-means*

First, we review the basic steps of the K-means algorithm:

1. Randomly select K points as the initial centroids;
2. Repeat the iterations as follows until convergence:
   (a) assign each sample to its nearest centroid, forming K clusters;
   (b) recalculate the centroid of each cluster;

Each iteration step can use a MapReduce job to complete the calculation. The (a) step corresponds to the map stage, and the (b) step corresponds to the reduce stage. Assuming the current centroids are $C = (c_1, c_2, \ldots, c_K)$, the corresponding map and reduce functions are:

1. **Map function:** Input a subset of the dataset. For each sample $\boldsymbol{x}_i$, calculate the distance between $\boldsymbol{x}_i$ and each current centroid. $\boldsymbol{x}_i$ is assigned to the closest cluster represented by its centroid (assumed to be cluster k), and the output key-value pair is $\langle k, \boldsymbol{x}_i \rangle$;
2. **Reduce function:** The samples with the same key in the input belong to the same cluster. Perform the summation operation and divide by the sample size in each key to get the updated centroid. Output key-value pairs $\langle k, \frac{\sum_{key=k} \boldsymbol{x}_i}{\sum_{key=k} 1} \rangle$.

15.2.5. *MapReduce Implementation of PageRank*

Suppose the graph G contains n nodes, and the entire graph is stored as an adjacency list. We use $N^-(v)$ to represent the set of nodes pointing to the node v, and $N^+(v)$ the set of nodes that v points to. As shown in Eq. (13.15), the updated formula of the PageRank value of the node in the PageRank algorithm is

$$\mathrm{PR}(v_j) = (1-\beta)\frac{1}{n} + \beta \sum_{v_i \in N^-(v_j)} \frac{\mathrm{PR}(v_i)}{|N^+(v_i)|}. \tag{15.8}$$

The iterative steps of the PageRank algorithm are: (1) each node allocates its score evenly to the nodes it points to; (2) each node sums all received scores, and then update its value according to Eq. (15.8). The above two steps can be implemented using the map and reduce functions respectively. It should be noted that we need to use the network structure information in each iteration. Therefore, in the reduce function, in addition to the score assigned by the current node v_i, it is also necessary to output its neighbor node list $N^+(v_i)$. The map and reduce functions that implement the PageRank algorithm are:

1. **Map function:** Input the node information $\langle v_i, (\mathrm{PR}(v_i), N^+(v_i))\rangle$, where $\mathrm{PR}(v_i)$ is the current score of the node. First, output the node's structure information as a key-value pair $\langle v_i, N^+(v_i)\rangle$. For each neighbor node v_j in $N^+(v_i)$, output a key-value pair $\langle v_j, \frac{\mathrm{PR}(v_i)}{|N^+(v_i)|}\rangle$;

2. **Reduce function:** For each key-value pair whose key is v_j, if its value is list type, then denote it as $N^+(v_j)$; if value is numeric, then all the values is summed to get $\sum_{v_i \in N^-(v_j)} \frac{\text{PR}(v_i)}{|N^+(v_i)|}$. Then update the value of node v_j according to Eq. (15.8) to get $\text{PR}(v_j)$. Output the key-value pair $\langle v_j, (\text{PR}(v_j), N^+(v_j))\rangle$.

It should be pointed out that the above single MapReduce job can only complete one round of score update. We need multiple MapReduce jobs to run serially to get the final result.

15.2.6. *Summary*

For more MapReduce implementations of data analysis models, interested readers can refer to the paper by Chu *et al.*[229] It should be noted that there may be multiple MapReduce implementations for the same model. This section just introduces one possible implementation. The Mahout project of the Apache Software Foundation focuses on how to implement the common machine learning and data mining algorithms in MapReduce.

Although most models can be implemented with MapReduce, MapReduce is not a good choice for distributed model training. This is because the training of most models requires multiple iterations, which means we need to iterate over the same dataset many times. In MapReduce, each iteration corresponds to one or more MapReduce jobs, and disks are used for data sharing between different jobs. In practice, the number of iterations is often hundreds and frequent disk reads can greatly reduce performance.

In the next section, we will introduce a memory-based distributed computing system called Spark. It designs a distributed dataset structure that uses memory instead of disk to store data. Since data is stored in memory, data sharing can be more efficient. Spark is more suitable for iterative data analysis tasks. In addition, MapReduce does not support well for interactive tasks that need to return results in milliseconds or seconds, streaming analytics of streaming change data.

15.3. Spark: Distributed Data Analysis

For many models in data science, the model training is done in an iterative form. In this scenario, we need to repeatedly access the same training set and calculate related functions on the training set to optimize the model parameters. In the previous section, we have discussed that each iteration step can be implemented using a MapReduce job. However, each MapReduce job needs to load data from disk, which causes serious performance problems.

For interactive data analysis (such as performing some SQL-like query operations in a massive dataset), it is often necessary to perform query operations repeatedly to understand and analyze the data. However, in MapReduce, each query needs to run a separate MapReduce job to go through the entire dataset from disk.

Hadoop is not a good solution for the above two types of tasks. The main bottleneck is that MapReduce uses the disk as a data storage in the calculation process. In this section, we will introduce a new distributed data processing platform Spark.[230] Spark is an efficient distributed computing system originally proposed by AMP Labs of the University of California, Berkeley. In 2010, it became the top project of Apache.

The core idea of Spark is to use memory instead of disk as the data storage. The data can be stored in the memory all the time, which can greatly improve the data processing speed.

The main differences between Spark and Hadoop include:

1. In Spark the intermediate output and results of jobs can be stored in memory, one no longer needs to read and write HDFS, greatly improving the efficiency of MapReduce;
2. Spark provides more dataset manipulation methods, giving data analysts more flexibility, unlike Hadoop which only provides map and reduce operations;
3. Spark has good support for machine learning algorithms and graph algorithms.

The Spark system is based on an innovative data abstraction type: resilient distributed dataset (RDD).[231] In this section, we first

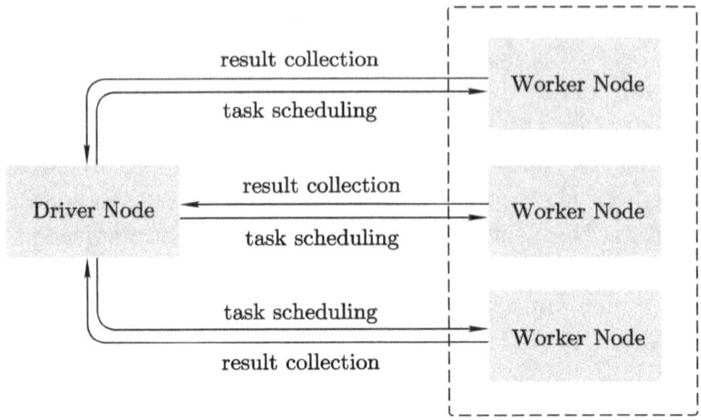

Figure 15.7 Data analysis task in Spark.

introduce RDD, then introduce the execution process of the data analysis task in Spark.

15.3.1. *Resilient Distributed Dataset*

The resilient distributed dataset (RDD) is the main data abstraction type in Spark. RDD is distributed. It abstracts a dataset physically distributed across multiple nodes into a logically complete dataset, so that data analysts can process a massive dataset just like a single-machine dataset. RDD is also resilient. The dataset is distributed and stored in the memory of the cluster nodes. When the memory size of the node is not enough, the data can be stored in the hard disk. Although this will reduce the performance, it can avoid the data processing task failure due to insufficient memory of the cluster node.

As shown in Figure 15.7, the dataset is initially stored in the disk of multiple data nodes of the distributed file system (such as HDFS). The data analyst completes the corresponding analysis program, mainly defining one or more RDDs, and then calling the operations on the RDD for data processing tasks. The data analysis program is submitted to Spark's *Driver*, and the Driver connects to the *Worker*s of the Spark cluster. The initial data are stored on the disk of the Worker. By defining RDD, Spark can load data into memory as RDD *partitions*. RDD partition can reside in memory throughout the data analysis process, greatly speeding up the analysis.

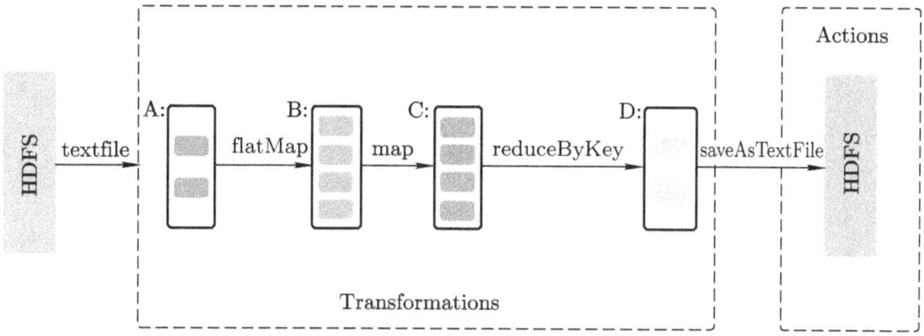

Figure 15.8 The data analysis task in Spark is implemented by constructing a transformation DAG between multiple RDDs.

There are three main ways to create a RDD. The first way is to parallelize the dataset in the existing single-machine environment and convert it into a RDD in the Spark cluster. The second way is to read file from the distributed file system, such as HDFS. The third way is to convert an existing RDD to a new RDD.

In Spark, data analysts do their work on data processing by using two types of operations defined on RDD: *transformations* and *actions*. Transformation operations include *map*, *filter*, *join*, and *flatMap*, which are used to construct logical dependencies between RDDs to define the entire data analysis process. Actions include *collect*, *reduce*, *count*, and *save*, which are used to collect results, save the results to an external storage system, or return them to the Driver. A complete list of RDD transformation operations and action operations can be found in the official documentation.

Note that all RDD transformation operations are delayed. The RDD transformation operation will generate a new RDD. The new RDD depends on the original RDD, and each RDD contains multiple partitions. The data analysis program actually constructs a directed acyclic graph (DAG) consisting of multiple RDDs as shown in Figure 15.8. Finally, the directed acyclic graph is submitted to Spark as a data analysis job.

15.3.2. *Execution Process of Spark Program*

A Spark program consists of two parts: the driver program and the worker program. The execution process of the Spark program is

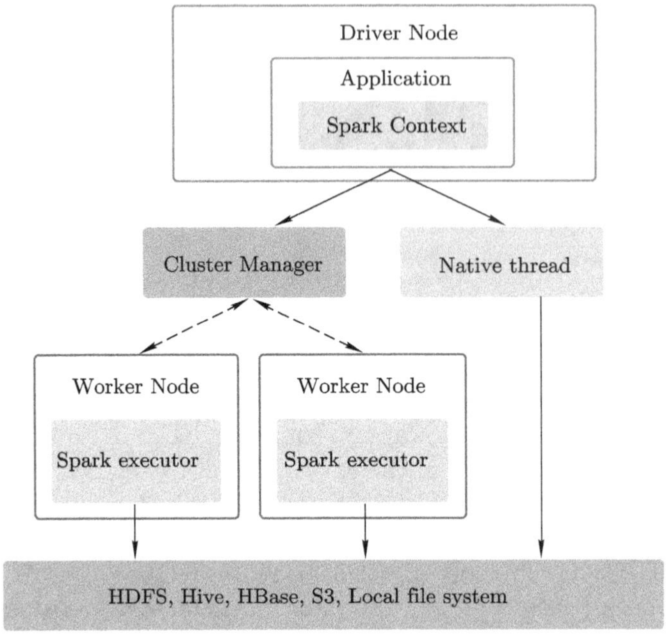

Figure 15.9 Spark program running diagram.

shown in Figure 15.9. The driver program is the entrance. A SparkContext is created to prepare the required running environment. The driver also performs the cluster resource application and task assignment. The worker program is a process running in the Worker node, and is responsible for running data processing tasks on the RDD partition stored in that node. Spark allows a pseudo-distributed environment to be configured on a single-machine, in which case the worker program uses one thread of the machine for simulation.

Below, we show how to use Spark to perform data processing with an example. Suppose we have a data file named *weibo.csv* on HDFS, and each line of the file represents a Chinese microblog. Now we need to perform the word segmentation task, and split each *microblog* into a collection of words, separated by spaces. The data analysis code implemented by Python is shown in Figure 15.10.

A SparkContext object *sc* is created first, where "*spark://bibdr-n1*" is the path of the Spark Driver, and "*testAPP*" is the name of this data analysis task. Then transform the *weibo.csv* file stored in

```
from pyspark import SparkContext
import jieba

sc = SparkContext("spark://bibdr-n1", 'testAPP')

weiboRDD = sc.textFile('hdfs://bibdr-n1:9000/test/weibo.csv')

def segment(text):
      return ' '.join(list(jieba.cut(text)))

segmentedWeibo = weiboRDD.map(segment) // transformation
top100 = segmentWeibo.top(100)// action
for text in top100:
    print(text,"n')
```

Figure 15.10 Chinese word segmentation using Spark.

HDFS to a RDD named *weiboRDD* in Spark via the *textFile* function. We implement a *segment* function to perform word segmentation. By invoking the *map* transformation of the *weiboRDD* object, the *segment* function is used as the input of the *map* function to generate a RDD named *segmentedWeibo*. Finally, the first 100 words are obtained *top* action, and the result *top100* is returned to the Driver node.

15.3.3. *Spark vs. Hadoop*

Compared with MapReduce, Spark's main advantage is to improve the performance of data processing. For example, researchers used Hadoop and Spark to train logistic regression models using the same training set, and found that Spark is more than 100 times more efficient.[229] The main reason is that data in Spark is shared through memory, while through disk in Hadoop.

Spark is not a substitute for Hadoop. Hadoop and Spark are complementary. Hadoop mainly provides a distributed storage system named HDFS for large-scale data. The data source of Spark may come from HDFS, and the data analysis results of Spark can be saved into HDFS. Hadoop also provides a computing framework named MapReduce. When data analysts perform data processing and analysis, they can choose either MapReduce or Spark. The main difference is that Spark performs better than MapReduce on some iterative and interactive tasks.

Spark also provides several data analysis tools, as shown in Figure 15.11. Spark SQL provides structured query and processing support, primarily for structured tabular data in databases. Sparking Streaming mainly supports the processing of stream data. MLlib is based on the Spark core module and implements many common machine learning algorithms. GraphX provides processing and analysis tools for graph data.

15.4. Other Distributed Systems

Based on Hadoop, there are a series of tools and projects that focus on specific tasks and scenarios. This section provides a brief introduction to some common projects.

Zookeeper aims to solve data management problems in distributed environments, including cluster management, universal naming and configuration synchronization, etc.[232]

HDFS was originally designed to store and manage text files. Hive and HBase are structured data processing tools for distributed clusters. Hive was originally open sourced by Facebook to handle massively structured log data. It is now a data warehouse based on Hadoop. Hive defines a SQL-like query language, HQL, which converts SQL into MapReduce tasks. HBase is a distributed column storage database, which is the open source implementation of Google BigTable.[233]

Mahout is a Hadoop-based data mining algorithm library. Mahout's main objective is to use MapReduce to implement classic

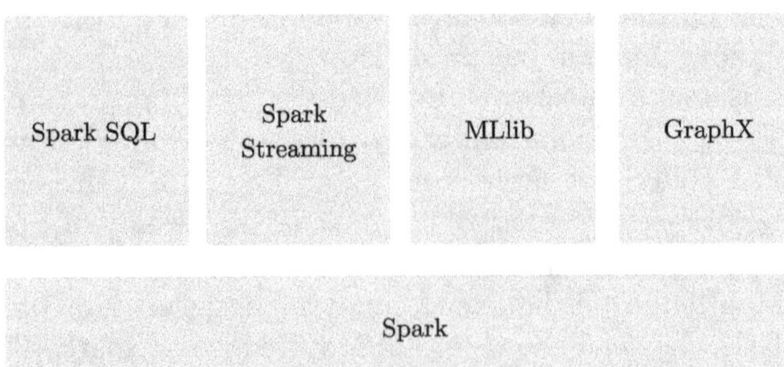

Figure 15.11 Spark data analysis tool.

algorithms in machine learning, facilitating data analysts to build models in a distributed environment. Pig is a data flow system based on Hadoop. Sqoop is a data synchronization tool which is mainly used to transfer data between traditional databases and Hadoop.

Ambari helps data analysts quickly configure and deploy Hadoop ecosystem-related platforms, and provides cluster maintenance and monitoring capabilities.

Appendix A

Matrix Operation

This appendix lists some matrix related concepts, arithmetic rules, and matrix decomposition.

A.1. Basic Concepts

Suppose $\boldsymbol{A}$ is an $m \times n$ matrix, and a_{ij} denotes the elements of the i-th row and j-th column of $\boldsymbol{A}$. If the number of rows and columns of the matrix are equal, $\boldsymbol{A}$ is called a square matrix. If the elements on the main diagonal in a square matrix are 1 and other elements are 0, it is called the unit matrix, generally referred to as $\boldsymbol{I}$.

A.1.1. *Transpose of Matrix*

The transposed matrix of $\boldsymbol{A}$ is usually denoted as $\boldsymbol{A}^{\mathrm{T}}$, which is an $n \times m$ matrix whose element in the i-th column and the j-th row are equal to the element of the i-th row and the j-th column of $\boldsymbol{A}$. Assuming that $\boldsymbol{A}$ and $\boldsymbol{B}$ are two matrices, c is a constant, then the transpose operation of the matrix satisfies the following properties:

1. $(\boldsymbol{A} \pm \boldsymbol{B})^{\mathrm{T}} = \boldsymbol{A}^{\mathrm{T}} \pm \boldsymbol{B}^{\mathrm{T}}$;
2. $(c\boldsymbol{A})^{\mathrm{T}} = c\boldsymbol{A}^{\mathrm{T}}$;
3. $(\boldsymbol{A}\boldsymbol{B})^{\mathrm{T}} = \boldsymbol{B}^{\mathrm{T}}\boldsymbol{A}^{\mathrm{T}}$.

A.1.2. *Inverse of Matrix*

For $n \times n$ square matrix $\boldsymbol{A}$, if there is $n \times n$ square matrix $\boldsymbol{B}$ such that $\boldsymbol{AB} = \boldsymbol{BA} = \boldsymbol{I}$, then $\boldsymbol{B}$ is the inverse matrix of $\boldsymbol{A}$. If matrix $\boldsymbol{A}$ has an inverse matrix, then $\boldsymbol{A}$ is called an invertible matrix. If $\boldsymbol{A}$ is an invertible matrix, its inverse matrix will be unique. The inverse matrix of $\boldsymbol{A}$ is also denoted as $\boldsymbol{A}^{-1}$. The inverse of matrix satisfies the following properties:

1. $(\boldsymbol{A}^{-1})^{-1} = \boldsymbol{A}$;
2. $(c\boldsymbol{A})^{-1} = \frac{1}{c}\boldsymbol{A}^{-1}(c \neq 0)$;
3. $(\boldsymbol{AB})^{-1} = \boldsymbol{B}^{-1}\boldsymbol{A}^{-1}$;
4. $(\boldsymbol{A}^{\mathrm{T}})^{-1} = (\boldsymbol{A}^{-1})^{\mathrm{T}}$.

A.1.3. *Rank of Matrix*

The rank of a matrix is the maximum number of linearly independent vectors in its row vectors (or column vectors). The rank of matrix $\boldsymbol{A}$ is usually denoted as $\mathrm{rank}(\boldsymbol{A})$. Assuming that $\boldsymbol{A}$ is a $m \times n$ matrix, it satisfies $\mathrm{rank}(\boldsymbol{A}) \leqslant \min\{m, n\}$. If the inequality takes the equal sign, i.e., $\mathrm{rank}(\boldsymbol{A}) = \min\{m, n\}$, then $\boldsymbol{A}$ is called a full rank matrix. For $n \times n$ square matrix $\boldsymbol{A}$, it is a full rank matrix, if and only if $\boldsymbol{A}$ is an invertible matrix. If $\boldsymbol{A}$ is not a full rank matrix, then $\boldsymbol{A}$ is also called a singular matrix.

A.1.4. *Trace of Matrix*

The sum of the main diagonal elements in the $n \times n$ square matrix $\boldsymbol{A}$ is called the trace of the matrix $\boldsymbol{A}$. The trace is generally denoted as $\mathrm{tr}(\boldsymbol{A})$. It is obvious by the definition that $\mathrm{tr}(\boldsymbol{A}) = \sum_{i=1}^{n} a_{ii}$. Assuming that both $\boldsymbol{A}$ and $\boldsymbol{B}$ are $n \times n$ square matrices, c and d are constants, the trace of matrix satisfies the following properties:

1. $\mathrm{tr}(\boldsymbol{A}) = \mathrm{tr}(\boldsymbol{A}^{\mathrm{T}})$;
2. $\mathrm{tr}(\boldsymbol{A} + \boldsymbol{B}) = \mathrm{tr}(\boldsymbol{A}) + \mathrm{tr}(\boldsymbol{B})$;
3. $\mathrm{tr}(c\boldsymbol{A} + d\boldsymbol{B}) = c\mathrm{tr}(\boldsymbol{A}) + d\mathrm{tr}(\boldsymbol{B})$;
4. $\mathrm{tr}(\boldsymbol{AB}) = \mathrm{tr}(\boldsymbol{BA})$.

A.1.5. *Vector Norm and Matrix Norm*

Norm is a measure of the magnitude of the elements in a vector or matrix. Commonly used vector norms include 1-norm, 2-norm and p-norm. Let $\boldsymbol{x}$ be a n-dimensional column vector, the 1-norm of $\boldsymbol{x}$ is the sum of the absolute values of its elements:

$$\|\boldsymbol{x}\|_1 = \sum_i |x_i|. \tag{A.1}$$

The 2-norm of a vector is the square root of the sum of the squares of all the elements in the vector, i.e.,

$$\|\boldsymbol{x}\|_2 = \sqrt{\sum_i x_i^2}. \tag{A.2}$$

The p-norm of a vector is the $1/p$ power of the sum of p power of the element's absolute value, i.e.,

$$\|\boldsymbol{x}\|_p = \left(\sum_i |x_i|^p\right)^{1/p}. \tag{A.3}$$

Commonly used matrix norms include 1-norm, ∞-norm and F-norm. The 1-norm of a matrix refers to the maximum of the sum of the matrix column vector elements' absolute values, i.e.,

$$\|\boldsymbol{A}\|_1 = \max_j \sum_i |a_{ij}|. \tag{A.4}$$

The ∞-norm of a matrix is the maximum of the sum of the matrix row vector element's absolute values, i.e.,

$$\|\boldsymbol{A}\|_\infty = \max_i \sum_j |a_{ij}|. \tag{A.5}$$

F-norm, also called Frobenius norm, refers to the square root of the sum of all element's squares in the matrix, i.e.,

$$\|\boldsymbol{A}\|_\mathrm{F} = \sqrt{\sum_i \sum_j a_{ij}^2}. \tag{A.6}$$

A.1.6. *Positive Definiteness of the Matrix*

For $n \times n$ square matrix $\boldsymbol{A}$, if $\boldsymbol{w}^{\mathrm{T}}\boldsymbol{A}\boldsymbol{w} > 0$ is satisfied for any non-zero vector $\boldsymbol{w}$, then $\boldsymbol{A}$ is a positive definite matrix. If $\boldsymbol{w}^{\mathrm{T}}\boldsymbol{A}\boldsymbol{w} \geqslant 0$ is satisfied for any vector $\boldsymbol{w}$, then $\boldsymbol{A}$ is a positive semidefinite matrix. We can also determine the positive definiteness of the matrix according to the eigenvalues of the matrix: if all eigenvalues of matrix $\boldsymbol{A}$ are greater than 0, then $\boldsymbol{A}$ is a positive definite matrix.

A.2. Matrix Derivation

Suppose $\boldsymbol{x}$, $\boldsymbol{w}$, $\boldsymbol{a}$ and $\boldsymbol{b}$ are column vectors, and $\boldsymbol{X}$, $\boldsymbol{A}$ and $\boldsymbol{B}$ are matrices. Some common matrix derivation formulas are as follows:

$$\frac{\partial \boldsymbol{x}^{\mathrm{T}}\boldsymbol{w}}{\partial \boldsymbol{x}} = \boldsymbol{w};$$

$$\frac{\partial \boldsymbol{x}^{\mathrm{T}}\boldsymbol{x}}{\partial \boldsymbol{x}} = 2\boldsymbol{x};$$

$$\frac{\partial \|\boldsymbol{x}\|_2^2}{\partial \boldsymbol{x}} = 2\boldsymbol{x};$$

$$\frac{\partial \boldsymbol{x}^{\mathrm{T}}\boldsymbol{A}\boldsymbol{x}}{\partial \boldsymbol{x}} = (\boldsymbol{A} + \boldsymbol{A}^{\mathrm{T}})\boldsymbol{x};$$

$$\frac{\partial \boldsymbol{w}^{\mathrm{T}}\boldsymbol{X}\boldsymbol{w}}{\partial \boldsymbol{X}} = \boldsymbol{w}\boldsymbol{w}^{\mathrm{T}};$$

$$\frac{\partial \boldsymbol{w}^{\mathrm{T}}\boldsymbol{X}\boldsymbol{a}}{\partial \boldsymbol{X}} = \boldsymbol{w}\boldsymbol{a}^{\mathrm{T}};$$

$$\frac{\partial \boldsymbol{w}^{\mathrm{T}}\boldsymbol{X}^{\mathrm{T}}\boldsymbol{X}\boldsymbol{a}}{\partial \boldsymbol{X}} = \boldsymbol{X}(\boldsymbol{w}\boldsymbol{a}^{\mathrm{T}} + \boldsymbol{a}\boldsymbol{w}^{\mathrm{T}});$$

$$\frac{\partial \mathrm{tr}(\boldsymbol{X})}{\partial \boldsymbol{X}} = \boldsymbol{I};$$

$$\frac{\partial \mathrm{tr}(\boldsymbol{X}\boldsymbol{A})}{\partial \boldsymbol{X}} = \boldsymbol{A}^{\mathrm{T}};$$

$$\frac{\partial \mathrm{tr}(\boldsymbol{X}^{\mathrm{T}}\boldsymbol{A})}{\partial \boldsymbol{X}} = \boldsymbol{A};$$

$$
\begin{aligned}
\frac{\partial \mathrm{tr}(\boldsymbol{A}\boldsymbol{X}^{\mathrm{T}})}{\partial \boldsymbol{X}} &= \boldsymbol{A};\\
\frac{\partial \mathrm{tr}(\boldsymbol{A}\boldsymbol{X}\boldsymbol{B})}{\partial \boldsymbol{X}} &= \boldsymbol{A}^{\mathrm{T}}\boldsymbol{B}^{\mathrm{T}};\\
\frac{\partial \mathrm{tr}(\boldsymbol{A}\boldsymbol{X}^{\mathrm{T}}\boldsymbol{B})}{\partial \boldsymbol{X}} &= \boldsymbol{A}\boldsymbol{B}.
\end{aligned}
$$

A.3. Matrix Decomposition

A.3.1. *Eigenvalue Decomposition*

Suppose $\boldsymbol{A}$ is a $n \times n$ square matrix. The eigenvalue of $\boldsymbol{A}$ and the corresponding eigenvector are scalar λ and non-zero vector $\boldsymbol{u}$ that satisfy

$$\boldsymbol{A}\boldsymbol{u} = \lambda \boldsymbol{u}. \tag{A.7}$$

The eigenvalues of $\boldsymbol{A}$ are usually written as a diagonal matrix $\boldsymbol{\Sigma}$, i.e., $\sigma_{ii} = \lambda_i$. If we combine the corresponding eigenvectors $\boldsymbol{u}_i$ into a matrix $\boldsymbol{U}$ by column, according to Eq. (A.7), we get

$$\boldsymbol{A}\boldsymbol{U} = \boldsymbol{U}\boldsymbol{\Sigma}. \tag{A.8}$$

If matrix $\boldsymbol{A}$ contains n linearly independent eigenvectors, i.e., $\boldsymbol{U}$ is a $n \times n$ full rank matrix. Therefore, $\boldsymbol{U}$ is also an invertible matrix, then Eq. (A.8) can be rewritten as

$$\boldsymbol{A} = \boldsymbol{U}\boldsymbol{\Sigma}\boldsymbol{U}^{-1}. \tag{A.9}$$

The process of decomposing the $n \times n$ full-rank square matrix $\boldsymbol{A}$ into the matrix multiplication in Eq. (A.9) is called eigenvalue decomposition. In particular, when $\boldsymbol{A}$ is a symmetric square matrix, it can be decomposed into

$$\boldsymbol{A} = \boldsymbol{U}\boldsymbol{\Sigma}\boldsymbol{U}^{\mathrm{T}}, \tag{A.10}$$

where $\boldsymbol{U}$ is an orthogonal matrix, and $\boldsymbol{\Sigma}$ is a diagonal matrix.

The sum of all the eigenvalues of $\boldsymbol{A}$ is equal to the trace of $\boldsymbol{A}$, i.e., $\mathrm{tr}(\boldsymbol{A}) = \sum_i^n \lambda_i$. Denote the eigenvalue of matrix $\boldsymbol{X}$ as $\mathrm{eig}(\boldsymbol{X})$, the following property satisfies:

$$\mathrm{eig}(\boldsymbol{A}\boldsymbol{B}) = \mathrm{eig}(\boldsymbol{B}\boldsymbol{A}). \tag{A.11}$$

A.3.2. *Singular Value Decomposition*

The eigenvalue decomposition requires that the matrix be a square matrix. In practice, most of the matrices we encounter are not square matrices. For $m \times n$ matrix $\boldsymbol{A}$, we can perform Singular Value Decomposition (SVD):

$$\boldsymbol{A} = \boldsymbol{U}\boldsymbol{\Sigma}\boldsymbol{V}^{\mathrm{T}}, \tag{A.12}$$

where $\boldsymbol{U}$ is an $m \times m$ square matrix, and its column vectors satisfy $\boldsymbol{U}^{\mathrm{T}}\boldsymbol{U} = \boldsymbol{I}$. $\boldsymbol{U}$ is usually called the left singular vector of $\boldsymbol{A}$. $\boldsymbol{V}$ is an $n \times n$ square matrix satisfying $\boldsymbol{V}^{\mathrm{T}}\boldsymbol{V} = \boldsymbol{I}$. The column vector of $\boldsymbol{V}$ is usually called the right singular vector of $\boldsymbol{A}$. $\boldsymbol{\Sigma}$ is a $m \times n$ rectangular diagonal matrix whose diagonal element σ_i is called the singular value of matrix $\boldsymbol{A}$.

In fact, the singular value decomposition and the eigenvalue decomposition have the following correspondence:

1. $\boldsymbol{U}$ is composed of the eigenvectors of the matrix $\boldsymbol{A}\boldsymbol{A}^{\mathrm{T}}$;
2. $\boldsymbol{V}$ is composed of the eigenvectors of the matrix $\boldsymbol{A}^{\mathrm{T}}\boldsymbol{A}$;
3. The square of the singular value of matrix $\boldsymbol{A}$ is both the eigenvalue of $\boldsymbol{A}\boldsymbol{A}^{\mathrm{T}}$ and $\boldsymbol{A}^{\mathrm{T}}\boldsymbol{A}$.

The above correspondence can be explained by the following two formulas:

$$\boldsymbol{A}\boldsymbol{A}^{\mathrm{T}} = \boldsymbol{U}\boldsymbol{\Sigma}\boldsymbol{V}^{\mathrm{T}}\boldsymbol{V}\boldsymbol{\Sigma}^{\mathrm{T}}\boldsymbol{U}^{\mathrm{T}} = \boldsymbol{U}(\boldsymbol{\Sigma}\boldsymbol{\Sigma}^{\mathrm{T}})\boldsymbol{U}^{\mathrm{T}}, \tag{A.13}$$

$$\boldsymbol{A}^{\mathrm{T}}\boldsymbol{A} = \boldsymbol{V}\boldsymbol{\Sigma}^{\mathrm{T}}\boldsymbol{U}^{\mathrm{T}}\boldsymbol{U}\boldsymbol{\Sigma}\boldsymbol{V}^{\mathrm{T}} = \boldsymbol{V}(\boldsymbol{\Sigma}^{\mathrm{T}}\boldsymbol{\Sigma})\boldsymbol{V}^{\mathrm{T}}. \tag{A.14}$$

Appendix B

Probability Basis

B.1. Basic Concepts

Probability measures the occurrence of an event. It is a number between 0 and 1. For example, $p(\mathrm{A})$ represents the probability of occurrence of event A. Joint probability is the probability that multiple events occur simultaneously. For example, $p(\mathrm{AB})$ represents the probability that event A and event B occur simultaneously. Conditional probability is used to measure the possibility of an event occurring given that another event has occurred. Usually $p(\mathrm{B}|\mathrm{A})$ is used to indicate that the probability of occurrence of B given A has occurred and $p(\mathrm{A}|\mathrm{B})$ represents the probability of occurrence of A given B has occurred. Joint probability can usually be written as

$$p(\mathrm{AB}) = p(\mathrm{A}|\mathrm{B})p(\mathrm{B}), \tag{B.1}$$

or

$$p(\mathrm{AB}) = p(\mathrm{B}|\mathrm{A})p(\mathrm{A}). \tag{B.2}$$

Combining the right terms of the equations (B.1) and (B.2), we can get

$$p(\mathrm{A}|\mathrm{B}) = \frac{p(\mathrm{B}|\mathrm{A})p(\mathrm{A})}{p(\mathrm{B})}. \tag{B.3}$$

Equation (B.3) is called Bayes' theorem. Usually we call $p(\mathrm{A})$ as a prior distribution, $p(\mathrm{B})$ as evidence, and $p(\mathrm{B}|\mathrm{A})$ as a likelihood. $p(\mathrm{A}|\mathrm{B})$ is the posterior distribution.

B.2. Common Probability Distribution

B.2.1. *Gaussian Distribution*

The Gaussian distribution is the most commonly used distribution, also known as normal distribution. In data science, we usually assume that model error follows the Gaussian distribution. The probability density function of the one-dimensional Gaussian distribution is

$$f(x|\mu,\sigma) = \frac{1}{\sqrt{2\pi}\sigma}\mathrm{e}^{-\frac{(x-\mu)^2}{2\sigma^2}}, \tag{B.4}$$

where μ is the mean, σ is the standard deviation. Usually we denote the Gaussian distribution as $N(\mu,\sigma^2)$. For example, $N(0,1)$ represents the standard normal distribution (that is, the Gaussian distribution with the mean of 0 and standard deviation of 1). Figure B.1 shows the probability density functions of the Gaussian distribution under different parameters.

The probability density function of the d-dimensional Gaussian distribution is

$$f(\boldsymbol{x}|\boldsymbol{\mu},\boldsymbol{\Sigma}) = \frac{1}{(2\pi)^{d/2}|\boldsymbol{\Sigma}|^{2/d}}\mathrm{e}^{-\frac{1}{2}(\boldsymbol{x}-\boldsymbol{\mu})^{\mathrm{T}}\boldsymbol{\Sigma}^{-1}(\boldsymbol{x}-\boldsymbol{\mu})}, \tag{B.5}$$

where $\boldsymbol{\mu}$ is the mean, $\boldsymbol{\Sigma}$ is the covariance matrix, and $|\boldsymbol{\Sigma}|$ is the determinant of $\boldsymbol{\Sigma}$.

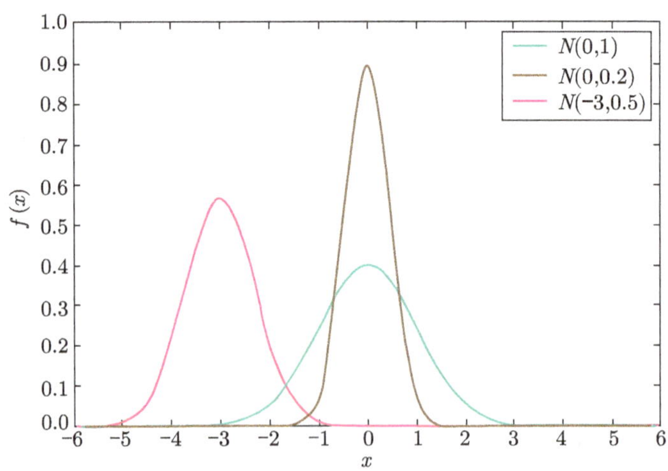

Figure B.1 Probability density functions of Gaussian distribution of three different parameters.

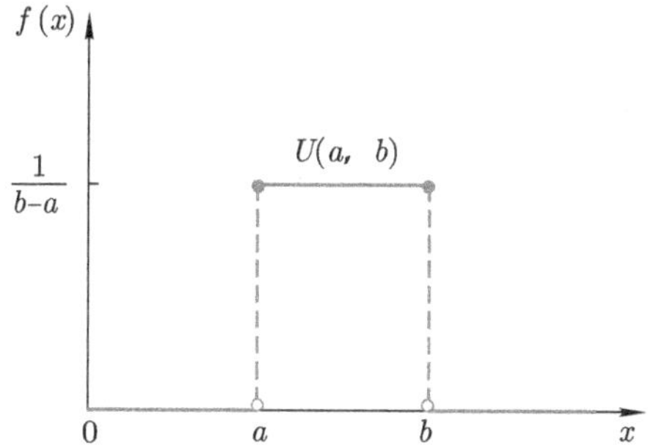

Figure B.2 Probability density function of continuous uniform distribution.

B.2.2. *Uniform Distribution*

Uniform distribution is a simple probability distribution, which can be divided into two types: discrete uniform distribution and continuous uniform distribution. For discrete random variable x with d values $\{1, 2, \ldots, d\}$, every one of d values has equal probability

$$p(x = i) = \frac{1}{d}. \tag{B.6}$$

Continuous uniform distribution on the interval $[a, b]$ is usually denoted as $U(a, b)$. Its probability density function is

$$f(x) = \begin{cases} \dfrac{1}{b-a}, & a \leqslant x \leqslant b; \\ 0, & \text{otherwise}, \end{cases} \tag{B.7}$$

with mean $\frac{a+b}{2}$ and variance $\frac{(b-a)^2}{12}$.

B.2.3. *Bernoulli Distribution*

The Bernoulli distribution is a probability distribution of a binary discrete variable. For random variable $x \in \{0, 1\}$, if it takes value 1 with probability μ, then the probability of taking value 0 is $1 - \mu$. The probability distribution of the Bernoulli distribution is

$$\text{Bern}(x|\mu) = \mu^x (1 - \mu)^{1-x}, \tag{B.8}$$

with mean μ and variance $\mu(1 - \mu)$.

B.2.4. *Binomial Distribution*

If we take n samples independently from a Bernoulli distribution with the mean of μ, what is the probability that 1 appears k times in these n samples? The above probability distribution is called the Binomial distribution. The probability distribution of the Binomial distribution is

$$\mathrm{Bin}(k|n,\mu) = \mathrm{C}_n^k \mu^k (1-\mu)^{n-k}, \tag{B.9}$$

where $\mathrm{C}_n^k = \frac{n!}{k!(n-k)!}$ is the number of combinations, the mean is $n\mu$ and the variance is $n\mu(1-\mu)$.

B.2.5. *Multinomial Distribution*

Suppose we have a d-dimensional random variable $\boldsymbol{x} = \{x_1, x_2, \ldots, x_d\}$. The expectation of each dimension of $\boldsymbol{x}$ is $\boldsymbol{\mu} = \{\mu_1, \mu_2, \ldots, \mu_d\}$, satisfying $\sum_{i=1}^d \mu_i = 1$. Each variable x_i is a Bernoulli distribution with the expected value μ_i, then

$$p(\boldsymbol{x}|\boldsymbol{\mu}) = \prod_{i=1}^{d} \mu_i^{x_i}. \tag{B.10}$$

If we take n samples independently from the above distribution, the probability of number of times that each dimension takes value 1 is $\boldsymbol{m} = \{m_1, m_2, \ldots, m_d\}$. The probability distribution of $\boldsymbol{m}$ can be characterized by multinomial distribution. Multinomial distribution is a generalization of binomial distribution from two-dimensional variable to multi-dimensional variable, and its probability distribution is

$$\mathrm{Mult}(\boldsymbol{m}|n,\boldsymbol{\mu}) = \frac{n!}{\prod_{i=1}^{d}(m_i!)} \prod_{i=1}^{d} \mu_i^{m_i}, \tag{B.11}$$

with mean $n\boldsymbol{\mu}$ and variance $n\boldsymbol{\mu}^{\mathrm{T}}\boldsymbol{\mu}$.

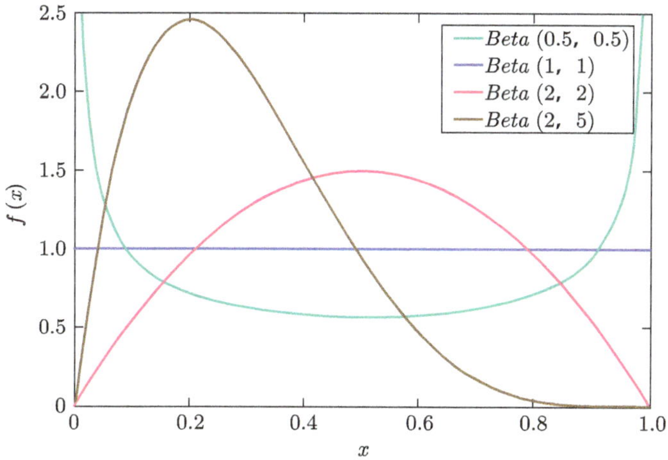

Figure B.3 Probability density function of Beta distribution.

B.2.6. *Beta Distribution*

Beta distribution is a probability distribution of continuous random variables on the interval [0, 1]. The probability density function is

$$\mathrm{Beta}(x|\alpha,\beta) = \frac{1}{\mathrm{B}(\alpha,\beta)} x^{\alpha-1}(1-x)^{\beta-1}, \tag{B.12}$$

where α and β are positive parameters. $\mathrm{B}(\alpha,\beta) = \frac{\Gamma(\alpha)\Gamma(\beta)}{\Gamma(\alpha+\beta)}$ is the normalization factor, where $\Gamma(\cdot)$ is the Gamma function. The mean of Beta distribution is $\frac{\alpha}{\alpha+\beta}$ and variance is $\frac{\alpha\beta}{(\alpha+\beta)^2(\alpha+\beta+1)}$. The probability density functions of Beta distribution with different α and β values are shown in Figure B.3.

B.2.7. *Dirichlet Distribution*

Dirichlet distribution is a multivariate generalization of Beta distribution. Suppose we have a d-dimensional random variable $\boldsymbol{x} = \{x_1, x_2, \ldots, x_d\}$. $\boldsymbol{x}$ is a $(d-1)$-simplex satisfying $x_i \in [0,1]$ and $\sum_{i=1}^{d} x_i = 1$. Dirichlet distribution has a d-dimensional parameter

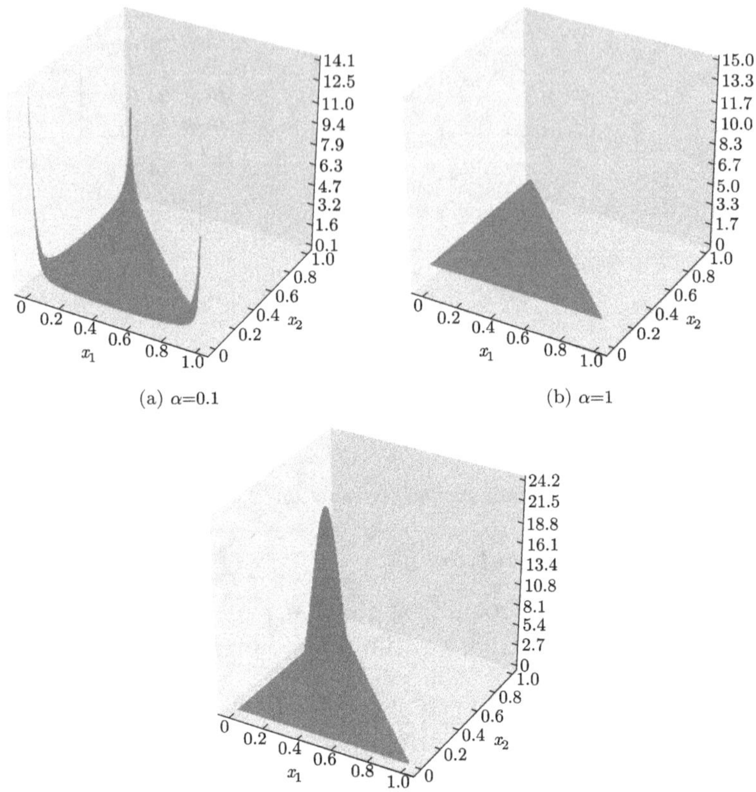

(a) α=0.1

(b) α=1

(c) α=10

Figure B.4 Probability density function of symmetric Dirichlet distribution.

$\boldsymbol{\alpha} = (\alpha_1, \alpha_2, \ldots, \alpha_d)^{\mathrm{T}}$ satisfying $\alpha_i > 0$. The probability density function of Dirichlet distribution is

$$\mathrm{Dir}(\boldsymbol{x}|\boldsymbol{\alpha}) = \frac{1}{\mathrm{B}(\boldsymbol{\alpha})} \prod_{i=1}^{d} x_i^{\alpha_i - 1}, \tag{B.13}$$

where $\mathrm{B}(\boldsymbol{\alpha}) = \frac{\prod_{i=1}^{d} \Gamma(\alpha_i)}{\Gamma\left(\sum_{i=1}^{d} \alpha_i\right)}$ is the normalization factor. The mean of the Dirichlet distribution is $\boldsymbol{\alpha} / \sum_{i=1}^{d} \alpha_i$.

In practice, each dimension of the parameter $\boldsymbol{\alpha}$ usually has the same value, which is called the symmetric Dirichlet distribution. In the 3-dimensional case, the density functions of symmetric Dirichlet distribution when $\alpha \in \{0.1, 1, 10\}$ are shown in Figure B.4.

Appendix C

Optimization Algorithm

In data science, model training is essentially solving an optimization problem. For example, in linear regression, our goal is to minimize the mean square error function, and the parameter is the coefficient vector. In this appendix, we will first introduce some basic concepts of optimization, and then introduce two types of optimization algorithms: gradient descent and Lagrange multiplier method.

C.1. Basic Concepts

C.1.1. *Convex Function*

For a unary function $f(x)$, if it satisfies for any $t \in [0, 1]$ that

$$f(tx_1 + (1-t)x_2) \leqslant tf(x_1) + (1-t)f(x_2), \tag{C.1}$$

then we call $f(x)$ a convex function. If it satisfies for any $t \in (0, 1)$ that

$$f(tx_1 + (1-t)x_2) < tf(x_1) + (1-t)f(x_2), \tag{C.2}$$

then $f(x)$ is a strict convex function. We can understand the characteristics of the convex function visually. The secant of the convex function is above the function curve, as shown in Figure C.1. The definition of the convex function can be directly generalized to the

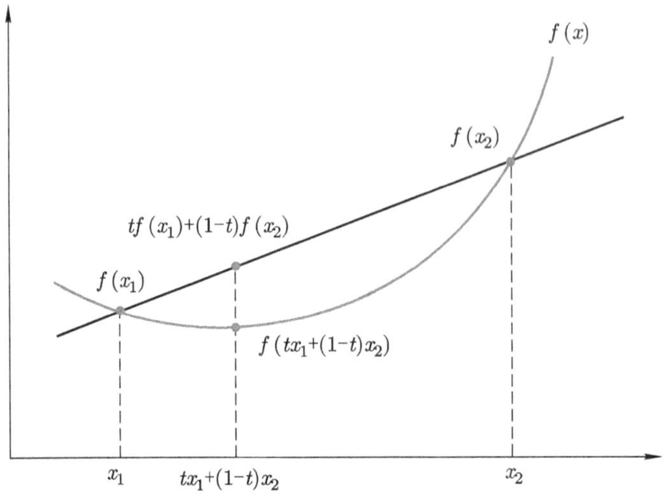

Figure C.1 The secant of the convex function is above the function curve.

multivariate function. Assuming that $f(\boldsymbol{x})$ is a multivariate function, if it satisfies for any $t \in [0,1]$ that

$$f(t\boldsymbol{x}_1 + (1-t)\boldsymbol{x}_2) \leqslant tf(\boldsymbol{x}_1) + (1-t)f(\boldsymbol{x}_2), \tag{C.3}$$

then $f(\boldsymbol{x})$ is a convex function.

In the training of data science models, if the objective function to optimization is a convex function, the local minimum is also the global minimum. This also means that the model we obtain is globally optimal and will not fall into the local optimum. For example, the objective function of the support vector machine is $\frac{1}{2}\|\boldsymbol{w}\|_2^2$, which is a convex function. How to determine whether a function is a convex function or not? For the unary function $f(x)$, we can determine by the sign of its second derivative $f''(x)$. If the second derivative of the function is always non-negative, i.e., $f''(x) \geqslant 0$, then $f(x)$ is a convex function. For multivariate function $f(\boldsymbol{x})$, we can determine by the positive definiteness of its Hessian matrix.[a] If the Hessian matrix is a semi-positive definite matrix, then $f(\boldsymbol{x})$ is a convex function.

[a]Hessian matrix is a square matrix composed of the second derivative of a multivariate function $f(x)$, whose i-th row and j-th column element is $\frac{\partial^2 f(\boldsymbol{x})}{\partial x_i \partial x_j}$.

Some common convex functions are listed below:

1. The quadratic function is usually a convex function (the quadratic coefficient is bigger than zero);
2. The exponential function is a convex function;
3. The negative logarithmic function is a convex function;
4. Commonly used matrix norm functions and vector norm functions are convex functions;
5. The linear combination (with non-negative weights) of convex functions is also a convex function.

C.1.2. *Jensen's Inequality*

For convex functions, we can generalize an important inequality called Jensen's inequality. If $f(\cdot)$ is a convex function and X is a random variable, then

$$f(E(X)) \leqslant E(f(X)). \tag{C.4}$$

(C.4) is the general form of Jensen's inequality. We introduce another form of the *Jensen's inequality as follows*. Suppose there are n samples $\{x_1, x_2, \ldots, x_n\}$ and their corresponding weights $\{\alpha_1, \alpha_2, \ldots, \alpha_n\}$ satisfying $\alpha_1 \geqslant 0$ and $\sum_{i=1}^{n} \alpha_i = 1$. For a convex function $f(\cdot)$, the following inequality holds:

$$f\left(\sum_{i=1}^{n} \alpha_i x_i\right) \leqslant \sum_{i=1}^{n} \alpha_i f(x_i). \tag{C.5}$$

Using Jensen's inequality, we can usually find a simple lower bound of a complex function. If it is difficult to directly optimize a complex function, we can continuously optimize its lower bounds.

C.2. Gradient Descent

Gradient descent, also known as steepest descent, is a first-order algorithm for solving unconstrained optimization problems. The gradient descent method is based on the following observations: if a real-valued function $f(\boldsymbol{x})$ is differentiable and defined at point a, then $f(\boldsymbol{x})$ decreases fastest along the direction of negative gradient $-\nabla f(\boldsymbol{x})$ at point a.

Assuming that $f(\boldsymbol{x})$ is a function with a first-order continuous partial derivative, the unconstrained optimization problem solved by the gradient descent method is

$$\min_{\boldsymbol{x}} \quad f(\boldsymbol{x}). \tag{C.6}$$

The gradient descent method is an iterative algorithm. Starting from the initial value $\boldsymbol{x}^{(0)}$, it is updated by the following iterative formula:

$$\boldsymbol{x}^{(t+1)} = \boldsymbol{x}^{(t)} - \lambda_t \nabla f(\boldsymbol{x}^{(t)}), \tag{C.7}$$

where λ_t is the step size. Through the iteration process, we can continuously decrease the value of the objective function until convergence. The value of $\boldsymbol{x}$ is updated in the negative gradient direction in each iteration. When the objective function is a convex function, the solution of the gradient descent method is globally optimal.

If the objective function $f(\boldsymbol{x})$ is second-order continuous and differentiable, it can be solved by Newton's method. The number of iterations of Newton's method is much smaller than the gradient descent method. However, solving the inverse of the Hessian matrix of the objective function is computationally expensive. Especially for high-dimensional problems, the computation of the inverse of the Hessian matrix is almost infeasible. In this case, we can approximate the inverse of the Hessian matrix to reduce computational load. This method is called the quasi-Newton method.

C.3. Lagrangian Multiplier Method

As we mentioned earlier, many problems in data science are essentially optimization problems. Usually there are constraints, so we often need to solve a constrained optimization problem. Lagrange multiplier is the most classic method to solve constrained optimization problems. It rewrites the objective function by introducing a set of Lagrange multipliers, which can transform a constrained optimization problem into an unconstrained one.

Suppose we need to solve the following optimize problem:

$$\begin{aligned} \min_{\boldsymbol{x}} \quad & f(\boldsymbol{x}), \\ \text{s.t.} \quad & h_i(\boldsymbol{x}) = 0, \ (i = 1, \ldots, m), \\ & g_j(\boldsymbol{x}) \leqslant 0, \ (j = 1, \ldots, n). \end{aligned} \tag{C.8}$$

We can introduce the Lagrange multipliers $\boldsymbol{\lambda} = \{\lambda_1, \lambda_2, \ldots, \lambda_m\}$ and $\boldsymbol{\mu} = \{\mu_1, \mu_2, \ldots, \mu_n\}$ to form the corresponding Lagrangian function

$$L(\boldsymbol{x}, \boldsymbol{\lambda}, \boldsymbol{\mu}) = f(\boldsymbol{x}) + \sum_{i=1}^{m} \lambda_i h_i(\boldsymbol{x}) + \sum_{j=1}^{n} \mu_j g_j(\boldsymbol{x}), \tag{C.9}$$

where $\mu_j \geqslant 0$. Now, our problem is equivalent to the following problem:

$$\min_{\boldsymbol{x}} \max_{\boldsymbol{\lambda}, \boldsymbol{\mu}} L(\boldsymbol{x}, \boldsymbol{\lambda}, \boldsymbol{\mu}). \tag{C.10}$$

This problem is often called the primal problem. We can define the corresponding dual problem:

$$\max_{\boldsymbol{\lambda}, \boldsymbol{\mu}} \min_{\boldsymbol{x}} L(\boldsymbol{x}, \boldsymbol{\lambda}, \boldsymbol{\mu}). \tag{C.11}$$

Usually the extremum of the primal problem is greater than the extremum of the dual problem. When the objective function and the constraint functions satisfy certain condition, the primal problem and the dual problem are equivalent. In this case, we can include that the strong duality holds. The condition is called the Slater condition and its contents are as follows:

1. $f(\boldsymbol{x})$ and $g(\boldsymbol{x})$ are convex functions;
2. $h(\boldsymbol{x})$ is an affine function;
3. There exits $\boldsymbol{x}_0$, for $j \in \{1, 2, \ldots, n\}$, $g_j(\boldsymbol{x}_0) < 0$.

The Slater condition is a sufficient condition for strong duality. When the optimization problem satisfies the strong duality, we can obtain the optimal solution through a set of conditions. These conditions are called Karush–Kuhn–Tucker conditions (referred to as KKT conditions).

KKT conditions are as follows:

$$\begin{cases} \dfrac{\partial L(\boldsymbol{x}, \boldsymbol{\lambda}, \boldsymbol{\mu})}{\partial \boldsymbol{x}} = 0; \\ \dfrac{\partial L(\boldsymbol{x}, \boldsymbol{\lambda}, \boldsymbol{\mu})}{\partial \boldsymbol{\lambda}} = 0; \\ \dfrac{\partial L(\boldsymbol{x}, \boldsymbol{\lambda}, \boldsymbol{\mu})}{\partial \boldsymbol{\mu}} = 0; \\ g_j(\boldsymbol{x}) \leqslant 0, & j = 1, 2, \ldots, n; \\ \mu_j \geqslant 0, & j = 1, 2, \ldots, n; \\ \mu_j g_j(\boldsymbol{x}) = 0, & j = 1, 2, \ldots, n. \end{cases} \tag{C.12}$$

When the strong duality is satisfied, the KKT conditions are necessary conditions for the optimal solution.

Appendix D

Distance

In data science, the distance or similarity of the data needs to be calculated in many scenarios, such as the K-nearest neighbor algorithm and most clustering models. Assume that $d(\cdot)$ is a distance function, the following properties need to be satisfied:

1. It is usually non-negative: $d(x, y) \geqslant 0$;
2. The distance between a sample and itself is zero: $d(x, x) = 0$;
3. The distance is usually symmetric: $d(x, y) = d(y, x)$;
4. It satisfies the triangle inequality: $d(x, z) \leqslant d(x, y) + d(y, z)$.

This appendix introduces several distance metrics, including Euclidean distance, Manhattan distance, Mahalanobis distance, Hamming distance, cosine similarity, Pearson correlation coefficient, Jaccard distance and KL divergence. Distance measures play an important role in data analysis. The choice of distance metrics depends on both the data types and the problem we are solving.

D.1. Euclidean Distance

Euclidean distance is a frequently distance metric. It measures the absolute distance between two samples in multidimensional space. Suppose $\boldsymbol{x}_1 = \{x_{11}, x_{12}, \dots, x_{1d}\}$ and $\boldsymbol{x}_2 = \{x_{21}, x_{22}, \dots, x_{2d}\}$ are two d-dimensional samples, then the Euclidean distance between $\boldsymbol{x}_1$

and $\boldsymbol{x}_2$ is

$$d(\boldsymbol{x}_1, \boldsymbol{x}_2) = \sqrt{\sum_{i=1}^{d}(x_{1i} - x_{2i})^2}. \tag{D.1}$$

It can be seen that Euclidean distance is determined by the square sum of the difference of each dimension between two samples. Euclidean distance is sensitive to scale of features and easily dominated by those features with large range of values. Thus data standardization are usually needed before using Euclidean distance.

D.2. Manhattan Distance

Suppose $\boldsymbol{x}_1 = \{x_{11}, x_{12}, \ldots, x_{1d}\}$ and $\boldsymbol{x}_2 = \{x_{21}, x_{22}, \ldots, x_{2d}\}$ are two d-dimensional samples, the Manhattan distance between $\boldsymbol{x}_1$ and $\boldsymbol{x}_2$ is

$$d(\boldsymbol{x}_1, \boldsymbol{x}_2) = \sum_{i=1}^{d} |x_{1i} - x_{2i}|. \tag{D.2}$$

Manhattan distance is equal to the sum of absolute values of the difference between two samples of each dimension. The meaning of Manhattan distance can correspond to the shortest taxi distance between two locations in a planned square building city (such as Manhattan). When using Manhattan distance, the scale of features should also be considered. Compared with Euclidean distance, Manhattan distance is more robust and less sensitive to outliers.

D.3. Mahalanobis Distance

Mahalanobis distance is another metric to measure the distance between two samples. It needs not only the information of the sample itself, but also the information of the whole dataset (the covariance matrix $\boldsymbol{\Sigma}$). The calculation of Mahalanobis distance is shown in Eq. (D.3).

$$d(\boldsymbol{x}_1, \boldsymbol{x}_2) = \sqrt{(\boldsymbol{x}_1 - \boldsymbol{x}_2)^{\mathrm{T}} \boldsymbol{\Sigma}^{-1} (\boldsymbol{x}_1 - \boldsymbol{x}_2)}. \tag{D.3}$$

Unlike Euclidean distance or Manhattan distance, the correlation between different features are taken into account in Mahalanobis distance. For example, height can bring information about weight. Another advantage of Mahalanobis distance is that it is scale *invariant*, that is, the scale of features will not affect the distance of samples. Thus it can be directly applied to a dataset without standardization.

D.4. Hamming Distance

Hamming distance is mainly used to measure the similarity between two strings of the same length. It is defined as the number of different characters at the corresponding positions of two strings. Hamming distance indicates the number of characters needed to be replaced to convert one string into another. Assuming that the length of the string is d, the Hamming distance between string $\boldsymbol{x}_1 = \{x_{11}, x_{12}, \dots, x_{1d}\}$ and string $\boldsymbol{x}_2 = \{x_{21}, x_{22}, \dots, x_{2d}\}$ is

$$\text{Hamming}(\boldsymbol{x}_1, \boldsymbol{x}_2) = d - \sum_{i=1}^{d} \text{I}(x_{1i}, x_{2i}), \tag{D.4}$$

where $\text{I}(\cdot)$ is the indicator function, and when its value equals to 1, the characters corresponding to a certain position are the same. In practice, strings can be English words, signal sequences and DNA sequences.

D.5. Cosine Similarity

Cosine similarity is used to measure the similarity of two samples in multidimensional space. It is calculated by the angle between two sample vectors. The range of cosine similarity is $[-1, 1]$, which is defined as

$$\cos(\boldsymbol{x}_1, \boldsymbol{x}_2) = \frac{\boldsymbol{x}_1^{\text{T}} \boldsymbol{x}_2}{\|\boldsymbol{x}_1\|_2 \|\boldsymbol{x}_2\|_2} = \frac{\sum_{i=1}^{d} x_{1i} x_{2i}}{\sqrt{\sum_{i=1}^{d} x_{1i}^2} \sqrt{\sum_{i=1}^{d} x_{2i}^2}}. \tag{D.5}$$

The larger the cosine similarity is, the smaller the angle will be between the two sample vectors. When the directions of two sample vectors coincide completely, the cosine similarity will be 1; when the directions of two sample vectors are completely opposite, the cosine similarity will be -1. The cosine similarity is usually used when measuring the similarity of textual data.

D.6. Pearson Correlation Coefficient

Pearson correlation coefficient can be used to measure the similarity between two samples. Suppose $\boldsymbol{x}_1 = \{x_{11}, x_{12}, \ldots, x_{1d}\}$ and $\boldsymbol{x}_2 = \{x_{21}, x_{22}, \ldots, x_{2d}\}$ are two d-dimensional samples, then Pearson correlation coefficient is defined as

$$\mathrm{Corr}(\boldsymbol{x}_1, \boldsymbol{x}_2) = \frac{\sum_{i=1}^{d}(x_{1i} - \overline{x_1})(x_{2i} - \overline{x_2})}{\sqrt{\sum_{i=1}^{d}(x_{1i} - \overline{x_1})^2}\sqrt{\sum_{i=1}^{d}(x_{2i} - \overline{x_2})^2}}, \tag{D.6}$$

where $\overline{x_1}$ and $\overline{x_2}$ are average values of sample $\boldsymbol{x}_1$ and $\boldsymbol{x}_2$ respectively. Pearson correlation coefficient ranges from $[-1, 1]$. It is position invariant and scale invariant.

D.7. Jaccard Similarity

Jaccard similarity is usually used to measure the similarity of two sets. For example, if an article is regarded as a set of words, the similarity between two articles can be calculated by the number of words that appear in both articles and the total number of different words in two articles. Assuming that A and B are two sets, the Jaccard similarity between them is

$$\mathrm{Jaccard}(\mathrm{A}, \mathrm{B}) = \frac{|\mathrm{A} \cap \mathrm{B}|}{|\mathrm{A} \cup \mathrm{B}|}. \tag{D.7}$$

D.8. KL Divergence

Kullback–Leibler divergence (KL divergence) is a method used to measure the distance between two distributions. Suppose there are two probability distributions P and Q. If P and Q are discrete probability distributions and the random variable takes d values, the

KL divergence is defined as

$$\mathrm{KL}(P\|Q) = \sum_{i=1}^{d} P(x=i) \ln \frac{P(x=i)}{Q(x=i)}. \tag{D.8}$$

If both P and Q are distributions on continuous random variable, then the KL divergence is defined as

$$\mathrm{KL}(P\|Q) = \int P(x) \ln \frac{P(x)}{Q(x)} \mathrm{d}(x). \tag{D.9}$$

Strictly speaking, KL divergence does not satisfy the definition of distance. Firstly, KL divergence does not satisfy the property of symmetry, that is, $\mathrm{KL}(P\|Q)$ and $\mathrm{KL}(P\|Q)$ are not always equal. Secondly, KL divergence does not satisfy the triangular inequality of distance.

Appendix E

Model Evaluation

When we use various statistical models or machine learning models for a prediction task, the most concerned problems are whether the model can offer a "good" prediction, which model performs better and, which model can better explain the data. These problems all involve how to measure the model performance through model evaluation, based on which we can perform model selection.

In this appendix, we first introduce some basic concepts of model evaluation, then introduce several dataset splitting and selection methods, and finally introduce some performance metrics commonly used in regression, classification and clustering problems.

E.1. Basic Concepts

E.1.1. *Independent and Identically Distributed Data*

Given a dataset $D = \{(\boldsymbol{x}_1, y_1), (\boldsymbol{x}_2, y_2), \ldots, (\boldsymbol{x}_n, y_n)\}$, where y_i is the label of the corresponding sample $\boldsymbol{x}_i$. We usually assume that the samples are independent and identically distributed, that is, all samples come from the same probability distribution and are statistically independent form each other. However, in practice, not all datasets satisfy the above assumption, such as time series data.

E.1.2. *Bias and Variance*

Assuming that for a prediction problem, the real model is $y = f(\boldsymbol{x})$. Usually we learn a prediction model $y = h(\boldsymbol{x})$ from the training set $D = \{(\boldsymbol{x}_1, y_1), (\boldsymbol{x}_2, y_2), \ldots, (\boldsymbol{x}_n, y_n)\}$ by a learning algorithm. Assuming $(\boldsymbol{x}, y)$ is a specific sample, bias measures the deviation between $\bar{h}(\boldsymbol{x})$ (the expectation of the prediction model $h(\boldsymbol{x})$) and the real model $f(\boldsymbol{x})$, that is, $\bar{h}(\boldsymbol{x}) - f(\boldsymbol{x})$. Variance measures how the performance of the prediction model changes in different training sets, that is $E((h(\boldsymbol{x}) - \bar{h}(\boldsymbol{x}))^2)$. Noise measures the distance between the observed labels and the real model, that is $E((y - f(\boldsymbol{x}))^2)$. The expected prediction error of the model can be expressed as $E((y - h(\boldsymbol{x}))^2)$, which indicates the deviation between the prediction result of the model and the sample label. The following relationship exists between the expected prediction error, bias, variance and noise of the model:

$$\underbrace{E((y - h(\boldsymbol{x}))^2)}_{\text{prediction error}} = \underbrace{E((h(\boldsymbol{x}) - \bar{h}(\boldsymbol{x}))^2)}_{\text{variance}} + \underbrace{(\bar{h}(\boldsymbol{x}) - f(\boldsymbol{x}))^2}_{\text{bias}^2} + \underbrace{E((y - f(\boldsymbol{x}))^2)}_{\text{noise}}. \tag{E.1}$$

E.1.3. *Hyper-parameter and Parameter Tuning*

Hyper-parameter generally refers to the tunable parameters in machine learning models, such as the number of hidden layer units in neural networks and the regularization coefficient in linear models. Hyper-parameters need to be set before model training. Different configuration of hyper-parameters often results in significant differences in model performance. Therefore, in model selection and evaluation, in addition to choosing the suitable model, the process of setting the hyper-parameters of the selected model is also required. This process is generally called parameter tuning.

Parameter tuning is a common problem in solving practical problems by machine learning models. Hyper-parameter setting is the key to good prediction results, especially in the case of deep learning models. Parameter tuning is essentially model selection. We can choose the optimal hyper-parameter by setting different hyper-parameter configurations, and then using the hold-out method or cross-validation method to compute the generalization ability of the model under different hyper-parameter configurations.

Usually, we need to have a preliminary preset range of hyper-parameters, and then set up several possible values within this range to evaluate the model performance using the hold-out method or cross-validation method for each hyper-parameter configuration. This process is called grid search. Unfortunately, when there are many hyper-parameters in the model, grid search is easy to fall into the curse of dimensionality. Assume there are four hyper-parameters, with 10 grid points for each hyper-parameter. There are 10^4 possible parameter configurations, and 10^4 models need to be trained. The amount of computation is huge. Therefore, when adjusting multiple hyper-parameters, more knowledge is needed beforehand to reduce the scope of grid search. We can also choose models or learners insensitive to hyper-parameters without compromising prediction accuracy.

E.1.4. *Overfitting and Underfitting*

When training a learner with training data, the error of the learner on the training set is called training error, or empirical error. When the model is trained, the key is whether the model can have small prediction error on new samples (test set), which is called generalization error. However, test samples can not be seen in advance, so we can only use the training set to fit the model in the process of model training. Generally speaking, if we use a complex model to fit the training data, we will get a learner with minimum training error. However, the generalization ability will be poor. This phenomenon is called overfitting. The main reason for overfitting is that complex learners learn "too well" from training set, and it is very likely that some unique characteristics of training set will be regarded as the characteristics of new samples. Figure E.1 shows the trend of training error and generalization error with the increase of model complexity. It can be seen that with the increase of model complexity, overfitting becomes more and more obvious. Underfitting means that the model is too simple or does not fit the model to the optimum, which results in poor performance in both training set and test set, as shown in the beginning part of Figure E.1.

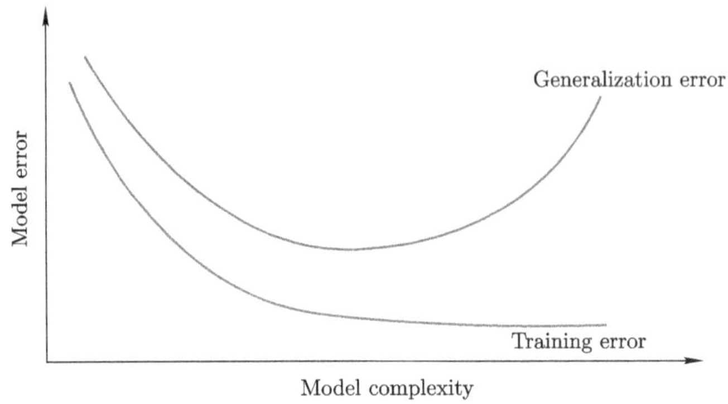

Figure E.1 The relationship between training error, generation error and model complexity.

E.2. Dataset Splitting

Given a dataset $D = \{(\boldsymbol{x}_1, y_1), \ldots, (\boldsymbol{x}_n, y_n)\}$, we want to find the model with best generalization ability. The general method is to split the dataset into training set and test set. Then fit the model or learner using training set and evaluate the performance using test set. Below we will introduce several dataset splitting methods.

E.2.1. *Hold-out*

The simplest method is to randomly split a dataset into two mutually exclusive sets: training set and test set. Randomly splitting is to keep the training set and the test set roughly the same distribution, which is also the basic idea of hold-out. In terms of classification, we need to do stratified sampling to ensure that the proportion of samples in each category is roughly the same for training set and test set. For example, in binary classification, two-thirds of the samples are randomly selected from all the positive samples as the positive part of the training set, then two-thirds of the samples are randomly selected from the negative samples as the negative part of the training set, and the remaining samples are used as the test set.

In practice, we often need to split the dataset randomly several times. After each split, we perform model training and model evaluation. Finally the mean and variance of generalization error are calculated. The mean can be used as the result of final model evaluation, and the variance can reflect the stability of the model for different data splits. The size of training set is also important for model performance. If the training set is too large, the model will fit well, but it will lead to poor generalization, and vice versa. In practical, $2/3 \sim 4/5$ samples are usually taken as training set and the rest as test set.

E.2.2. *Cross Validation*

Cross validation is probably the most commonly used method of model evaluation in practice. Its basic idea is to split the dataset into k mutually exclusive subsets randomly and evenly (stratified sampling can be carried out if necessary, so that the proportions of different categories are roughly the same in each subset). Assuming that the original dataset is denoted as D, we randomly split it into k subsets $\{D_1, D_2, \ldots, D_k\}$. After obtaining k subsets, we conduct k-round model training and model evaluation. In each round, one subset is selected as the test set, and the remaining subsets are training set. The model is trained with the training set and evaluated with the test set. Suppose we use a performance evaluation metric and calculate the performance of the i-round as f_i. Finally, we take the average performance metric of k rounds as the final evaluation result. Such a process is called k-fold cross validation. Figure E.2 shows the process of 5-fold cross validation. When necessary, the dataset can be split randomly several times, and the results of each k-fold cross validation can be averaged to obtain more accurate estimation of model performance. An extreme case of k-fold cross validation is when k equals n (the number of samples in dataset D). In that case, only one sample is selected as test set and the other $n-1$ samples are used as training set. This method is called leave-one-out cross validation. Leave-one-out cross validation can generally estimate the model performance accurate, because it is not affected by the randomly splitting process of the dataset. However, model training needs to be conducted for n times in this method, which is computationally intensive and usually only applicable for small datasets.

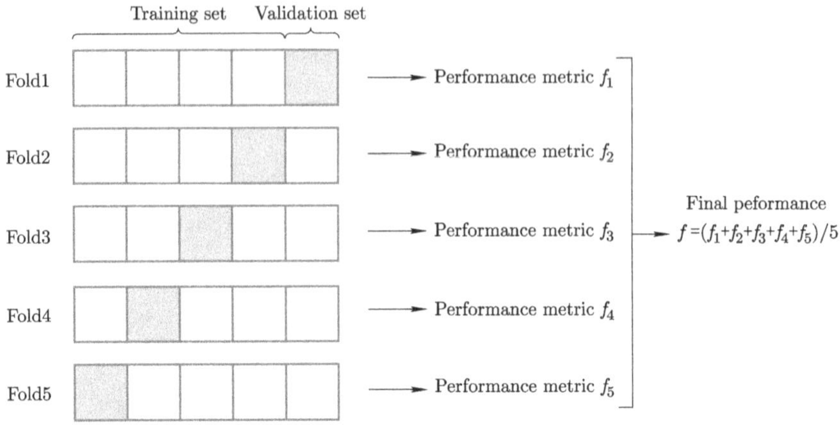

Figure E.2 The process of 5-fold cross validation.

E.2.3. *Bootstrapping*

In statistics, bootstrapping generally refers to random sampling with replacement.[234] That is to say, selecting samples from a dataset D of size n repeatedly with replacement to form a new dataset D_{bs} of any size. Thus, we can use bootstrapping to get a new dataset D_{bs} with the same size as the original dataset. Then we can use D_{bs} as training set to train the model. Finally we can evaluate the model on test set $D\backslash D_{bs}$.

In the new dataset (with size n) obtained by bootstrapping, some samples in the original dataset will appear more than once, while some samples will not appear. We can estimate the proportion of samples that do not appear in the new dataset. The probability that a sample will not be selected is:

$$p(\text{not chosen}) = \left(1 - \frac{1}{n}\right)^n. \tag{E.2}$$

When $n \to \infty$, $p(\text{not chosen})$ equals to $\frac{1}{\mathrm{e}} \approx 0.368$. That is, about 36.8% of the samples do not appear in the new dataset. In order to reduce the errors caused by such randomness, we need to repeat bootstrapping b rounds, train and test the model separately, and then calculate the mean of the model performance as the final estimation of generalization performance.

We can use bootstrapping to generate several different datasets from the original dataset. These datasets can be used as training sets to train base models in ensemble model (refer to Chapter 5). Bootstrapping is a good choice when the dataset is small and it is difficult to use hold-out and cross validation.

E.3. Model Evaluation Metrics

E.3.1. *Regression Metrics*

Assume that the test set consists of n samples. The true value of the sample is $\boldsymbol{y}$, and the true value of sample i is y_i. The model prediction value is $\hat{\boldsymbol{y}}$. The model prediction of sample i is $\hat{y}_i$. The most commonly used metric for regression is the mean squared error (MSE), which can be calculated as

$$\text{MSE}(\boldsymbol{y}, \hat{\boldsymbol{y}}) = \frac{1}{n}\sum_{i=1}^{n}(y_i - \hat{y}_i)^2. \tag{E.3}$$

It can be seen that the mean square error estimates the expectation of the square difference between the true value of the sample and the predicted value. Root mean square error (RMSE) is the arithmetic square root of the mean square error, which is calculated as

$$\text{RMSE}(\boldsymbol{y}, \hat{\boldsymbol{y}}) = \sqrt{\text{MSE}(\boldsymbol{y}, \hat{\boldsymbol{y}})} = \sqrt{\frac{1}{n}\sum_{i=1}^{n}(y_i - \hat{y}_i)^2}. \tag{E.4}$$

Mean square error and root mean square error usually magnify the influence of outliers on model evaluation results. One way to overcome this problem is to replace the square with the absolute value of the difference between the true value and the predicted value.

This metric is called mean absolute error (MAE), and is calculated as

$$\text{MAE}(\boldsymbol{y}, \hat{\boldsymbol{y}}) = \frac{1}{n}\sum_{i=1}^{n}|y_i - \hat{y}_i|. \tag{E.5}$$

The ranges of mean square error, root mean square error and mean absolute error are $[0, +\infty)$. They are all affected by the scale of the target feature, so it is difficult to compare models between different tasks. In statistics, the coefficient of determination R^2 is usually used to evaluate regression models. If the average of the true value of the samples is $\bar{y} = \frac{1}{n}\sum_{i=1}^{n} y_i$, the sum of the total squares is

$$\mathrm{SS}_{\mathrm{tot}} = \sum_{i=1}^{n}(y_i - \bar{y})^2. \tag{E.6}$$

The sum of residual squares is

$$\mathrm{SS}_{\mathrm{res}} = \sum_{i=1}^{n}(y_i - \hat{y}_i)^2, \tag{E.7}$$

and the coefficient of determination R^2 is

$$\mathrm{R}^2(\boldsymbol{y}, \hat{\boldsymbol{y}}) = 1 - \frac{\mathrm{SS}_{\mathrm{res}}}{\mathrm{SS}_{\mathrm{tot}}}. \tag{E.8}$$

Usually, the range of R^2 is $[0, 1]$, and the closer R^2 approaches 1, the better the regression performance will be; the closer R^2 approaches 0, the worse the regression performance is. In some cases, the sum of residual squares may be bigger than the sum of the total squares, and R^2 may be less than 0.

E.3.2. *Classification Metrics*

In the classification problem, there are two kinds of labels for each sample: one is the true label y of the sample, the other is prediction label $\hat{y}$. According to the two labels of each sample, a confusion matrix can be obtained. As shown in Table E.1, each column of the confusion matrix represents the predicted category, and the sum of each column represents the number of samples predicted to be in that category. Each row represents the true category, and sum of each row represents the number of samples for that category. Most classification metrics can be calculated based on the confusion matrix. Below we will introduce metrics for binary classification and multi-classification problems respectively. For the binary

Table E.1 Confusion matrix for binary classification problem.

	Predicted category	
True category	1	0
1	TP (number of true positive samples)	FN (number of false negative samples)
0	FP (number of false positive samples)	TN (number of true negative samples)

classification problem, each sample can be divided into the following four types:

1. True positive sample (TP): the true category and predicted category of the sample are both positive.
2. True negative sample (TN): the true category and predicted category of the sample are both negative.
3. False positive sample (FP): The true category of the sample is negative, while the predicted category is positive.
4. False negative sample (FN): The true category of the sample is positive, while the predicted category is negative.

The corresponding confusion matrix is shown in Table E.1.

The most common classification metric is accuracy, which indicates the proportion of correctly predicted samples. Given the confusion matrix, accuracy is calculated as

$$\text{Accuracy} = \frac{\text{TP} + \text{TN}}{\text{TN} + \text{FN} + \text{FP} + \text{TP}}. \tag{E.9}$$

When the sample categories are imbalanced, accuracy is not a good metric for classification. For example, in the text sentiment classification task, assuming that positive samples accounts for 80% and negative samples only 20%. A classification model which predicts all samples to be positive is obviously useless, but its accuracy rate can reach 80%. For imbalanced dataset, precision and recall are better metrics than accuracy. Precision refers to the proportion of true positive samples in the predicted positive samples:

$$\text{Precision} = \frac{\text{TP}}{\text{TP} + \text{FP}}. \tag{E.10}$$

Recall, also known as sensitivity or hit rate, refers to the proportion of true positive samples in the positive samples:

$$\text{Recall} = \frac{\text{TP}}{\text{TP} + \text{FN}}. \tag{E.11}$$

From the above formulas, we can see that the numerator of precision is the same as that of the recall, but the denominator is different. Referring to the confusion matrix, it can be seen that the calculation of precision can be understood as first extracting samples from predicted positive samples (TP + FP), and then counting the number of true positive samples (TP) in these samples. The recall is calculated as counting how many samples are correctly predicted (TP) among the positive samples (TP + FN). Usually, precision and recall are negatively correlated. High precision often corresponds to low recall, and vice versa. In practice, we often need to make a trade-off between precision and recall.

Considering the precision and recall is one-sided when separated, it is necessary to consider both of them. In the binary classification problem, F-measure is a metric that considers the precision and recall rate comprehensively, which is calculated is as follows:

$$\text{F}_\beta = \frac{(1 + \beta^2)\text{Precision} \times \text{Recall}}{\beta^2 \times \text{Precision} + \text{Recall}}, \tag{E.12}$$

where β is a positive number, which is used to adjust the weight of the precision and recall. The greater the β, the greater the weight of recall will be; the smaller the β, the greater the weight of precision will be. The weight of precision and recall is the same when $\beta = 1$, which is called F_1 value. It is the harmonic average of precision and recall:

$$F_1 = \frac{2 \times \text{Precision} \times \text{Recall}}{\text{Precision} + \text{Recall}}. \tag{E.13}$$

Specificity indicates the proportion of true negative samples in negative samples, which can be understood as the recall of negative category:

$$\text{Specificity} = \frac{\text{TN}}{\text{TN} + \text{FP}}. \tag{E.14}$$

In practice, the class imbalance problem makes the classification performance inappropriately reflected by traditional evaluation metrics such as precision. For example, suppose that there are 90 samples in category A and 10 samples in category B. Classifier C_1 predicts all test samples into category A. Classifier C_2 correctly predicts 70 samples in category A and 5 samples in category B. Accuracy of C_1 is 90%, and 75% of C_2. However, C_2 is obviously more useful. In cost-sensitive learning, the cost will be different for mistakes in different categories. Thus, the default classification threshold of 0.5 seems inappropriate.

To solve the above problem, we will introduce another model evaluation method for classification problem: receiver operating characteristic (ROC). The main analysis tool is a ROC curve drawn on two-dimensional plane. The abscissa of the curve is false positive rate (FPR), and the ordinate is true positive rate (TPR), and the calculation formulas are

$$\mathrm{FPR} = \frac{\mathrm{FP}}{\mathrm{FP} + \mathrm{TN}}, \tag{E.15}$$

$$\mathrm{TPR} = \frac{\mathrm{TP}}{\mathrm{TP} + \mathrm{FN}}. \tag{E.16}$$

For a classifier, we can get a pair of points consisting of TPR and FPR according to its performance on test samples. Thus, the classifier can be mapped to a point on the ROC plane. By adjusting the classification threshold of this classifier, we can get a curve passing through point $(0, 0)$ and point $(1, 1)$, which is the ROC curve of this classifier. Generally, the ROC curve should be above the line between $(0, 0)$ and $(1, 1)$. The ROC curve formed by linking point $(0, 0)$ and point $(1, 1)$ actually represents a random classifier.

Although it is useful to use ROC curve to represent the performance of classifier, we always expect a numerical value. This value is called the Area Under ROC Curve (AUC). As the name implies, AUC is the size of the area under ROC curve. The AUC of a random model is 0.5, and the AUC of actual model is between 0.5 and 1.0. Larger AUC represents better performance. Generally speaking, AUC is more than 0.75, which can be considered as a good model. Figure E.3 shows several different ROC curves. The closer the curve is to the upper left point (0,1), the better the performance of the corresponding classifier will be. ROC and AUC are not suitable for multi-classification problem. Although some researchers have proposed ROC curves and AUC under multi-classification, they are less

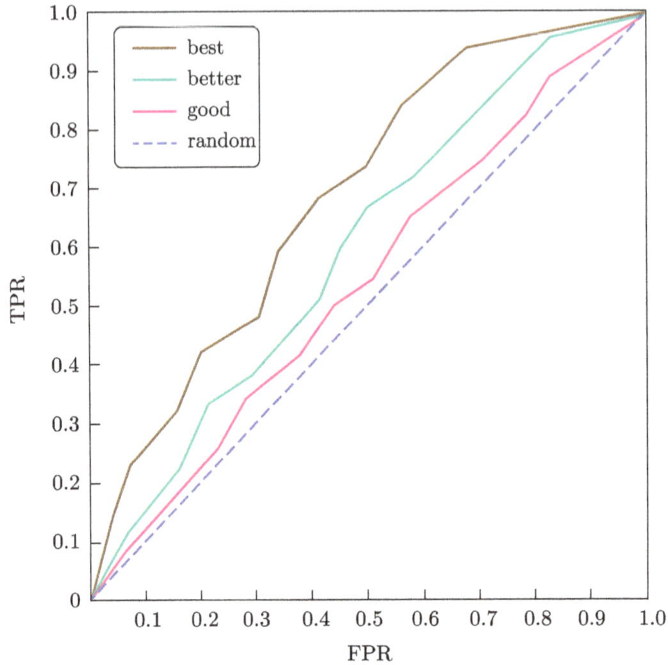

Figure E.3 ROC curve.

practical. Compared with the binary classification problem, the evaluation metrics that can be used directly by the multi-classification problem is accuracy, while other metrics need to be adjusted.

The size of confusion matrix for multi-classification problem is generalized from 2×2 to $C \times C$, in which $C > 2$ is the number of categories. Each category can be regarded as positive, while all other categories are negative. A multi-classification problem can be transformed into multiple binary classification problems. The precision and recall of each binary classification problem are calculated, and the average precision and recall is used to evaluate the multi-classification problem.

E.3.3. *Clustering Metrics*

Clustering is a typical unsupervised learning task. It is difficult to evaluate the performance of clustering models due to the lack of sample labels. Intuitively, clustering is to group similar samples into

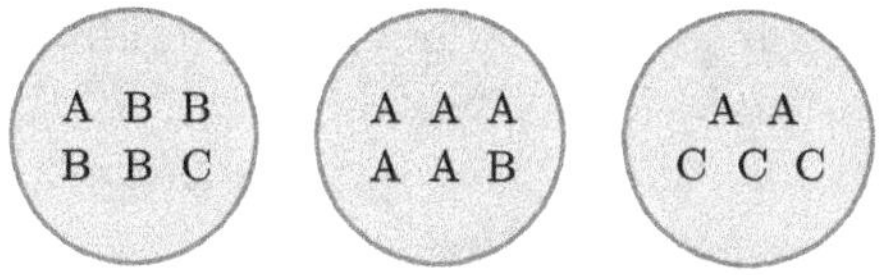

Figure E.4 Samples are grouped into three categories by clustering algorithm.

the same cluster and dissimilar samples into different clusters. From this point of view, researchers have developed some metrics to evaluate clustering results.[234,235] However, such "internal" evaluation metrics have great limitations and uncertainties. In practice, we often rely on external data to evaluate clustering results. Each sample has a label, which is equivalent to giving a real reference clustering result as shown in Figure E.4. We can evaluate the clustering results according to the sample labels.

Purity is a metric which depicts the proportion of correctly classified samples. Before calculate purity, we need to group the sample labels. Assuming that n_{ij} represents the number of samples clustered into cluster i but belonging to label category j, and $n_i = \sum_{j=1}^{C} n_{ij}$ is the total number of samples in cluster i, then $p_{ij} = \frac{n_{ij}}{n_i}$ represents the label distribution of samples in cluster i. We define the purity of cluster i as $p_i \triangleq \max_j p_{ij}$, then the purity of clustering results are as follows:

$$\text{purity} = \sum_i \frac{n_i}{n} p_i. \tag{E.17}$$

In Figure E.4, the purity metric can be calculated as

$$\text{purity} = \frac{6}{17} \times \frac{4}{6} + \frac{6}{17} \times \frac{5}{6} + \frac{5}{17} \times \frac{3}{5} = 0.71.$$

The range of purity is [0, 1]. Generally, higher purity means better clustering performance. However, if we group each sample into a single cluster, the purity is 1, which is obviously a meaningless clustering result.

E.3.4. *Rand Index*

Rand index is a measure of similarity between two clustering results. Let $U = \{u_1, u_2, \ldots, u_R\}$ and $V = \{v_1, v_2, \ldots, v_C\}$ be two different clustering results for n samples. U is the result obtained by clustering algorithm and V is the result obtained by sample labels. Then, we can define a 2×2 confusion matrix, which contains the following numbers: TP (true positive) represents the number of pairs of samples that appear in the same cluster in both U and V; TN (true negative) represents the number of pairs of samples that appear in different clusters in both U and V; FN (false negative) represents the number of pairs of samples that appear in different clusters in U but in the same cluster in V; FP (false positive) represents the number of pairs of samples that appear in the same cluster in U but in different clusters in V. The rand index is

$$\mathrm{RI} = \frac{\mathrm{TP} + \mathrm{TN}}{\mathrm{TP} + \mathrm{FP} + \mathrm{FN} + \mathrm{TN}}. \tag{E.18}$$

Rand index can measure the proportion of making correct clustering decisions. Obviously rand index satisfies $0 \leqslant \mathrm{RI} \leqslant 1$, and larger value means better result. For example, the three clusters in Figure E.4 contain 6, 6 and 5 samples. The number of pairs of samples in the same cluster is

$$\mathrm{TP} + \mathrm{FP} = \mathrm{C}_6^2 + \mathrm{C}_6^2 + \mathrm{C}_5^2 = 40.$$

TP is

$$\mathrm{TP} = \mathrm{C}_4^2 + \mathrm{C}_5^2 + \mathrm{C}_3^2 + \mathrm{C}_2^2 = 20,$$

where the last two items come from the third cluster, that is, there are C_2^3 sample pairs with the label C, and there are C_2^2 sample pairs with the label A. So $\mathrm{FP} = 40 - 20 = 20$. Similarly, we can calculate $\mathrm{FN} = \mathrm{C}_4^1\mathrm{C}_1^1 + \mathrm{C}_1^1\mathrm{C}_7^1 + \mathrm{C}_1^1\mathrm{C}_3^1 + \mathrm{C}_5^1\mathrm{C}_2^1 = 24$, $\mathrm{TN} + \mathrm{FN} = \mathrm{C}_6^1\mathrm{C}_6^1 + \mathrm{C}_5^1\mathrm{C}_6^1 + \mathrm{C}_5^1\mathrm{C}_6^1 = 96$. So $\mathrm{TN} = 72$, $\mathrm{RI} = \frac{20+72}{20+20+24+72} = 0.68$.

Mutual information measures the uncertainty when a random variable decreases given another random variable. It can also be used to measure the clustering performance. That is, to calculate the mutual information between two clustering results U and V. The probability that a randomly selected sample belongs to both cluster u_i and cluster v_j is

$$p_{UV}(i, j) = \frac{|u_i \cap v_j|}{n}. \tag{E.19}$$

The probability that a randomly selected sample belongs to the cluster u_i is $p_U(i) = \frac{|u_i|}{n}$. Similarly the probability of belonging to the cluster V_j is $p_V(j) = \frac{|v_j|}{n}$. Therefore, we can define mutual information as

$$\mathrm{MI}(U,V) = \sum_{i=1}^{R}\sum_{j=1}^{C} p_{UV}(i,j) \ln \frac{p_{UV}(i,j)}{p_U(i)p_V(j)}. \tag{E.20}$$

The range of mutual information is from 0 to $\min\{H(U), H(V)\}$, in which $H(\cdot)$ is the entropy of the corresponding clustering. It should be noted that mutual information can reach its maximum value when there are many small clusters. To prevent it, the normalized mutual information is often used:

$$\mathrm{NMI}(U,V) = \frac{\mathrm{MI}(U,V)}{(H(U)+H(V))/2}. \tag{E.21}$$

It can be seen that the range of NMI is $[0,1]$. Larger NMI means better clustering result.

References

[1] DIRAC P A M. The quantum theory of the electron [C]. Proceedings of the Royal Society of London A: Mathematical, Physical and Engineering Sciences, 1928, 117(778): 610–624.

[2] LI J Z, ABSHER D M, Tang H, *et al.* Worldwide human relationships inferred from genome-wide patterns of variation [J]. Science, 2008, 319(5866): 1100–1104.

[3] GHRIST R. Barcodes: the persistent topology of data [J]. Bulletin of the American Mathematical Society, 2008, 45(1): 61–75.

[4] DOUGHERTY J, KOHAVI R, SAHAMI M. Supervised and unsupervised discretization of continuous features [C]. Proceedings of the 12th International Conference on Machine Learning, Tahoe City, California. San Francisco: Morgan Kaufmann, 1995: 194–202.

[5] WONG A K C, CHIU D K Y. Synthesizing statistical knowledge from incomplete mixed-mode data [J]. IEEE Transactions on Pattern Analysis and Machine Intelligence, 1987, 6: 796–805.

[6] LIU H, SETIONO R. Feature selection via discretization of numeric attributes [J]. IEEE Transactions on Knowledge and Data Engineering, 1997, 9(4): 642–645.

[7] TAY F E H, SHEN L X. A modified chi2 algorithm for discretization [J]. IEEE Transactions on Knowledge and Data Engineering, 2002, 14(3): 666–670.

[8] KERBER R. Chimerge: Discretization of numeric attributes [C]. Proceedings of the 10th National Conference on Artificial Intelligence, June 12–16, 1992, San Jose, California. Palo Alto: AAAI Press, 1992: 123–128.

[9] KURGAN L A, CIOS K J. CAIM discretization algorithm [J]. IEEE Transactions on Knowledge and Data Engineering, 2004, 16(2): 145–153.

[10] GARCIA S, LUENGO J, SAEZ J A, *et al.* A survey of discretization techniques: Taxonomy and empirical analysis in supervised learning [J]. IEEE Transactions on Knowledge and Data Engineering, 2013, 25(4): 734–750.

[11] HAWKINS D M. Identification of outliers [M]. Heidelberg: Springer, 1980.

[12] BREUNIG M, KRIEGEL H P, NG R T, *et al.* LOF: Identifying density-based local outliers [C]. Proceedings of the 2000 ACM SIGMOD International Conference on Management of Data, May 5–8, 2000, Dallas, Texas. New York: ACM, 2000: 93–104.

[13] CHANDOLA V, BANERJEE A, KUMAR V. Anomaly detection: A survey [J]. ACM Computing Surveys, 2009, 41(3): 15.

[14] HODGE V, AUSTIN J. A survey of outlier detection methodologies [J]. Artificial Intelligence Review, 2004, 22(2): 85–126.

[15] AGGARWAL C C, YU P S. Outlier detection for high dimensional data [C]. Proceedings of the 2001 ACM SIGMOD International Conference on Management of Data, May 21–24, Santa Barbara, California. New York: ACM, 2001: 37–46.

[16] LENZERINI M. Data integration: A theoretical perspective [C]. Proceedings of the 21st ACM SIGMOD-SIGACT-SIGART Symposium on Principles of Database Systems, June 3-5, Madison, Wiscosin. New York: ACM 2002: 233–246.

[17] YEH I C. Modeling of strength of high-performance concrete using artificial neural networks [J]. Cement and Concrete Research, 1998, 28(12): 1797–1808.

[18] GUVENIR A H, DEMIROZ G, ILTER N. Learning differential diagnosis of erythemato-squamous diseases using voting feature intervals [J]. Artificial Intelligence in Medicine, 1998, 13(3): 147–165.

[19] KUSA A, DANECHOVA Z, FINDRA S, *et al.* Gender differences in purchase decision-making styles [J]. European Journal of Science and Theology, 2014, 10(5): 113–123.

[20] HOERL A E, KENNARD R W. Ridge regression: Biased estimation for nonorthogonal problems [J]. Technometrics, 1970, 12(1): 55–67.

[21] TIBSHIRANI R. Regression shrinkage and selection via the lasso [J]. Journal of the Royal Statistical Society, Series B, 1996, 58(1): 267–288.

[22] FRIEDMAN J, HASTIE T, TIBSHIRANI R. The elements of statistical learning [M]. Berlin: Springer, 2010.

[23] FU W J. Penalized regressions: the Bridge versus the LASSO [J]. Journal of Computational and Graphical Statistics, 1998, 7(3): 397–416.

[24] WU T T, LANGE K. Coordinate descent algorithms for lasso penalized regression [J]. The Annals of Applied Statistics, 2008, 2(1): 224–244.

[25] EFRON B, HASTIE T, JOHNSTONE I, *et al.* Least angle Regression [J]. The Annals of Statistics, 2004, 32(2): 407–499.

[26] PARIKH N, BOYD S. Proximal algorithms [J]. Foundations and Trends in Optimization, 2014, 1(3): 127–239.

[27] BECK A, TEBOULLE M. A fast iterative shrinkage-thresholding algorithm for linear inverse problems [J]. SIAM Journal on Imaging Sciences, 2009, 2(1): 183–202.

[28] STAMEY T A, KABALIN J N, MCNEAL J E, *et al.* Prostate specific antigen in the diagnosis and treatment of adenocarcinoma of the prostate. II. Radical prostatectomy treated patients [J]. The Journal of Urology, 1989, 141(5): 1076–1083.

[29] ZOU H, HASTIE T. Regularization and variable selection via the elastic net [J]. Journal of the Royal Statistical Society, Series B, 2005, 67(2): 301–320.

[30] YUAN M, LIN Y. Model selection and estimation in regression with grouped variables [J]. Journal of the Royal Statistical Society, Series B, 2006, 68(1): 49–67.

[31] FRIEDMAN J H. Multivariate adaptive regression splines [J]. The Annals of Statistics, 1991, 19(1): 1–67.

[32] HAYKIN S. Neural networks: A comprehensive foundation [M]. 2nd ed. Upper Saddle River: Prentice Hall PTR, 1998.

[33] CRISTIANINI N, SHAWE-TAYLOR J. An introduction to support vector machines and other kernel-based learning methods [M]. New York: Cambridge University Press, 2000.

[34] RASMUSSEN C E, WILLIAMS C K I. Gaussian processes for machine learning (adaptive computation and machine learning) [M]. Cambridge: The MIT Press, 2005.

[35] CORTEZ P, MORAIS A J R. A data mining approach to predict forest fires using meteorological data [C]. Proceedings of the EPIA 2007-Portuguese conference on artificial intelligence, December 3–7, 2007. Guimarates, Portugal. Heidelberg: Springer, 2007: 512–523.

[36] FANAEE-T H, GAMA J. Event labeling combining ensemble detectors and background knowledge [J]. Progress in Artificial Intelligence, 2014, 2(2-3): 113–127.

[37] ZHOU F, CLAIRE Q, KING R D. Predicting the geographical origin of music [C]. Proceedings of 2014 IEEE International Conference on Data Mining, December 14–17, 2014, Shenzhen, China. New York: IEEE, 2015: 1115–1120.

[38] COVER T, HART P. Nearest neighbor pattern classification [J]. IEEE Transactions on Information Theory, 1967, 13(1): 21–27.
[39] GINI C. Variabilita e mutabilita [M]. Bologna: Tipografia di Paolo Cuppini, 1912.
[40] QUINLAN J R. Induction of decision trees [J]. Machine Learning, 1986, 1(1): 81–106.
[41] QUINLAN J R. C4. 5: Programs for machine learning [M]. San Francisco: Morgan Kaufmann, 1993.
[42] BREIMAN L, FRIEDMAN J, STONE C J, *et al.* Classification and regression trees [M]. Chapman and Hall/CRC, 1984.
[43] BREIMAN L, FRIEDMAN J, Olshen R, *et al.* Classification and regression trees [M]. Monterrey: Wardswirth and Brooks, 1984.
[44] PLATT J. Sequential minimal optimization: A fast algorithm for training support vector machines [J]. Advances in Kernel Methods-support Vector Learning, 1998, 208(1): 212–223.
[45] EVETT I W, SPIEHLER E J. Rule induction in forensic science [J]. Knowledge Based Systems, 1989, 2(3): 152–160.
[46] FREY P W, SLATE D J. Letter recognition using Holland-style adaptive classifiers [J]. Machine Learning, 1991, 6(2): 161–182.
[47] LICHMAN M. UCI machine learning repository [DB/OL].
[48] STREET N, WOLBERG W H, MANGASARIAN O L, *et al.* Nuclear feature extraction for breast tumor diagnosis [C]. Proceedings of SPIE, July 29, San Jose, California, 1993. Bellingham: SPIE Press, 1993: 861–870.
[49] BREIMAN L. Bagging predictors [J]. Machine Learning, 1996, 24(2): 123–140.
[50] WOLPERT D H. Stacked generalization [J]. Neural Networks, 1992, 5: 241–259.
[51] BREIMAN L. Random forests [J]. Machine Learning, 2001, 45(1): 5–32.
[52] CARUANA R, KARAMPATZIAKIS N, YESSENALINA A. An empirical evaluation of supervised learning in high dimensions [C]. Proceedings of the 25th International Conference on Machine Learning, June 5–9, 2008, Helsinki, Finland. San Francisco: Morgan Kaufmann, 2008: 96–103.
[53] FREUND Y, SCHAPIRE R E. A decision-theoretic generalization of online learning and an application to boosting [J]. Journal of Computer and System Sciences, 1997: 119–139.
[54] ER O, TANRIKULU C, ABAKAY A, *et al.* An approach based on probabilistic neural network for diagnosis of Mesothelioma's disease [J]. Computers and Electrical Engineering, 2012, 38(1): 75–81.

[55] CORTEZ P, CERDEIRA A, ALMEIDA F, *et al.* Modeling wine preferences by data mining from physicochemical properties [J]. Decision Support Systems, 2009, 47(4): 547–553.

[56] PELLEG D, MOORE A W. X-means: Extending k-means with efficient estimation of the number of clusters [C]. Proceedings of the 17th International Conference on Machine Learning, June 29–July 2, 2000, Stanford, California. San Francisco: Morgan Kaufmann, 2000: 727–734.

[57] HANSEN M H, YU B. Model selection and the principle of minimum description length [J]. Journal of the American Statistical Association, 2001, 96(454): 746–774.

[58] HAMERLY G, ELKAN C. Learning the k in k-means [C]. Proceedings of the 16th International Conference on Neural Information Processing Systems, December 1–5, 2009, Bangkok, Thailand. Cambridge: MIT Press, 2003: 281–288.

[59] ARTHUR D, VASSILVITSKII S. K-means++: The advantages of careful seeding [C]. Proceedings of the 18th Annual ACM-SIAM Symposium on Discrete Algorithms, January 7–9, 2007, New Orleans, Louisiana. New York: SIAM, 2007: 1027–1035.

[60] LINGRAS P, PETERS G. Rough clustering [J]. Wiley Interdisciplinary Reviews: Data Mining and Knowledge Discovery, 2011, 1(1): 64–72.

[61] MANNING C D, RAGHAVAN P, SCHUTZE H. Introduction to information retrieval [M]. New York: Cambridge University Press, 2008.

[62] SHI J B, MALIK J. Normalized cuts and image segmentation [J]. IEEE Transactions on Pattern Analysis and Machine Intelligence, 2000, 22(8): 888–905.

[63] LUXBURG U V. A tutorial on spectral clustering [J]. Statistics and Computing, 2007, 17(4): 395–416.

[64] ESTER M, KRIEGEL H P, SANDER J, *et al.* A density-based algorithm for discovering clusters in large spatial databases with noise [C]. Proceedings of the 2nd International Conference on Knowledge Discovery and Data Mining, August 2–4, 1996, Portland, Oregon. Cambridge: AAAI Press, 1996: 226–231.

[65] ORBANZ P, TEH Y W. Bayesian nonparametric models [M]. New York: Springer, 2011: 81–89.

[66] TEH Y W, JORDAN M I. Hierarchical Bayesian nonparametric models with applications [M]. Hjort N L, Holmes C, Muller P. Bayesian Nonparametrics. Cambridge: Cambridge University Press, 2010: 158–207.

[67] HJORT N L, HOLMES C, MULLER P, *et al.* Bayesian nonparametrics [M]. Cambridge: Cambridge University Press, 2010.

[68] QUINLAN J R. Combining instance-based and model-based learning [C]. Proceedings of the 10th International Conference on Machine Learning, June 27–29, 1993, University of Massachusetts, Amherst. San Francisco: Morgan Kaufmann, 1993: 236–243.

[69] AGRAWAL R, SRIKANT R. Fast algorithms for mining association rules in large databases [C]. Proceedings of the 20th International Conference on Very Large Data Bases, September 12–15, 1994, Santiago de Chile, Chile. San Francisco: Morgan Kaufmann, 1994: 487–499.

[70] HAN J W, PEI J, YIN Y W, *et al.* Mining frequent patterns without candidate generation: A frequent-pattern tree approach [J]. Data Mining and Knowledge Discovery, 2004, 8(1): 53–87.

[71] HARPER F M, KONSTAN J A. The movielens datasets: History and context [J]. ACM Transactions on Interactive Intelligent Systems, 2016, 5(4): 19.

[72] FREDRIKSON M, ANNAS P, FISCHER H, *et al.* Gender and age differences in the prevalence of specific fears and phobias [J]. Behaviour Research and Therapy, 1996, 34(1): 33–39.

[73] HAHSLER M, HORNIK K, REUTTERER T. Implications of probabilistic data modeling for mining association rules [M]. Berlin: Springer, 2006: 598–605.

[74] PEARSON K F R S. LIII. On lines and planes of closest fit to systems of points in space [J]. Philosophical Magazine and Journal of Science, Series 6, 1901, 2(11): 559–572.

[75] FISHER R A. The use of multiple measurements in taxonomic problems [J]. Annals of Eugenics, 1936, 7(2): 179–188.

[76] AHDESMAKI M, STRIMMER K. Feature selection in omics prediction problems using cat scores and false nondiscovery rate control [J]. The Annals of Applied Statistics, 2010, 4(1): 503–519.

[77] KRUSKAL J B, WISH M. Multidimensional scaling [M]. London: Sage, 1978.

[78] TORGERSON W S. Theory and methods of scaling [M]. New York: John Wiley and Sons, 1958.

[79] KRUSKAL J B. Nonmetric of multidimensional scaling: A numerical method [J]. Psychometrika, 1964, 29(2): 115–129.

[80] PICH C. Applications of multidimensional scaling to graph drawing [D]. State College: The Pennsylvania State University, 2009.

[81] TROSSET M W. Applications of multidimensional scaling to molecular conformation [J]. Computing Science and Statistics, 1998, 29: 148–152.

[82] ROWEIS S T, SAUL L K. Nonlinear dimensionality reduction by locally linear embedding [J]. Science, 2000, 290(5500): 2323–2326.

[83] SAUL L K, ROWEIS S T. Think globally, fit locally: Unsupervised learning of low dimensional manifolds [J]. Journal of Machine Learning Research, 2003, 4(2): 119–155.

[84] SCHOLKOPF B, SMOLA A, MULLER K R. Nonlinear component analysis as a kernel eigenvalue problem [J]. Neural Computation, 1998, 10(5): 1299–1319.

[85] MIKA S, SCHOLKOPF B, SMOLA A J, *et al.* Kernel PCA and denoising in feature spaces [C]. Proceedings of the 11th conference on Advances in Neural Information Processing Systems, 1998. Denver, Colorado. Cambridge: MIT Press, 1998: 536–542.

[86] WANG Q. Kernel principal component analysis and its applications in face recognition and active shape models [EB/OL].

[87] TENENBAUM J B, SILVA V D, LANGFORD J C. A global geometric framework for nonlinear dimensionality reduction [J]. Science, 2000, 290(5500): 2319–2323.

[88] Coifman R R, LAFON S, LEE A B, *et al.* Geometric diffusions as a tool for harmonic analysis and structure definition of data: Diffusion maps [C]. Proceedings of the National Academy of Sciences of the United States of America, October 8-10, 2004, Irvine, California. Pittsburgh: National Academies Press, 2005, 102(21): 7426–7431.

[89] COIFMAN R R, LAFON S. Diffusion maps [J]. Applied and Computational Harmonic Analysis, 2016, 21(1): 5–30.

[90] MAATEN L V D, HINTON G. Visualizing data using t-SNE [J]. Journal of Machine Learning Research, 2008, 9(11): 2579–2605.

[91] MAATEN L V D. Accelerating t-SNE using tree-based algorithms [J]. Journal of Machine Learning Research, 2014, 15(1): 3221–3245.

[92] HINTON G E, SALAKHUTDINOV R R. Reducing the dimensionality of data with neural networks [J]. Science, 2006, 313(5786): 504–507.

[93] SAMMON J W. A nonlinear mapping for data structure analysis [J]. IEEE Transactions on computers, 1969, 100(5): 401–409.

[94] BELKIN M, NIYOGI P. Laplacian eigenmaps and spectral techniques for embedding and clustering [C]. Proceedings of the 14th conference on Advances in Neural Information Processing Systems, 2001. Denver, Colorado. Cambridge: MIT Press, 2001: 585–591.

[95] MAATEN L V D, POSTMA E, HERIK J V D. Dimensionality reduction: a comparative view [J]. Journal of Machine Learning Research, 2009, 10(1): 66–71.

[96] HUANG G B, JAIN V, LEARNED-MILLER E. Unsupervised Joint Alignment of Complex Images [C]. Proceedings of the 11th IEEE

International Conference on Computer Vision, October 14-20, 2007, Rio de Janeiro, Brazil, 2007: 1–8.

[97] DASH M, LIU H. Feature selection for classification [J]. Intelligent Data Analysis, 1997, 1(3): 131–156.

[98] HE X F, CAI D, NIYOGI P. Laplacian score for feature selection [C]. Proceedings of the 18th conference on Advances in Neural Information Process Systems, 2005. Cambridge: MIT Press, 2006: 507–514.

[99] ZHAO Z, LIU H. Spectral feature selection for supervised and unsupervised learning [C]. Proceedings of the 24th International Conference on Machine Learning, June 20–24, 2007, Corvalis, Oregon. New York: ACM, 2007: 1151–1157.

[100] CAI D, ZHANG C Y, HE X F. Unsupervised feature selection for multi-cluster data [C]. Proceedings of the 16th ACM SIGKDD International Conference on Knowledge Discovery and Data Mining, July 25–28, 2010, Washington D. New York: ACM, 2010: 333–342.

[101] MORO S, CORTEZ P, RITA P. A data-driven approach to predict the success of bank telemarketing [J]. Decision Support Systems, 2014, 62(1): 22–31.

[102] YEH I C, LIEN C H. The comparisons of data mining techniques for the predictive accuracy of probability of default of credit card clients [J]. Expert Systems with Applications, 2009, 36(2): 2473–2480.

[103] TSANAS A, XIFARA A. Accurate quantitative estimation of energy performance of residential buildings using statistical machine learning tools [J]. Energy and Buildings, 2012, 49(49): 560–567.

[104] LIM T T, LOH W Y, SHIH Y S. A comparison of prediction accuracy, complexity, and training time of thirty-three old and new classification algorithms [J]. Machine Learning, 2000, 40(3): 203–228.

[105] HAMMERSLEY J M, CLIFFORD P. Markov fields on finite graphs and lattices, 1971. Unpublished manuscript.

[106] CLIFFORD P. Markov random fields in statistics [M]. Oxford: Clarendon Press, 1990: 19–32.

[107] JURAFSKY D, MARTIN J H. Speech and language processing [M]. 2nd ed. London: Pearson, 2014.

[108] SZELISKI R. Computer vision: algorithms and applications [M]. London: Springer, 2010.

[109] SCHWEIKERT G, ZIEN A, ZELLER G, *et al.* mGene: accurate SVM-based gene finding with an application to nematode genomes [J]. Genome Research, 2009, 19(11): 2133.

[110] NILLSON D, GOLDBERGER J. Sequentially finding the n-best list in hidden Markov models [C]. Proceedings of the 17th International Joint Conference on Artificial Intelligence, August 4–10, 2001, Seattle, Washington. San Francisco: Morgan Kaufmann, 2001: 1280–1285.

[111] HAMILTON J D. Analysis of time series subject to changes in regime [J]. Journal of Econometrics, 1990, 45(1): 39–70.

[112] SHANNON M, BYRNE W. A formulation of the autoregressive HMM for speech synthesis [J]. American Journal of Physiology Endocrinology and Metabolism, 2009, 283(5): 725–726.

[113] BERCHTOLD A. The double chain Markov model [J]. Communications in Statistics-Theory and Methods, 1999, 28(11): 2569–2589.

[114] GHAHRAMANI Z, JORDAN M I. Factorial hidden Markov models [J]. Machine Learning, 1997, 29(2-3): 245–273.

[115] KOLTER J Z, JAAKOLA T S. Approximate inference in additive factorial HMMs with application to energy disaggregation [C]. Proceedings of the 15th International Conference on Artificial Intelligence and Statistics, 2012, 22: 1472–1482.

[116] ZHONG M J, GODDARD N, SUTTON C. Signal aggregate constraints in additive factorial HMMs, with application to energy disaggregation [C]. Proceedings of the 27th International Conference on Neural Information Processing Systems, December 8–13, 2014, Montreal, Canada. Cambridge: MIT Press, 2014: 3590–3598.

[117] LAFFERTY J, MCCALLUM A, PEREIRA F. Conditional random fields: Probabilistic models for segmenting and labeling sequence data [C]. Proceedings of the 18th International Conference on Machine Learning, June 28–July 1, 2001, Williamstown, Massachusetts. San Francisco: Morgan Kaufmann, 2001: 282–289.

[118] BERNAL A, CRAMMER K, HATZIGEORGIOU A, *et al.* Global discriminative learning for higher-accuracy computational gene prediction [J]. PLOS Computational Biology, 2007, 3(3): 488–497.

[119] HINTON G. A practical guide to training restricted Boltzmann machines [M]. Montavon G, Orr G B, Muller K R. Neural Networks: Tricks of the Trade, 2nd ed. Berlin: Springer, 2012: 599–619.

[120] SALTON G, WONG A, YANG C S. A vector space model for automatic indexing [J]. Communications of the ACM, 1975, 18(11): 613–620.

[121] DEERWESTER S, DUMAIS S T, FURNAS G W, *et al.* Indexing by latent semantic analysis [J]. Journal of the American Society for Information Science, 1990, 41(6): 391.

[122] XU Z J, WANG J, LIU Y, *et al.* A review of computer vision development and trends [J]. Journal of XI'AN University of Posts and Telecommunications, 2012, 12(6): 1–8.

[123] LI Z H, YANG Z H, LIN H F. Semantic output output-based disease-protein knowledge extraction [J]. Journal of Shandong University (Natural Science), 2016, 51(3): 104–110.

[124] HOFMANN T. Probabilistic latent semantic indexing [C]. Proceedings of the 22nd Annual International ACM SIGIR Conference on Research and Development in Information Retrieval, August 15–19, 1999, Berkeley, California. New York: ACM, 1999: 50–57.

[125] BLEI D, NG A, JORDAN M. Latent Dirichlet allocation [J]. Journal of Machine Learning Research, 2003, 3(7): 993–1022.

[126] BLEI D, LAFFERTY J. Correlated topic models [C]. Proceedings of the 18th conference on Advances in neural information processing systems, 2005. Cambridge: MIT Press, 2005: 147–154.

[127] TEH Y W, JORDAN M I, BEAL M J, *et al.* Hierarchical Dirichlet processes [J]. Journal of the American Statistical Association, 2006, 101(476): 1566–1581.

[128] BLEI D M, MCAULIFFE J D. Supervised topic models [C]. Proceedings of the 20th conference on Advances in neural information processing systems, December 3–6, 2007, Vancouver, Canada. Cambridge: MIT Press, 2008: 327–332.

[129] TURNEY P D. Thumbs up or thumbs down? Semantic orientation applied to unsupervised classification of reviews [C]. Proceedings of the 40th Annual Meeting on Association for Computational Linguistics, July 7–12, 2002, Philadelphia, Pennsylvania. Stroudsburg: ACL, 2002. 417–422.

[130] NASUKAWA T, YI J. Sentiment analysis: Capturing favorability using natural language processing [C]. Proceedings of the 2nd International Conference on Knowledge Capture, October 23–25, 2003. Sanibel Island, Florida. New York: ACM, 2003: 70–77.

[131] PANG B, LEE L, VAITHYANATHAN S. Thumbs up? Sentiment classification using machine learning techniques. EMNLP, 2002: 79–86.

[132] PANG B, LEE L, VAITHYANATHAN S. Thumbs up? Sentiment classification using machine learning techniques [C]. Proceedings of the ACL-02 conference on Empirical methods in natural language processing, 2002. Stroudsburg: ACL, 2002, 10: 79–86.

[133] KENNEDY A, INKPEN D. Sentiment classification of movie reviews using contextual valence shifters [J]. Computational Intelligence, 2006, 22(2): 110–125.

[134] WHITELAW C, GARG N, ARGAMON S. Using appraisal groups for sentiment analysis [C]. Proceedings of the 14th ACM International Conference on Information and Knowledge Management, October 31–November 5, 2005, Bremen, Germany. New York: ACM, 2005: 625–631.

[135] NG V, DASGUPTA S, ARIFIN S M. Examining the role of linguistic knowledge sources in the automatic identification and classification of

reviews [C]. Proceedings of the COLING/ACL on Main Conference Poster Sessions. Association for Computational Linguistics, July 17–18, 2006, Sydney, Australia. Stroudsburg: ACL, 2006: 611–618.

[136] YU J X, ZHA Z J, WANG M, *et al.* Aspect ranking: identifying important product aspects from online consumer reviews [C]. Proceedings of the 49th Annual Meeting of the Association for Computational Linguistics: Human Language Technologies, June 19, 2011, Portland, Oregon. Stroudsburg: ACL, 2011: 1496–1505.

[137] CARENINI G, NG R T, ZWART E. Extracting knowledge from evaluative text [C]. Proceedings of the 3rd International Conference on Knowledge Capture, October 2–5, Banff, Canada. New York: ACM, 2005: 11–18.

[138] SU Q, XU X Y, GUO H L, *et al.* Hidden sentiment association in Chinese web opinion mining [C]. Proceedings of the 17th International Conference on World Wide Web. April 21–25, Beijing, China. New York: ACM, 2008: 959–968.

[139] DING X W, LIU B, YU P S. A holistic lexicon-based approach to opinion mining [C]. Proceedings of the 2008 International Conference on Web Search and Data Mining, February 11–12, Palo Alto, California. New York: ACM, 2008: 231–240.

[140] KIM S M, HOVY E. Determining the sentiment of opinions [C]. Proceedings of the 20th international conference on Computational Linguistics, August 23–27, 2004, University of Geneva, Switzerland. Stroudsburg: ACL, 2004: 1367.

[141] CHOI Y, BRECK E, CARDIE C. Joint extraction of entities and relations for opinion recognition [C]. Proceedings of the 2006 Conference on Empirical Methods in Natural Language Processing, July 22–23, 2006, Sydney, Australia. Stroudsburg: ACL, 2006: 431–439.

[142] JOHANSSON R, MOSCHITTI A. Reranking models in finegrained opinion analysis [C]. Proceedings of the 23rd International Conference on Computational linguistics, August 23–27, Beijing, China. Stroudsburg: ACL, 2010: 519–527.

[143] RUPPENHOFER J, SOMASUNDARAN S, WIEBE J. Finding the sources and targets of subjective expression [C/CD]. Proceedings of the 6th International Conference on Language Resources and Evaluation, May 28–30, Marrakech, Morocco. LREC, 2008, 24(1): 2781–2788.

[144] KIM S M, HOVY E. Extracting opinions, opinion holders, and topics expressed in online news media text [C]. Proceedings of the Workshop on Sentiment and Subjectivity in Text, July 22, 2006, Sydney, Australia. Stroudsburg: ACL, 2006: 1–8.

[145] PANG B, LEE L. Opinion mining and sentiment analysis [J]. Foundations and Trends in Information Retrieval. 2008, 2(1-2): 1–135.

[146] LIU B. Sentiment analysis and opinion mining [J]. Synthesis Lectures on Human Language Technologies, 2012, 5(1): 1–167.

[147] PANG B, LEE L. A Sentimental Education: Sentiment Analysis Using Subjectivity Summarization Based on Minimum Cuts [C]. Proceedings of the 42nd Annual Meeting in Association for Computational Linguistics, July 21–26, 2004, Barcelona, Spain. Stroudsburg: ACL, 2004: 271–278.

[148] ALMEIDA T A, HIDALGO J M G, YAMAKAMI A. Contributions to the study of SMS spam filtering: new collection and results [C]. Proceedings of the 11th ACM symposium on Document Engineering, September 19–22, 2011, Mountain View, California. New York: ACM, 2011: 259–262.

[149] ZACHARY W W. An information flow model for conflict and fission in small groups [J]. Journal of Anthropological Research, 1977, 33(4): 452–473.

[150] ERDOS P, RENYI A. On random graphs I [J]. Publicationes Mathematicae, 1959, 6: 290–297.

[151] WATTS D J, STROGATZ S H. Collective dynamics of 'small-world' networks [J]. Nature, 1998, 393(6684): 440–442.

[152] MILGRAM S. The small world problem [J]. Psychology Today, 1967, 2(1): 185–195.

[153] BARABASI A L, ALBERT R. Emergence of scaling in random networks [J]. Science, 1999, 286(5439): 509–512.

[154] FREEMAN L C. A set of measures of centrality based on betweenness [J]. Sociometry, 1977, 40(1): 35–41.

[155] NEWMAN M E J. Fast algorithm for detecting community structure in networks [J]. Physical Review E, 2004, 69(6): 066133.

[156] BRIN S, PAGE L. Reprint of: The anatomy of a large-scale hypertextual web search engine [J]. Computer Networks, 2012, 56(18): 3825–3833.

[157] HAVELIWALA T H. Topic-sensitive pagerank [C]. Proceedings of the 11th International Conference on World Wide Web, May 7–11, 2002, Honolulu, Hawaii. New York: ACM, 2002: 517–526.

[158] KLEINBERG J M. Authoritative sources in a hyperlinked environment [J]. Journal of the ACM, 1999, 46(5): 604–632.

[159] GIRVAN M, NEWMAN M E J. Community structure in social and biological networks [C]. Proceedings of the National Academy of Sciences of the United States of America, May 22, 2001. Pittsburgh: National Academies Press, 2001, 99(12): 7821–7826.

[160] NEWMAN M E J, GIRVAN M. Finding and evaluating community structure in networks [J]. Physical Review E, 2004, 69(2): 026113.

[161] BLONDEL V D, GUILLAUME J L, LAMBIOTTE R, *et al.* Fast unfolding of communities in large networks [J]. Journal of Statistical Mechanics: Theory and Experiment, 2008, 10: 10008.

[162] LEHMANN J, ISELE R, JAKOB M, *et al.* DBpedia-A large-scale, multilingual knowledge base extracted from Wikipedia [J]. Semantic Web journal, 2014, 6(2): 167–195.

[163] ZOU L, MO J H, CHEN L, *et al.* gStore: Answering SPARQL queries via subgraph matching [J]. Proceedings of the VLDB Endowment, 2011, 4(8): 482–493.

[164] ZOU L, OZSU M T, CHEN L, *et al.* gStore: a graph-based SPARQL query engine [J]. The VLDB Journal, 2014, 23(4): 565–590.

[165] ZENG K, YANG J C, WANG H X, *et al.* A distributed graph engine for web scale RDF data [C]. Proceedings of the 39th International Conference on Very Large Data Bases, August 26, Trento, Italy. Berlin: Springer, 2013, 265–276.

[166] FADER A, SODERLAND S, ETZIONI O. Identifying relations for open information extraction [C]. Proceedings of the Conference on Empirical Methods in Natural Language Processing, July 27–31, 2011, Edinburgh, United Kingdom. Stroudsburg: ACL, 2011: 1535–1545.

[167] LIANG P. Learning executable semantic parsers for natural language understanding [J]. Communications of the ACM, 2016, 59(9): 68–76.

[168] REBELE T, SUCHANEK F M, HOFFART J, *et al.* YAGO: A multilingual knowledge base from Wikipedia, Wordnet, and Geonames [C]. Proceedings of the 15th International Semantic Web Conference, October 17–21, Kobe, Japan. New York: Springer, 2016: 177–185.

[169] BORDES A, USUNIER N, GARCIA-DURAN A, *et al.* Translating embeddings for modeling multi-relational data [C]. Proceedings of the 27th International Conference on Neural Information Processing Systems, December 5–10, 2013, Lake Tahoe, Nevada. New York: Curran Associates Inc, 2013: 2787–2795.

[170] LESKOVEC J, KREVL A. SNAP datasets: Stanford large network dataset collection [DB/OL].

[171] GEHRKE J, GINSPARG P, KLEINBERG J M. Overview of the 2003 KDD cup [J]. ACM SIGKDD Exploration Newsletter, 2003, 5(2): 149–151.

[172] GOODFELLOW I, BENGIO Y, COURVILLE A. Deep learning [M/OL]. Cambridge: MIT Press, 2016.

[173] HE K M, ZHANG X Y, REN S Q, *et al.* Deep residual learning for image recognition [C]. Proceedings of the IEEE Conference on Computer Vision and Pattern Recognition, June 26–July 1, 2016, Seattle, Washington. New York: Curran Associates Inc, 2016: 770–778.

[174] SILVER D, HUANG A, MADDISON C J, *et al.* Mastering the game of Go with deep neural networks and tree search [J]. Nature, 2016, 529(7587): 484–489.

[175] MCCONNELL R K. Method of and apparatus for pattern recognition; United States, 4567610 [P]. 1986.

[176] LOWE D G. Object recognition from local scale-invariant features [C]. Proceedings of the 7th IEEE International Conference on Computer Vision, September 20–25, 1999, Kerkyra, Greece. Washington DC: IEEE Computer Society, 1999: 1150–1157.

[177] HAHNLOSER R L T. On the piecewise analysis of networks of linear threshold neurons [J]. Neural Networks, 1998, 11(4): 691–697.

[178] GOODFELLOW I J, WARDE-FARLEY D, MIRZA M, *et al.* Maxout networks [C]. Proceedings of Machine Learning Research. International Conference on Machine Learning, June 17–19, 2013, Atlanta, Georgia. ICML, 28(3): 1319–1327.

[179] MAAS A L, HANNUN A Y, NG A Y. Rectifier nonlinearities improve neural network acoustic models [C/OL]. Proceedings of ICML Workshop on Deep Learning for Audio, Speech and Language Processing, 2013.

[180] NAIR V, HINTON G E. Rectified linear units improve restricted boltzmann machines [C]. Proceedings of the 27th International Conference on Machine Learning, June 21–24, 2010, Haifa, Israel. New York: Omnipress, 2010: 807–814.

[181] HORNIK K, STINCHCOMBE M, WHITE H. Multilayer feedforward networks are universal approximators [J]. Neural Networks, 1989, 2(5): 359–366.

[182] ELDAN R, SHAMIR O. The power of depth for feedforward neural networks [C]. Proceedings of the 29th Annual Conference on Learning Theory, June 23–26, 2016, New York, USA. Stroudsburg: ACL, 2016: 907–940.

[183] TELGARSKY M. Benefits of depth in neural networks [C]. Proceedings of the 29th Annual Conference on Learning Theory, June 23–26, 2016, New York, USA. Stroudsburg: ACL, 2016: 1517–1539.

[184] RUMELHART D E, HINTON G E, WILLIAMS R J. Learning representations by back-propagating errors [J]. Nature, 1986, 323(9): 533–536.

[185] ABADI M, Agarwal A, Barham P, *et al.* TensorFlow: Large-scale machine learning on heterogeneous distributed systems [EB/OL].

[186] BERGSTRA J, BREULEUX O, BASTIEN F, *et al.* Theano: A CPU and GPU math compiler in Python [C/OL]. Proceedings of the 9th Python in Science Conference, June 28–July 3, 2010, Austin, Texas: 3–10.

[187] CHEN T Q, LI M, LI Y T, *et al.* Mxnet: A flexible and efficient machine learning library for heterogeneous distributed systems [EB/OL].

[188] BOTTOU L. Online learning and stochastic approximations [J]. Online Learning in Neural Networks, 1998, 17(9): 142.

[189] POLYAK B T. Some methods of speeding up the convergence of iteration methods [J]. USSR Computational Mathematics and Mathematical Physics, 1964, 4(5): 1–7.

[190] SUTSKEVER I, MARTENS J, DAHL G E, *et al.* On the importance of initialization and momentum in deep learning [C]. Proceedings of Machine Learning Research. International Conference on Machine Learning, June 17–19, 2013, Atlanta, Georgia. ICML, 28(3): 1139–1147.

[191] NESTEROV Y. A method of solving a convex programming problem with convergence rate O(1/k2) [J]. Soviet Mathematics Doklady, 1983, 27(2): 372–376.

[192] DUCHI J, HAZAN E, SINGER Y. Adaptive subgradient methods for online learning and stochastic optimization [J]. Journal of Machine Learning Research, 2011, 12(7): 2121–2159.

[193] KINGMA D, BA J. Adam: A method for stochastic optimization [EB/OL].

[194] IOFFE S, SZEGEDY C. Batch normalization: accelerating deep network training by reducing internal covariate shift [C]. Proceedings of the 32nd International Conference on Machine Learning, July 6–11, 2015, Lille, France: ICML, 2015: 448–456.

[195] SCHAUL T, ANTONOGLOU I, SILVER D. Unit tests for stochastic optimization [C]. International Conference on Learning Representation, April 14–16, 2014, Banff, Canada.

[196] ZEILER M D. ADADELTA: an adaptive learning rate method [EB/OL].

[197] LI Q X, TAI C, WEINAN E. Dynamics of stochastic gradient algorithms [EB/OL].

[198] LECUN Y. Generalization and network design strategies [J]. Connectionism in Perspective, 1998: 143–155.

[199] BOUREAU Y L, PONCE J, LECUN Y. A theoretical analysis of feature pooling in visual recognition [C]. Proceedings of the 27th International Conference on Machine Learning, June 21–24, 2010, Haifa, Israel. New York: Omnipress, 2010: 111–118.

[200] BOUREAU Y L, ROUX N L, BACH F, *et al.* Ask the locals: multi-way local pooling for image recognition [C]. Proceedings of the 2011 IEEE International Conference on Computer Vision, November 6–13, 2011, Washington DC: IEEE, 2011: 2651–2658.

[201] JIA Y Q, HUANG C, DARRELL T. Beyond spatial pyramids: Receptive field learning for pooled image features [C]. Proceedings of the 2012 IEEE International Conference on Computer Vision and Pattern Recognition, June 16–21, 2012, Providence, Rhode Island. Washington DC: IEEE, 2012: 3370–3377.

[202] LECUN Y, BOTTOU L, BENGIO Y, *et al.* Gradient-based learning applied to document recognition [J]. Proceedings of the IEEE, 1998, 86(11): 2278–2324.

[203] KRIZHEVSKY A, SUTSKEVER I, HINTON G E. Imagenet classification with deep convolutional neural networks [J]. Advances in Neural Information Processing Systems, 2012: 1097–1105.

[204] SRIVASTAVA N, HINTON G E, KRIZHEVSKY A, *et al.* Dropout: a simple way to prevent neural networks from overfitting [J]. Journal of Machine Learning Research, 2014, 15(1): 1929–1958.

[205] SIMONYAN K, ZISSERMAN A. Very deep convolutional networks for large-scale image recognition [EB/OL].

[206] LIN M, CHEN Q, YAN S C. Network in network [EB/OL].

[207] SZEGEDY C, VANHOUCKE V, IOFFE S, *et al.* Rethinking the inception architecture for computer vision [C]. Proceedings of the 29th IEEE Conference on Computer Vision and Pattern Recognition, June 26–July 1, 2016, Las Vegas, Nevada. Washington DC: IEEE, 2016: 2818–2826.

[208] SUTSKEVER I, VINYALS O, LE Q V. Sequence to sequence learning with neural networks [J]. Advances in Neural Information Processing Systems, 2014: 3104–3112.

[209] GREGOR K, DANIHELKA I, GRAVES A, *et al.* DRAW: a recurrent neural network for image generation [C]. Proceedings of the 32nd International Conference Machine Learning, July 6–11, 2015, Lille, France. ICML, 2015: 1462–1471.

[210] CHAN W, JAITLY N, LE Q, *et al.* Listen, attend and spell: A neural network for large vocabulary conversational speech recognition [C]. Proceedings of 2016 IEEE International Conference on Acoustics, Speech and Signal Processing, March 20–25, 2016, Shanghai, China. Washington DC: IEEE, 2016: 4960–4964.

[211] HOCHREITER S, SCHMIDHUBER J. Long short-term memory [J]. Neural Computation, 1997, 9(8): 1735–1780.

[212] BENGIO Y, FRASCONI P, SIMARD P. The problem of learning long-term dependencies in recurrent networks [C]. Proceedings

of International Symposium on Neural Networks. Washington DC: IEEE, 1993: 1183–1188.

[213] BENGIO Y, SIMARD P, FRASCONI P. Learning long-term dependencies with gradient descent is difficult [J]. IEEE Transactions on Neural Networks, 1994, 5(2): 157–166.

[214] CHO K, MERRIENBOER B V, GULCEHRE C, *et al.* Learning phrase representations using RNN encoder-decoder for statistical machine translation [C]. Proceedings of the 2014 Conference on Empirical Methods in Natural Language Processing, October 25–29, 2014, Doha, Qatar. Stroudsburg: ACL, 2014: 1724–1734.

[215] KOUTNIK J, GREFF K, GOMEZ F, *et al.* A clockwork RNN [C]. Proceedings of the 31st International Conference on Machine Learning, June 21–26, 2014, Beijing, China. JMLR.org, 2014: II 1863-II 1871.

[216] SCHUSTER M, PALIWAL K K. Bidirectional recurrent neural networks [J]. IEEE Transactions on Signal Processing, 1997, 45(11): 2673–2681.

[217] GRAVES A. Generating sequences with recurrent neural networks [EB/OL].

[218] VINCENT P, LAROCHELLE H, LAJOIE I, *et al.* Stacked denoising autoencoders: Learning useful representations in a deep network with a local denoising criterion [J]. Journal of Machine Learning Research, 2010, 11(12): 3371–3408.

[219] SALAKHUTDINOV R, HINTON G E. Deep Boltzmann machines [C]. Proceedings of the 12th International Conference on Artificial Intelligence and Statistics, April 16-18, 2009, Clearwater Beach, Florida. JLMR.org, 2009: 3.

[220] GOODFELLOW I, POUGET-ABADIE J, MIRZA M, *et al.* Generative adversarial nets [J]. Advances in Neural Information Processing Systems, 2014: 2672–2680.

[221] MNIH V, KAVUKCUOGLU K, SILVER D, *et al.* Human-level control through deep reinforcement learning [J]. Nature, 2015, 518(7540): 529–533.

[222] COLLOBERT R, KAVUKCUOGLU K, FARABET C. Torch7: A Matlab-like environment for machine learning [C/OL]. Proceedings of NIPS Workshop BigLearn, 2011.

[223] JIA Y Q, SHELHAMER E, DONAHUE J, *et al.* Caffe: convolutional architecture for fast feature embedding [EB/OL].

[224] LECUN Y, BOTTOU L, BENGIO Y, *et al.* Gradient-based learning applied to document recognition [J]. Proceedings of the IEEE, 1998, 86(11): 2278–2324.

[225] SAMARIA F S, HARTER A C. Parameterisation of a stochastic model for human face identification [C]. Proceedings of the 2nd IEEE Workshop on Applications of Computer Vision, December 5–7, 1994, Sarasota, Florida. New York: IEEE, 1994: 138–142.

[226] GARCIA H, LUDU A. The Google file system [J]. ACM Sigops Operating Systems Review, 2003, 37(5): 29–43.

[227] DEAN J, GHEMAWAT S. MapReduce: Simplified data processing on large clusters [J]. Communications of the ACM, 2008, 51(1): 107–113.

[228] KEARNS M. Efficient noise-tolerant learning from statistical queries [J]. Journal of the ACM, 1998, 45(6): 983–1006.

[229] CHU C T, KIM S K, LIN Y A, *et al.* Map-Reduce for machine learning on multicore [C]. Proceedings of the 20th Annual Conference on Neural Information Processing Systems, December 4–7, 2006, Canada. Cambridge: MIT Press, 2006: 281–288.

[230] ZAHARIA M, CHOWDHURY M, FRANKLIN M J, *et al.* Spark: Cluster computing with working sets [C]. Proceedings of the 2nd USENIX Conference on Hot Topics in Cloud Computing, Boston: USENIX, 2010: 1765–1773.

[231] ZAHARIA M, CHOWDHURY M, DAS T, *et al.* Resilient distributed datasets: A fault-tolerant abstraction for in-memory cluster computing [C]. Proceedings of the 9th USENIX Conference on Networked Systems Design and Implementation, April 25–27, 2012, San Jose, California. Berkeley: USENIX Association, 2012: 141–146.

[232] BURROWS M. The Chubby lock service for loosely-coupled distributed systems [C]. Proceedings of the 7th Symposium on Operating Systems Design and Implementation, November 6–8, 2006, Seattle, Washington. Berkeley: USENIX Association, 2006: 335–350.

[233] CHANG F, DEAN J, GHEMAWAT S, *et al.* Bigtable: A distributed storage system for structured data [J]. ACM Transaction on Computer Systems, 2008, 26(2), 2006: 4.

[234] EFRON B, TIBSHIRANI R J. An introduction to the bootstrap [M]. London: Chapman and Hall/CRC Press, 1994.

[235] JAIN A K, DUBES R C. Algorithms for clustering data [M]. Upper Saddle River: Prentice Hall Inc, 1988.

[236] KAUFMAN L, ROUSSEEUW P J. Finding groups in data: an introduction to cluster analysis [M]. Hoboken: John Wiley and Sons Inc, 2009.

[237] TIELEMAN T, HINTON G. Lecture 6.5-rmsprop: Divide the gradient by a running average of its recent magnitude. Coursera: Neural Networks for Machine Learning 4.2, 2012: 26–31.

www.ingramcontent.com/pod-product-compliance
Lightning Source LLC
LaVergne TN
LVHW052022200926
848481LV00005B/1559